12th International Scientific Conference Civil Engineering Design and Construction (DCB 2022)

IOP Conference Series: Materials Science and Engineering
Volume 1276

Online
7 – 10 September 2022

ISBN: 978-1-7138-6983-2
ISSN: 1757-8981

Printed from e-media with permission by:

Curran Associates, Inc.
57 Morehouse Lane
Red Hook, NY 12571

Some format issues inherent in the e-media version may also appear in this print version.

Institute of Physics
Dirac House, Temple Back
Bristol BS1 6BE UK

Phone: 44 1 17 929 7481
Fax: 44 1 17 920 0979

techtracking@iop.org

Additional copies of this publication are available from:

Curran Associates, Inc.
57 Morehouse Lane
Red Hook, NY 12571 USA
Phone: 845-758-0400
Fax: 845-758-2633
Email: curran@proceedings.com
Web: www.proceedings.com

TABLE OF CONTENTS

IOP Conf. Series: Materials Science and Engineering 1276 (2023) 011001 doi:10.1088/1757-899X/1276/1/011001

Preface

The **International scientific conference "Civil Engineering Design and Construction" (DCB)** is the twelfth international conference organized by the Scientific and Technical Union of Civil Engineering in Bulgaria (NTSSB), strongly supported by the Chamber of Engineers in the Investment Design in Bulgaria, Bulgarian Construction Chamber, Federation of the Bulgarian Scientific and Technical Unions, the Varna Free University and "Evrika" Foundation. Special attention was paid to sustainable construction, digitalization in the construction industry and building information modeling (BIM).

The conference is held every two years, starting in 2000 as a national conference of the Union of Bulgarian Construction Engineers and the Bulgarian Scientific and Technical Union of Civil Engineering. This conference is of special importance for Bulgaria, where scientists, researchers and experts present their achievements and share their experience or innovative ideas in various fields of design, construction and architecture. Since 2006, the DCB conference has been held as an international scientific forum for in-depth discussions and fruitful meetings.

The conference was held in Varna Free University "Chernorizets Hrabar", Bulgaria on September 08-10, 2022. The conference was partially held virtually due to the COVID-19 pandemic. The presentations of the participants presented on-site and those of the online participants were screened in the conference halls. The conference was held successfully with Zoom videoconferencing software.

The conference program was organized in 5 sessions, in which 2 plenary and 58 specialized reports were presented. Apart from principal Bulgarian scientific organizers and the University of Architecture, Civil Engineering and Geodesy, University of Structural Engineering and Architecture "Lyuben Karavelov", Varna Free University, New Bulgarian University and Bulgarian Academy of Science (BAS), scientists from Slovakia, Lithuania and Estonia, took part in the scientific forum.
In addition to thematic discussions on the main topics, a special session for young scientists was organized, supported by the "Evrika" Foundation. Leading companies working in the building and construction sectors were actively presented.

The accepted papers are published in the IOP Conference Series – Materials Science and Engineering and separate volume of Conference Proceedings and copyrighted with an ISSN.

Editors DCB'2022:
Assoc. Prof. Tony Venelinov, PhD – Editor in Chief
Assoc. Prof. Lachezar Hrischev, PhD, Eng

IOP Conf. Series: Materials Science and Engineering 1276 (2023) 011001 doi:10.1088/1757-899X/1276/1/011001

SCIENTIFIC COMMITTEES

International Scientific Committee

Chairman:
Acad. Yachko Ivanov – Bulgaria

Vice Chairmans:
Prof. Vlastimir Radonjanin – Serbia
Prof. Rao Surampalli – USA

Members:
Prof. Aleksandrs Korjakins - Litva
Prof. Azariy Lapidus – Russia
Prof. Andrzej Garbacz – Poland
Prof. Anna Nedyalkova – Bulgaria
Prof. Asterios Liolios – Greece
Prof. Bojidar Janev- USA
Prof. Damian Kachlakev - USA
Prof. Faruk Karadoğan – Turkey
Prof. Heinz Brandl – Austria
Prof. Ivan Vaniček – Czech Republic
Prof. Jurai Kralik – Slovakia
Prof. Liberato Ferrara – Italy
Prof. Mindaugas Dauksys – Lithuania
Prof. Petar Hristov - Bulgaria
Prof. Radomir Folić – Serbia
Prof. Victor Hristevski – North Macedonia
Assoc. Prof. Asli Gurgun – Turkey
Assoc. Prof. Ewa Blazik-Borowa – Poland
Assoc. Prof. Nabi Ibadov - Poland
Assoc. Prof. Robert Eadie – UK
Assoc. Prof. Rositsa Petkova-Slipets - Bulgaria
Eng. Iliyan Terziev – Bulgaria

National Scientific Committee

Co-Chairman's:
Assoc. Prof. Lachezar Hrischev
Prof. Dimitar Nazarski
Prof. Konstantin Kazakov

Members:
Prof. Bogomil Petrov
Prof. Chavdar Kolev
Prof. Dimitar Nazarski
Prof. Doncho Partov
Prof. Emil Burnazski
Prof. Fantina Rangelova
Prof. Ivelin Ivanov
Prof. Ivan Markov
Prof. Ivanka Paskaleva
Prof. Ivaylo Koprev
Prof. Krasimir Enimanev
Prof. Krasimir Krastanov
Prof. Marina Traykova
Prof. Stoyan Bratoev
Prof. Teodor Roshavelov
Prof. Todor Bulev
Assoc. Prof. Ana Yanakieva
Assoc. Prof. Anita Handruleva
Assoc. Prof. Daria Mihaleva
Assoc. Prof. Emad Abdulahad
Assoc. Prof. Georgi Linkov
Assoc. Prof. Ivan Pavlov
Assoc. Prof. Ivan Torodov
Assoc. Prof. Tony Venelinov
Assoc. Prof. Yourdanka Kandilkova
Assoc. Prof. Hristina Zayakova
Assoc. Prof. Ventseslav Stoyanov

IOP Conf. Series: Materials Science and Engineering 1276 (2023) 011002 doi:10.1088/1757-899X/1276/1/011002

Peer Review Statement

All papers published in this volume have been reviewed through processes administered by the Editors. Reviews were conducted by expert referees to the professional and scientific standards expected of a proceedings journal published by IOP Publishing Publishing.

- **Type of peer review:** Single Anonymous
- **Conference submission management system:** Morressier
- **Number of submissions received:** 32
- **Number of submissions sent for review:** 32
- **Number of submissions accepted:** 29
- **Acceptance Rate (Submissions Accepted / Submissions Received × 100):** 90.6
- **Average number of reviews per paper:** 1
- **Total number of reviewers involved:** 23
- **Contact person for queries:**
  Name: Assoc. Prof. Tony Venelinov
  Email: tvenelinov_fhe@uacg.bg
  Affiliation: University of Architecture, Civil engineering and Geodesy, Sofia, Bulgaria

Analysis and characterization of modern and ancient mortars

V Stoyanov[1,2], B Kostova[3*], B Dumanov[4]

[1] University of Structural Engineering and Architecture „Lyuben Karavelov",
175 Suhodolska Str., 1373 Sofia, Bulgaria
[2] Academy of the Ministry of Interior, Faculty of Fire Safety and Civil Protection,
171 Pirotska Str., 1309 Sofia, Bulgaria
[3] New Bulgarian University, Department of Natural Sciences,
21 Montevideo Blvd., 1618 Sofia, Bulgaria
[4] New Bulgarian University, Department of Archaeology,
21 Montevideo Blvd., 1618 Sofia, Bulgaria

* E-mail: bkostova@nbu.bg

Abstract. The analysis, characterization, and comparison of modern and ancient mortars provide fundamental (by tracing the change in the raw materials and recipes used from the past to the present) and practical (obtaining modern mortars compatible with the ancient) knowledge. This work aimed to characterize modern and ancient mortars by investigating their phase composition, binder-to-aggregate ratio, hydraulicity, and hygroscopic properties. The used methods for analyses were powder X-ray diffraction, Fourier transforms infrared spectroscopy, and thermal analysis. A modern mortar (sample A) preparation was with a binder of white Portland cement and an aggregate of river sand. Ancient mortar samples (sample B - cocciopesto and sample C - lime mortar) collection were from Bulgarian archaeological sites allocated within the Roman period. The results define investigated sample A as pozzolanic mortar with hygroscopic properties, sample B - as hydraulic lime mortar with hygroscopic properties, and sample C – as lime mortar without any hygroscopic properties. Obtained results with these, achieved for phase composition and binder-to-aggregate ratio, show that modern mortar is not compatible with the Romans and is not suitable for reconstruction and conservation of masonry from these archaeological sites.

1. Introduction

Modern binding composites increasingly use new and modified materials with improved functional properties. At the same time, to achieve the goals of construction's sustainable development, research is being done on recycled and environmentally friendly raw materials [1] to reach energy efficiency and reduce carbon emissions in binding composite production [2]. However, the basics of all modern achievements were on raw materials and recipes known since ancient times. Binding materials are one of the oldest composites - the first records of their use date back to 12000 BC.

The ancient mortar was composed of binder (lime, gypsum) and aggregate (sand, volcanic ash, crushed rocks, and ceramic). Preparation of the mortar's binder has been by quicklime and water. Water hydrates the quicklime to lime (portlandite). Over time, portlandite was carbonated and transformed into calcite by absorbing CO_2 from the air. Examples of ancient mortar with non-carbonated portlandite are also known in the literature [3]. The gypsum additive in mortar creates it rigid, strong, and durable [4]. The Egyptians widely used lime and gypsum mortars, and the Greeks made improvements to mortar technology [5]. Lime has been widely used by the Romans too. Besides lime mortar, the Romans also used hydraulic lime mortar. They produce it by adding to mortar the materials like volcanic ash. Given the limitations of the volcanic ash deposits, the Romans added ceramic fragments (cocciopesto mortar), which also gave hydraulic properties [5, 6]. The ceramics preparation was by mixing clay and additives of quartz, feldspar, etc., and subsequently firing. Firing at temperatures between 450 and 800°C results in ceramic amorphous pozzolanic phases-aluminosilicates. In mortars, they react with lime in the presence of water to produce calcium silicate hydrate (C-S-H) and/or calcium aluminate hydrate phases (C-A-H). Such phases improve the strength of mortars, are resistant to moisture (giving hydraulic properties to mortar), and make mortar suitable for use in humid environments. Firing at temperatures above 800°C produces ceramic high-temperature mullite and cristobalite instead of pozzolanic phases [7, 8].

Historically, the achievement for binding materials was the discovery of Portland cement in 1824 [5]. The hydraulic properties of Portland cement were obtained by adding clays. Mixing Portland cement with water rises exothermic reaction of hydration, accompanied by the formation of C-S-H phases (often amorphous), portlandite, and ettringite. The new-formed hydrated phases are poorly soluble in water, giving hydraulic properties to the mortar. The achievement of mortar's strength and durables was by adding gypsum. The presence of gypsum was responsible for the monosulphoaluminates formation. Such phases slow down the ettringite formation and reduce the portlandite amount in the mortar [9].

Following the history of mortar production, analysis, characterization, and comparison of modern and ancient mortars provide fundamental (by tracing the change in the raw materials and recipes used from the past to the present) and practical (obtaining modern mortars compatible with the ancient) knowledge. This work aimed to characterize one modern and two ancient mortars. The modern mortar sample was composed of a binder (white Portland cement) and aggregate (river sand). The ancient mortar samples were collected from two Bulgarian archaeological sites allocated within the Roman period. Macroscopically the Roman mortars differ in the content of ceramic fragments - one of them was recognized as cocciopesto, and the other – was mortar without crushed ceramic. The detection of the sample's phase composition was made by powder X-ray diffraction and Fourier transform infrared measurements. Thermal analysis was used to obtain the hydraulicity, hygroscopic properties, and binder-to-aggregate ratio (B:A) ratio.

2. Materials and methods

2.1. Materials

Modern mortar samples (named Sample A) were prepared of binder - white Portland cement, aggregate - sand, and distilled water. The B:A ratio was 1:3. According to the chemical composition provided by the white Portland cement- manufacturer Devnya Cement, Bulgaria (cement type CEM I 52.5 N), the following phase composition was calculated by the Bogue method [9]: belite (C_2S), alite (C_2S), tricalcium aluminate (C_3A), calcium aluminoferrite (C_4AF), and gypsum. The aggregate (river sand) was with over 85.0% content of SiO_2 – quartz, with a shape index – of 4.6% (EN 933-4 [10]). The prepared sample was with a composition satisfying the EN 196-1 (section 6) requirements [11]. Sample A was molded without any compacting treatment in 6 prisms (two prisms for each sample), 40´40´160 mm, for one day in a moist atmosphere (> 95% RH and 20°C). Then the unmolded sample was stored underwater (20°C) for 120 days.

Ancient mortars were from two Bulgarian archaeological sites. Sample B collection was by the area of the western part of the Upper Thracian lowland. This mortar was with ceramic fragments added to aggregate and recognized as cocciopesto. Sample C collection was by the southern foothills of Sarnena Sredna Gora Mountain (sample catalog number - №61 - Dimitrievo village, Emporion Pizos). This sample was without ceramic fragments.

2.2. *Methods*

The powder X-ray diffraction (PXRD) measurements were made by D2 Phaser BrukerAXS, CuKα radiation (λ = 0.15418 nm) (operating at 30 kV, 10 mA) from 3 to 70 °2θ with a step of 0.05°, 1 s/step (ground sample weight – 1.0±0.1 mg and particle size below 0.075 mm). The PDF (Powder Diffraction File, ICDD, 2001) database was used for determining the phases and minerals in the samples [12].

Fourier transform infrared (FTIR) measurements were performed by FTIR Spectrometer Nicolet 6700, covering the range of 400 - 4000 cm-1 with 100 scans and 1.928 cm^{-1} resolution. The samples were prepared as pellets with KBr.

Thermal analysis: simultaneous thermogravimetric/derivative thermogravimetric coupled with differential scanning calorimetric (TG/DTG-DSC) analysis was carried out on a Setline STA 1100, SETARAM, France, in the temperature range room temperature (RT) – 900°C; in the static air, with a heating rate of 10°C min^{-1}. The operational characteristics of the TG/DTG–DSC - system were: sample mass of 25.0±1.0 mg (mass resolution of 0.05 μg), temperature resolution of +/- 0.3°C, and alumina sample crucible with a volume of 100 μL.

3. Results and discussion

3.1. *PXRD*

Figure 1 shows the PXRD results of the samples investigation. The detected phases in Sample A were calcite ($CaCO_3$) PDF#06-6528, quartz (SiO_2) PDF #46-1045, feldspar ($CaAl_2Si_2O_8$) PDF #41-1486, alite - C_3S (Ca_3SiO_5) PDF#11-0593, belite - C_2S (Ca_2SiO_4) PDF#49-1673, portlandite ($Ca(OH)_2$) PDF#44-1481, ettringite ($6Ca(OH)_2.Al_2(SO_4)_3.26H_2O$) PDF#41-1451, and monosulphoaluminate ($Ca_4Al_2(OH)_{12}(SO_4).6H_2O$) PDF#50-1607. In the Sample B was registered calcite, quartz, feldspar ($NaAlSi_3O_8$) PDF #89-6426, illite ($KAl_2Si_3AlO_{10}(OH)_2$) PDF#02-0056, hematite (Fe_2O_3) PDF#33-0664, portlandite, and tobermorite ($Ca_5[(Si,Al)_6(O,OH)_{17}].5H_2O$) PDF#45-1481. The calcite and quartz were recognized only in Sample C.

3.2. *FTIR*

Figure 2 shows the results from the FTIR measurements. The results confirm the established phases by PXRD. The vibration of C-O in CO_2^{-3} of calcite were registered at following wavenumbers: 2881, 2608, and 2514 cm^{-1} (ν1+ ν3); 1780 cm^{-1} (ν1 + ν4); 1411 and 1423 cm^{-1} (ν3), 874 cm^{-1} (ν2), and 710 cm^{-1} (ν4) [13, 14]. The vibrations of Si-O in SiO_2 in quarts were detected at: 1160-1088 cm^{-1} (ν3), 798 cm^{-1} and 775 cm^{-1} (ν2), 692 cm^{-1} (ν1), and 472 cm^{-1} (ν4) [13]. The Si-Si, Si-Al(Si), Si(Al)-O, Si-O-Si, Al-O-Si, and O-Si(Al)-O bands were referred to feldspars as follows: 1032 cm^{-1} (ν Si-O stretching vibration), 646 cm^{-1} (ν Si-Al(Si) stretching), 585 cm^{-1} (ν O-Si(Al)-O bending vibrations), 532 cm^{-1} (ν Al-O-Si deformation vibrations), and 428 cm^{-1} (ν Si-O-Si deformation) [13, 15, 16]. FTIR registered feldspar, which PXRD was not detected in sample C. The Si-O stretching vibration at 1020 cm^{-1} was assigned to illite [13, 17]. The bands at: 3733 cm^{-1} could be assigned to Ca-O-H (ν3) in portlandite and ettringite; 3640 cm^{-1} – to Ca-O-H (ν3) in portlandite, ettringite, and monosulphoaluminate; 3428 cm^{-1} – to O-H structural (ν1) in portlandite and ettringite; 1621 cm^{-1} – to Ca-O-H (ν2) in portlandite and monosulphoaluminate; 1002 cm^{-1} – to S-O in SO_4^{2-} (ν1) in ettringite, and monosulphoaluminate; 865 cm^{-1} – to Al—O in AlO_6^{9-} in ettringite, and monosulphoaluminate; 458 cm^{-1} – to S-O in SO_4^{2-} in monosulphoaluminate; 452 cm^{-1} could be assigned to Si-O in SiO_3^{2-} (ν2) in tobermorite as well as to Fe-O vibration in hematite [9, 13].

FTIR registered bands of phases, which were not detected by PXRD. Bands at 2987 and 2970 cm^{-1} (for Samples B and C) were assigned to organic matter – stretching vibrations of C-H in CH_2 [6]. The ν O-H vibrations of H_2O molecules were registered at 3435 cm^{-1} (Sample C) [13, 17].

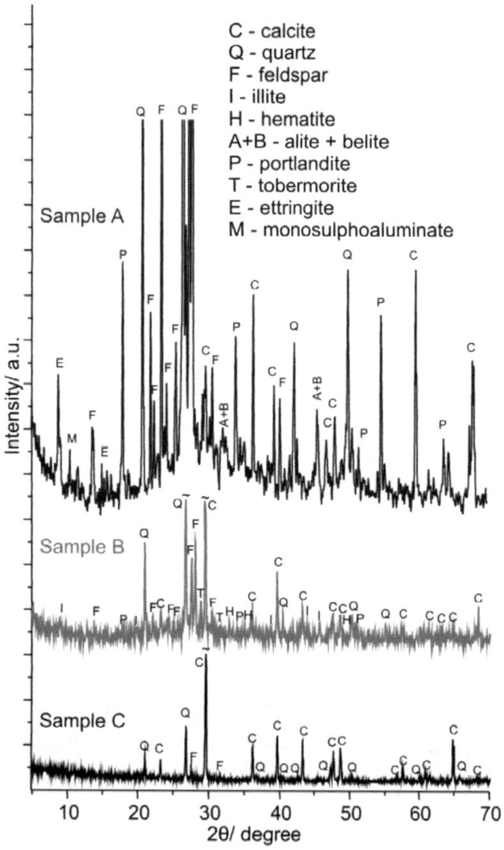

Figure 1. PXRD patterns of studied samples.

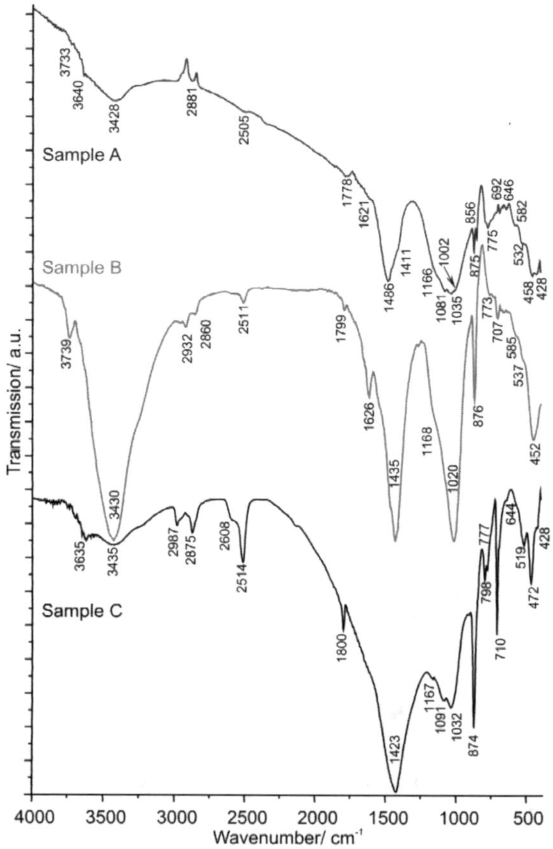

Figure 2. FTIR spectra of studied samples.

The PXRF and FTIR results show the sample's binder and aggregate composition. Sample A: the aggregate was composed of quartz and feldspar, and the binder of calcite, belite, alite, portlandite, ettringite, and monosulphoaluminate. The calcite, alite, and belite were Portland cement relict phases. Portlandite, ettringite, and monosulphoaluminate formation were during Portland cement hydration. Sample B: the aggregate was composed of quartz, feldspar, illite, and hematite. Illite and hematite were defined as ceramic phases. High-temperature ceramic phases were not detected therefore the hydraulic phase formation at the calcite–ceramic boundary occurs [7]. The binder of sample B was composed of calcite, portlandite, and a C-S-H phase's tobermorite. The detected hydraulic phases in samples A and B define them as hydraulic mortars. The aggregate of sample C was composed of quartz and feldspar, and the binder was only of calcite, which defies the sample as lime mortar.

3.3. Thermal analysis

Figures 3, 4, 5, and Table 1 present the thermal analysis results. The total mass loss (ML) of studied samples changed from 8.86% (sample A) up to 30.58% (sample C). The thermal decomposition of studies samples separates into five temperature ranges: RT - 220°C, 300 - 420°C, 450-550°C, 550 - 700°C, and 700 - 900°C.

RT - 220°C temperature range:

In the temperature range between RT - 220°C was measured ML from 0.15 to 1.16% (Table 1). This range can be divided into two subranges: RT-120°C and 120 - 220°C. Between RT and 120°C, the dehydration of hygroscopic water (humidity) [8, 19] occurs with ML change from 0.15 to 1.16%. The sum of ML up to 120°C defines mortar hygroscopic properties [20] (for sample A = 1.70%, for sample

B = 1.77%, and for sample C = 0.44). Samples A and B show hygroscopic properties, while sample C does not present hygroscopic properties, since it contains adsorbed water less than 1% during heating up to 120°C [20]. Between 100 and 220°C – the dehydration of C-S-H phases [21] (Figs. 3, 4, Table 1) and illite [22] for sample B (Fig. 4) occurs. For sample C, the ML between 100 and 220°C was not registered (Fig. 5, Table 1).

Table 1. Results of thermal analysis.

Sample	Temperature range									
	RT - 220°C		300 - 420°C		450 - 550°C		550 - 700°C		700 - 900°C	
	T_{infl} (°C)	ML (%)	T_{infl} (°C)	ML (%)	T_{infl} (°C)	ML (%)	T_{infl} (°C)	ML (%)	T_{infl} (°C)	ML (%)
A	43.0	0.54	-	-	467.9	0.66	577.7	1.08	703.5	2.11
	87.2	1.16					636.5	1.23		
	141.1	0.99								
B	48.4	0.26	382.2	0.62	462.6	0.39	698.4	0.92	802.5	7.98
	76.3	1.02								
	118.0	0.49								
	147.1	0.64								
	175.0	0.51								
	210.0	0.46								
C	47.8	0.15	324.2	0.19	-	-	-	-	857.1	28.9
	78.0	0.29								

T_{infl} – temperature of the point of inflection (°C).
ML – mass loss (%).

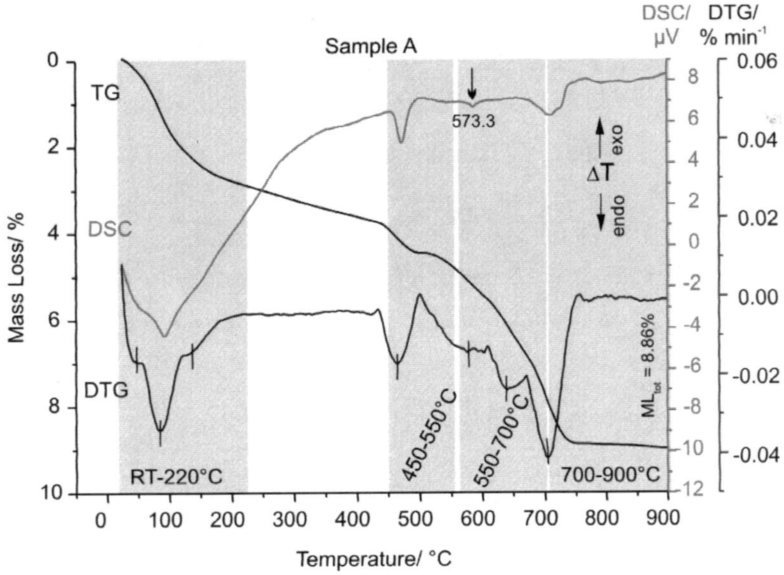

Figure 3. Thermal analysis of sample A.

300 - 420°C temperature range:

During the 300 - 420°C temperature range (with total ML from 0.14 up to 4.16%), an exothermic effect with temperature of the point of inflection (T_{infl}) at 382.2°C (sample B) and 324.2°C (sample C) with ML of 0.62 and 0.19%, respectively (Figs. 4, 5, Table 1). It is defined as organic components decomposition [20]. This effect was not registered for sample A (Fig. 3).

450 - 550°C temperature range:

In this temperature range, the portlandite dehydration occurs – at T_{infl} = 467.9°C for sample A and T_{infl} = 462.2°C (Sample B) (Figs. 3, 4, Table 1) [21].

550 - 700°C temperature range:

Two thermal reactions and effects occur in this range: (i) an endothermic effect at about 571°C without ML (Figs. 3-5), recognized as the α-β polymorph phase transition of quartz [20]; and (ii) endothermic reactions of hydraulic phases dehydroxylation [8, 20, 21], detected only for samples A and B (Figs. 3, 4, Table 1). The hydraulic component formation required chemically active phases such as clays, ceramic fragments, or volcanic ash [5, 6, 20]. Such components were not present in sample C.

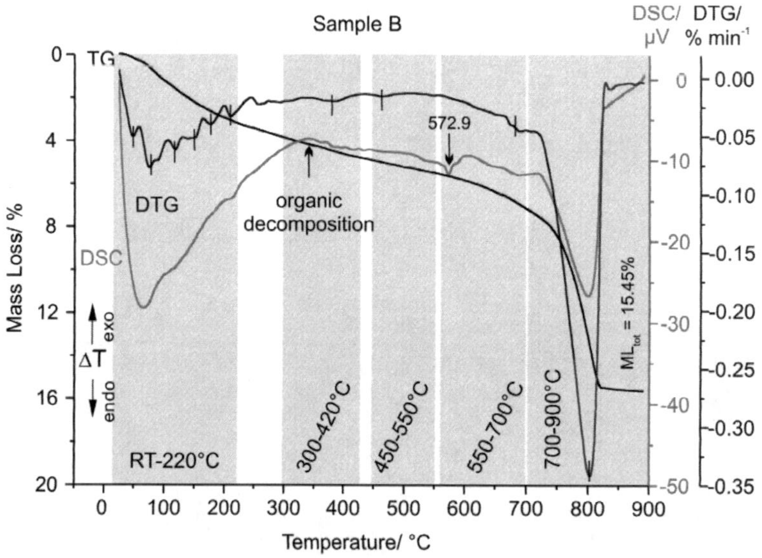

Figure 4. Thermal analysis of sample B.

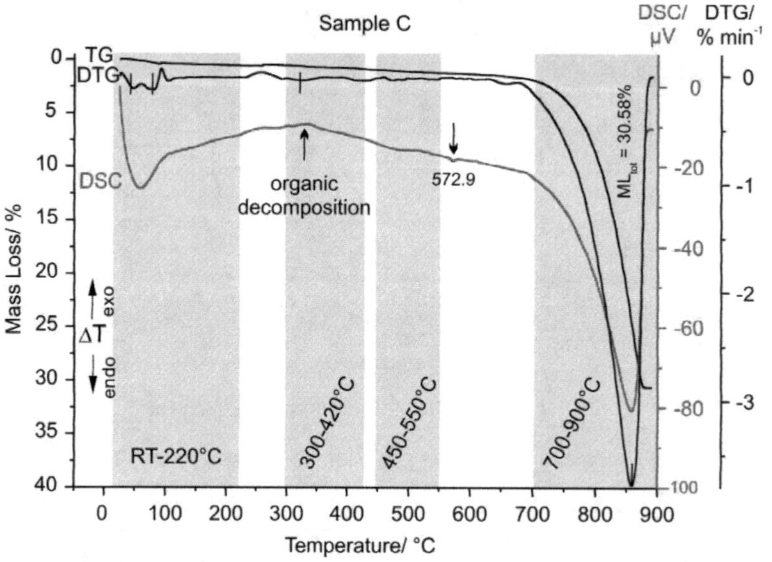

Figure 5. Thermal analysis of sample C.

700 - 900°C temperature range:

During this temperature range, a calcite decarbonation occurs [8, 9, 21] with ML of evolving CO_2 [23] (Table 1), registered for all studied samples. The ML of CO_2, measured for Roman mortars, was used to calculate the CaO and $CaCO_3$ content, and their B:A ratio [20, 24]. The results obtained were B:A = 1:4 for sample B and B:A = 2:1 for sample C.

The hydraulic properties of the mortars could be obtained by CO_2/OH ratio, calculated by ML in the 700 - 900°C temperature range and ML from structurally bonded water in hydraulic phases (i. e. ML in the 450 - 700°C temperature range) [25]. For sample A, the CO_2/OH ratio = 0.7, whereas for sample B = 6. Thus sample A was defined as pozzolanic mortar with CO2/OH ratio value of less than 3, sample B – as hydraulic lime mortar with CO_2/OH ratio value in the range of 4 – 10, and sample C – as lime mortar, having CO_2/OH ratio over 10 [25].

The results of the research characterize the investigated samples. Samples A and B show the same hygroscopic properties. Sample C was without any hygroscopic properties. The defined B:A ratios of Roman samples present their homogeneity. An increasing amount of aggregates decreases mortar's homogeneity and decreases its structure compactness. Sample B (with B:A ratio = 1:4) shows lower homogeneity in comparison to sample C (with B:A ratio = 2:1). The decrease in the structure compactness leads to a decrease in decarbonation temperature for sample B (Table 1). The described characteristics of sample B probably allow the existence of un-carbonated portlandite for a long time. The B:A ratio of sample A was lower than that of sample B, and the decarbonation temperature was lower than that of sample B (Table 1). Probably, this was a result of (i) lower structure compactness due to lower aging time in comparison to Roman mortar and (ii) the lower quantity of $CaCO_3$ in comparison to the number of hydraulic phases in the binder. The calculation of the $CaCO_3$ percentage was obtained from the ML of CO_2 (Table 1) [20, 24]. The $CaCO_3$ calculations show 4.8% in sample A, in sample B - 18.0%, and in sample C – 65.5%.

The excellent hydraulic properties of sample A defined it as pozzolanic mortar. Sample B also has hydraulic properties and which defined as hydraulic lime mortar. The use of samples A and B in a humid environment is suitable because they harden in contact with water. Sample C was impermeable for water, staying masonry dry. These results suggest that studied samples are replaceable in a dry environment, and sample C was not suitable for a wet environment. Nevertheless, the described above characteristics are not enough to determine the possibility of modern mortar application for the restoration and conservation of archaeological sites. To achieve compatible mortar composition with the ancient also to account should be taken the sulphate phase's content of the mortar. The introduction of sulphate phases to mortar, without it containing such phases initially, could accelerate its destruction due to sulphate/thaumasite attacks [26]. In both studied archaeological samples no sulphate phases have been detected. Therefore sample A, with such phases, could not be used to restore/conserve samples B and C.

4. Conclusion

Modern mortar production aims to achieve composites with uptake physical, chemical, and mechanical performance with architecture and construction applications. Additionally, manufacturing should be energy efficiencies and carbon emissions reduced. In the past, mortar production aimed to obtain bonding composites using natural resources available in the habitats. Accordingly, when carrying out conservation and restoration activities, it is important to follow the recipes applied to the specific place in the past to create a mortar compatible with the original.

Studies have shown similarities between modern (sample A) and ancient mortars (sample B) - both have hydraulic and hygroscopic properties and approximate close values of B:A ratios. Differences also have been identified. They reject the possibility of modern mortars' use for restoration and conservation due to sulfate phase presence, which over time would lead to sulphate and thaumasite attacks on studied ancient mortars. The use of modern mortar (sample A) instead of sample C would increase the moisture of the ancient mortar and, accordingly, weathering intensification.

The most suitable approach to masonry conservation and restoration is to analyze and characterize the ancient mortar and prepare a new one, which corresponds as much as possible to the original in composition, ratio B:A, hygroscopic and hydraulic properties. In modern cocciopesto preparation, it is necessary to monitor the firing temperature of the ceramic fragments used. When the original cocciopesto contains hydraulic phases, the ceramic fragments used must be fired at a temperature up to 800 °C and vice versa.

Acknowledgments

This work was funded by the National Science Fund of Bulgaria (grant KP-06-N39/9; all authors).

References

[1] Kechagia P, Koutroumpi D, Bartzas G, Peppas A, Samouhos M, Deligiannis S and Tsakiridis P E 2020 *Sci Total Environ* 143224, https://doi.org/10.1016/j.scitotenv.2020.143224
[2] Andrew R and Global M 2018 *Earth Syst Sc. Data Discuss* **10** 195–217 https://doi.org/10.5194/essd-10-195-2018.
[3] Wei G, Fang S and Yao Z 2021 *Arhaeometry* https://doi.org/10.1111/arcm.12601
[4] Goffer Z 2007 *Archaeological chemistry* (New Jersey: Wiley) p 602
[5] Klrca O 2005 Ancient binding materials, mortars and concrete technology: history and durability aspects *Structural Analysis of Historical Constructions, 1 Set: Possibilities of Numerical and Experimental Techniques*, ed P B Lourenco and P Roca (London: Taylor & Francis Group) pp 87-94
[6] Silva A, Wenk H R and Monteiro P J M 2005 *Thermochim Acta* **438** 35–40 https://doi.org/10.1016/j.tca.2005.03.003
[7] Böke H, Akkurt S, İpekoğlu B and Uğurlu E 2006 *Cement and Concrete Research* **36:6** 1115-1122 https://doi.org/10.1016/j.cemconres.2006.03.011
[8] Corti C, Rampazzi L, Bugini R, Sansonetti A, Biraghi M, Castelletti L, Nobile I and Orsenigo C 2013 *Thermochim Acta* **572** 71– 84 https://doi.org/10.1016/j.tca.2013.08.015
[9] Taylor H F W 1997 *Cement chemistry* (London: Thomas Telford Publishing) p 459
[10] EN 933-4:2008 Tests for geometrical properties of aggregates - Part 4: Determination of particle shape - Shape index
[11] EN 196-1:2006 Methods of testing cement – Part 1: Determination of strength
[12] Powder Diffraction File (PDF), copyright by International Centre for Diffraction Data (ICDD) Newtown Square; 12 Campus Blvd., Newtown Square, PA 19073-3273
[13] Chukanov N V 2014 *Infrared spectra of mineral species: Extended library* Dordrecht Heidelberg New York London: Springer) p 1726
[14] Stanienda K J 2016 *Carbonates Evaporites* https://doi.org/10.1007/s13146-016-0297-2
[15] Singha M and Singh L 2016 *Indian Journal of Pure & Applied Physics* **54** 116-122
[16] Theodosoglou E, Koroneos A, Soldatos T, Zorba T and Paraskevopoulos K M 2010 *Bull Geol Soc Greece* https://doi.org/10.12681/bgsg.11681
[17] Zviagina B B, Drits V A and Dorzhieva O V 2020 *Minerals* **10** 153 https://doi.org/10.3390/min10020153
[18] Genestar C, Pons C and Más A 2006 *Anal Chim Acta* **557** 373–379 https://doi.org/10.1016/j.aca.2005.10.058
[19] Borsoi G, Santos Silva A, Menezes P, Candeias A and Mirão J 2019 *Constr Build Mater* **204** 597–608 https://doi.org/10.1016/j.conbuildmat.2019.01.233
[20] Moropoulou A, Bakolas A and Bisbikou K 1995 *Thermochim Acta* **269/270** 779-795 https://doi.org/10.1016/0040-6031(95)02571-5
[21] Fares H, Rémond S, Noumowé A and De Schutter G. High 2011 Temperature Behavior of Self-Compacting Concretes *Self-Compacting Concrete* ed A. Loukili (UK, USA: ISTE Ltd and Wiley) chapter 5 pp 215-258
[22] Földvári M 2011 Handbook of the thermogravimetric system of minerals and its use in geological

practice *Occasional Papers of the Geological Institute of Hungary* (Vol. 213) ed G Maros (Budapest: Geological Institute of Hungary) p 180 https://doi.org/10.1556/ceugeol.56.2013.4.6
[23] Curulli A, Montesperelli G, Ronca S, Cavalagli N, Ubertini F, Padeletti G and Vecchio Ciprioti S V. 2020 *J Therm Anal Calorim* https://doi.org/10.1007/s10973-020-09937-9
[24] Cultrone G, Arizzi A, Sebastián E and Rodriguez-Navarro C 2008 *Environ Geol* **56** 741–752 https://doi.org/10.1007/s00254-008-1379-9
[25] Iordanidis A, Garcia-Guinea J, Strati A, Gkimourtzina A and Papoulidou A 2011 *J Therm Anal Calorim* **103** 577–586 https://doi.org/10.1007/s10973-010-1055-7
[26] van Hees R P J, Wijffels T J and van der Klugt L J A R 2003 *Cement & Concrete Composites* **25** 1165–1171 https://doi.org/.1016/j.cemconcomp.2003.07.003

IOP Conf. Series: Materials Science and Engineering 1276 (2023) 012002 doi:10.1088/1757-899X/1276/1/012002

End-of-life options for ceramic masonry units – environmental and performance-based perspective

Yana D Kancheva[1], Roumiana A Zaharieva[2]

[1]Dept. of Applied Geodesy, Faculty of Geodesy, University of Architecture, Civil Engineering and Geodesy (UACEG), 1 Hristo Smirnenski Blvd., 1046 Sofia, Bulgaria
[2]Dept. of Building Materials, Faculty of Civil Engineering, University of Architecture, Civil Engineering and Geodesy (UACEG), 1 Hristo Smirnenski Blvd., 1046 Sofia, Bulgaria

E-mail: kancheva_fgs@uacg.bg

Abstract. Ceramic bricks are among the most used building material throughout human history for their excellent technical properties, abundance of raw materials and easy production. Huge quantity of brick waste is generated from demolition works at the end-of-life stage of buildings. Several scenarios within waste management hierarchy can be considered: a) preparing for reuse of bricks into masonry units for similar purposes; b) mechanical recycling (crushing, screening) to crushed stone for various applications –in backfilling, pavement, drainage works, which are examples of down-cycling or as concrete aggregate and as a component in blended binders which are examples of upcycling; c) disposal at inert waste landfills or used as a separation layer at municipal waste landfills.
This study performs a life cycle assessment (LCA) of these three possible routes for ceramic waste and takes into account the relevant processes. The LCA results are calculated for the compulsory environmental indicators according to the methodology of EN 15804 along with the additional indicators assessing toxicity. The outcome of the LCA confirms that the most environmentally beneficial scenario is the reuse of bricks as it requires minimal activities for preparing. Technical feasibility issues and costs impact, associated to the three scenarios are also analysed. The conclusion is that an optimised and sustainable approach of bricks end-of-life shall be applied in order to boost the circularity in construction sector.

1. Introduction

Construction and demolition waste (CDW) forms more than one third of the total waste in EU [1]. According to the National Statistical Institute (NSI) [2] the total generated waste from construction in Bulgaria is 1773534 tons. According to the national legislation, at least 70% of the non-hazardous CDW must be directed for material recycling or other recovery [3], which complies to the provisions of the Waste Framework Directive (WFD) [4]. The dominating practices for utilization of CDW include backfilling, applications in road construction or as mineral aggregate for concrete. They have been explored in a number of studies [5]–[11] showing that CDW can successfully replace virgin raw materials. However, a significant amount of the CDW is still landfilled. When there is poor control on the CDW management, this leads to formation of unregulated landfills, pollution of soil and water areas as well lost benefits from the recovery of material [12], [13].

Waste from ceramic masonry units (WCM) is one of the CDW groups that can help for the achieving of the 70% target. In 2009, the share of ceramic CDW in Bulgaria was 14% of the total CDW but this share is expected to grow up to at least 25%, as most of the old buildings are made with ceramic masonry units [14]. Diverting this waste from landfilling can happen by utilization as secondary material and WCM has also a unique potential for reuse, provided the technical properties are well preserved during use.

Life cycle assessment (LCA) approach has been widely used over the last three decades to quantify various environmental impacts [15]. The LCA gives an overview of the impacts throughout the whole (or part of) the life cycle of a product. Though studies on production of ceramics bricks are available in the scientific literature [16]–[19], the end-of-life options are scarcely researched [12]. The aim of this paper is to propose the best recovery options of waste ceramic bricks by assessing the potential environmental impacts from several waste management scenarios and the technical performances of recovered material. The results are discussed in terms of significant contributions to the most relevant environmental indicators and of the potential prevention (saving) of impacts from acquisition and processing of virgin materials.

2. CDW from ceramic masonry units – characterisation and current waste management trends

2.1. Ceramic masonry units - general description

Ceramic materials, especially masonry bricks, are among the most used and oldest known building materials [20]. Traditional ceramic bricks are made from clay and have very good physical and mechanical properties. Depending on the type of the bricks, they can also provide certain thermal insulation quality. The uses of bricks in construction include multiple applications such as internal and external load-bearing walls, partition walls, chimneys, decks and arches, etc. Manufacturing results in durable products [21] but consumes large quantities of non-renewable resources [22]. High-temperature firing is part of the production process (i.e. intensive energy demand [23]) which in turn leads to significant emissions of greenhouse gases [24].

According to their gross dry density – smaller or bigger than 1000 km/m^3, the ceramic masonry units are divided into two groups: high density units and low-density units [25]. Bricks are characterized by a huge variety - with vertical or horizontal holes or without holes, having different shape and dimensions. In general, high-density bricks (with limited holes percentage) exhibit superior mechanical properties and offer many options at the end of life. In the oldest Bulgarian buildings, subject of demolition nowadays, the share of those dense bricks is predominant, because the production and the use of hollow bricks and big-size blocks started only in 70s of the past century.
Bricks have various applications as masonry units - for internal use as partition walls, or external use, protected by renders or exposed to atmospheric impacts, for load bearing walls or as self-bearing walls. Each use of bricks requires relevant properties. So, the type, size and properties of the bricks determine different options for their end-of-life treatment, which are briefly discussed below.

2.2. Ceramic waste

According to the Ordinance on construction waste management and on the use of recycled construction materials [26] bricks are classified as mineral waste which is "waste generated by the construction or removal of construction works consisting mainly of mineral materials such as bricks, concrete, mortars, natural stone, sand, ceramic building materials, concrete blocks and/or aerated concrete blocks, etc.". Non-hazardous ceramic masonry waste from CDW is classified in the European Waste Code system under code 17 01 02 (bricks). When the bricks waste contains a significant amount of concrete, tiles and other ceramics, it can be classified as 17 01 07. In the case of pollution of bricks by hazardous substances, e.g. during the use stage in industrial buildings or by some coatings, the bricks waste shall be classified as per code 17 01 06 (17 01 06 mixtures of, or separate fractions of concrete, bricks, tiles and ceramics containing hazardous substances). Usually, the brick masonry CDW belong to the first type of waste (17 01 02). This type is under consideration in that article, but a proper waste audit shall be performed prior to the demolition works.

2.3. Waste management practices for ceramic masonry bricks and blocks

2.3.1. Reuse. According to [14], the bricks that are suitable for reuse are usually of high density and higher strength (7.5-20 MPa) classes. These include single dense bricks and hollow bricks with vertical holes with smaller dimensions. Blocks with horizontal holes have smaller strengths (less than 5 MPa) and their integrity depends on the type of mortar and demolition method. In order to preserve the initial shape and size of the bricks, manual or mechanical dismantling is recommended when the bricks are intended for reuse. Machine demolition would lead to improved work efficiency but also to a high percentage of broken bricks. Before reuse, the bricks need to undergo a preparation process that includes activities that guarantee the purity and suitability for further use. The process starts with sorting (conglomerate crushing and block separation), cleaning (removal of adhered mortar, fine cleaning, emery rock treatment, and, if necessary, treatment with weak solution of acetic acid), characterisation (separation into brick lots and determination of performance characteristics related to the intended use) and preparation of the product documentation – figure 1. Bricks after preparing for reuse can be used in non-load bearing masonry walls, load-bearing walls, masonry with thermal insulating or fire-resistant functions, decorative facing masonry. It shall be noted the in cases, where reclaimed ceramic bricks are to be used as external masonry, frost resistance requirements may apply and they need to be protected from water saturation by cornices or appropriate covering layers. The preparing for reuse activities are covered by the recovery operation R12 or R5 – table 1.

Table 1. Recovery operations according to provisions of the Waste Framework Directive [4].

Operation code	Activity	Recovery operations allowed by Bulgarian CDW management legislation [27]
R 5	Recycling/reclamation of other inorganic materials	Preparing for reuse and Recycling into crushed fractions for various construction purposes
R 10	Land treatment resulting in benefit to agriculture or ecological improvement	Backfilling where the waste is used for reclamation purposes in excavated areas or for engineering purposes in landscaping
R 12	Exchange of wastes for submission to any of the operations numbered R1 to R11	Preparing for reuse and preparing for other recovery operations (i.e. crushing)
D 1	Deposit into or onto land, e.g. landfill	Disposal

Figure 1. Preparing for reuse: conglomerate separation, mortar removal, sorting, testing. (Photos by: Roumiana Zaharieva).

2.3.2. Recycling (recovery operation R5, table 1) of ceramic bricks can be done as lightweight aggregate for concrete and as a component in blended binders, or as fractioned crushed stone and sand for pavement and drainage works The first two processes can be considered as upcycling option, while the last ones are examples of downcycling. Partial replacement of natural aggregates in concrete is a promising scenario for utilization of crushed brick waste. The lower density and lower thermal conductivity of recycled brick aggregate encourages its use in lightweight concretes and concretes with improved thermal insulation capacity. However, the effects on the properties of concrete could be diverse. Some studies show that recycled ceramic aggregates have lower density, higher porosity and water absorption, lower strength [26]. These factors may influence the workability of the fresh concrete mix and the behaviour of hardened concrete (e.g. shrinkage, creep and durability).

2.3.3. Backfilling According to [28], "'backfilling' is a recovery operation where suitable waste is used for reclamation purposes in excavated areas or for engineering purposes in landscaping and where the waste is a substitute for non-waste materials". Though it is debatable whether backfilling by a crushed stone can be considered an actual recycling, the backfilling is included in the WFD [4] for achieving the target of 70% for re-use and recycling of CDW. This means that backfilling can be considered as a low-quality recovery as it replaces other materials that would be used for this purpose (e.g. natural crushed stone or rough soils), but does not meet the criteria for recycling Since WCM of codes 17 01 02 and 17 01 07 is inert, it can be successfully used for backfilling. Those activities are covered by operations R12 (the crushing of bricks) and R10 (table 1)

2.3.4. Landfill. According to the Bulgarian NSI [2], in 2020, around 12% of the CDW in Bulgaria has been landfilled. Mineral CDW of code 17 01 07 is inert, so it can be landfilled without potential hazards for the environment. However, with the vast quantities of generated CDW occupying significant area and not used recovery potential ot WCM, this is an option that needs to be avoided. Landfills for inert waste are limited and the mineral CDW are directed to landfills for municipal waste. Very often the mineral CDW is used as a separation layer and in case the landfill operator possesses permit for recovery operation R10, those quantities of waste are reported as "recovered", not as disposed, because they are used in backfilling operation.

3. Life cycle assessment (LCA) of the possible scenarios

3.1. Goal and scope of the study

The goal of the life cycle assessment is to calculate the end-of-life environmental impacts of ceramic brick waste from public and residential buildings considering several scenarios: preparing for reuse, recycling as lightweight aggregate for concrete, backfilling and landfill. The potential savings of emissions from substitution of virgin raw materials are also considered for a more thorough understanding of the benefits. The scope includes modules C2-C4 as per EN 15804 [29]: transport to recycling facility and landfill (C2), waste processing (C3) and disposal (C4). The same demolition method (module C1) is assumed and the influence of this process is not accessed. Benefits and loads (module D) are calculated based on average production processes for bricks and crushed stone as concrete aggregate and backfilling material.

3.2. System boundaries

This study considers high-density single brick waste from public and residential buildings. the functional referent unit is 1 ton of dense brick waste with dimensions 25x12x6.5 cm. Though the impact of the demolition method is not included, it is assumed that the ceramic waste is separated from concrete and impurities are removed prior to demolition of the wall (selective demolition). The boundary of the system with the previous system is set where the building ceases to fulfil its service and is subjected to demolition; the boundary with the successive system is set where the waste reaches the end of waste status and is transformed into a material that is ready for the next application (system) or is deposited at a landfill. Since the recovery processes are situated at that boundary, this would suggest dividing the burdens between the predecessor and descendant system. Regarding such allocation, ISO 14044 [30] states that impacts from one process to more than one systems (allocation) should be avoided by either dividing the process into sub-processes, or by expanding the system boundary to include additional functions. Available studies suggest varying approaches [31], [32] which can be grouped into the following options: 0:100 (substitution approach), 100:0 (cut-off) and

50:50 (equal share) [32] . The decision on whether to use the substitution approach (crediting material substitution) or the cut-off approach (no credits for the recyclable material as they are considered free of burdens), or equal distribution of the impacts is based on the specifics of the study and its goal.

For our case study, the approach of EN 15978 [33] and EN 15804 [29] is adopted. The allocation of end-of-life impacts is done by expanding the system and including the so-called module D (Benefits and loads beyond the system boundary). The purpose of module D is to evaluate the potential for reuse, recycling and recovery. This means that credits are assigned for the end-of-life operations (preparing for reuse, recycling, etc.) considering the secondary use of the bricks and the most likely material/process that is substituted, e.g. bricks produced from virgin clay materials or processing virgin rocks for backfilling use. The impacts from use of virgin materials in the manufacturing of a product are associated with the first use (cut-off approach) [34], i.e. secondary materials do not bring credits.

3.2.1. Scenario 1. Preparing for reuse

Scenario 1 considers that the brick wall would be demolished manually or mechanically in a way that preserves the integrity of most of the bricks, so that they can be reused. An investigation of technical properties of waste bricks having bulk density over 1000 kg/m^3 has been performed within [18] and it has been found that if the masonry wall is protected from water absorption and freeze-thaw attacks, the performance would be sufficient for secondary use. It can be expected that a certain share of the bricks would be broken during the demolition process. We assume this share to be 30% (i.e. 300 kg from 1 ton of WCM). The amount of cement-sand mortar on the remaining bricks is assumed to be on average around 20% by mass, i.e., 140 kg. So, the actual mass of bricks that is possible to be reused is around 560 kg from 1 ton CBW, the rest is usually send to landfill where it is actually used as material for backfilling or as separating mineral layer of municipal waste. Treatment operations include additional separation of bricks forming conglomerates of demolished blocks and additional cleaning of the masonry mortar, renders and plasters. All operations are performed manually, so no machinery associated emissions are considered. An average distance of 15 km is assumed for the transport of secondary bricks to the place of their reuse.

3.2.2. Scenario 2. Recycling

Recycling of bricks is considered with two options:

a) Recycling as crushed stone aggregate

This scenario includes crushing of the masonry wall to two major fractions of particle sizes: 25% by mass is assumed to be crushed sand 0/4 mm and 75% is the share of the coarse aggregate 4/22.4 mm. The crushed sand consists of a large share of mortar (around 80%) and much less bricks fraction. The coarse aggregate includes around 80% bricks and 20% mortar. The amounts for recycling as aggregates are presented in table 2 and the total amount derived from 1 ton is 650 kg.

Table 2. Fraction quantities for recycling as lightweight concrete aggregate.

Fraction	Composition	Dedicated use
0/4 mm (25% by mass)	Crushed bricks: 50 kg Mortar: 200 kg	For fine concrete aggregate Directed for backfilling
4/22.4 mm (75% by mass)	Crushed bricks: 600 kg Mortar: 150 kg	For coarse concrete aggregate Directed for backfilling

The ceramic fractions are suitable for substitution of lightweight aggregates, e.g. produced by expanding of clay materials (ceramsite, agloporite). Transportation distance to recycling facility is assumed to be 15 km.

b) Recovery as backfilling material

In this scenario, the whole waste quantity (1 ton of ceramic bricks and cement-sand mortar) is sent for backfilling after a treatment process of rough crushing without sieving of the debris to all-in fraction (0/63 mm). The crushing is usually performed in a mobile treatment plant at the demolition site. The distance to the backfilling site is assumed to be 15 km. Savings of emissions are calculated assuming the substituted material is natural crushed stone fraction 0/63 mm.

3.2.3. Scenario 3. Disposal

The whole amount of brick waste (1 ton) is sent to a landfill as inert waste, without being subjected to any preliminary processing. An average distance of 35 km corresponding to the Bulgarian situation is assumed to the landfilling site.

3.3. Life cycle inventory (LCI) and modelling

The scenarios are modelled with referent datasets from the ecoinvent database regarding the use of fuels and energy for the treatment operations. Savings from substitution of primary bricks production are calculated also with average referent dataset from the database. For substitution of virgin backfilling material, mining of crushed limestone is taken into account. The data is not geographically specific but represents the average current waste treatment and production technologies. Since the end-of-life processed are in general the same irrespective of the country, the technological and temporal coverage are more important in this case. Some variation can be expected in the results for the impacts savings from production of bricks with virgin clay as this is related to national electricity mixes, but this is not explored here. The modelling is done via the openLCA software by GreenDelta GmbH [35] and the ecoinvent database [36] using the impact assessment methods specified in EN 15804. Selected indicators are presented in the results section including total global warming potential (GWP), ozone depletion potential (ODP), photochemical ozone creation potential (POCP), abiotic depletion potential for elements (ADPE), use of non-renewable and renewable energy.

3.4. LCA results

It was mentioned that in some of the scenarios not all of the waste actually ends at the dedicated purpose as the ceramic CDW contains cement-sand mortar adhered to the bricks. So, at first, a calculation was done of the impacts in case 1 ton of brick waste is fully utilised as per single operation, i.e. preparing for reuse, recycling, backfilling and disposal. The results are presented into the following table:

Table 3. Impacts from waste processing of 1 ton brick waste.

Indicator	Unit	Preparing for reuse	Recycling as lightweight aggregate	Recovery as backfilling	Landfilling
Global warming potential	kg CO_2-eq.	2.611E+00	2.868E+00	1.215E+00	1.764E+01
ADPE	kg Sb-eq.	6.782E-05	6.821E-05	2.313E-05	2.800E-04
ODP	kg CFC11-eq.	5.962E-07	6.517E-07	2.732E-07	4.881E-06
POCP	kg NMVOC-eq.	4.490E-03	5.120E-03	2.340E-03	5.972E-02
Non-renewable energy	MJ-eq.	3.977E+01	4.329E+01	1.798E+01	3.317E+02
Renewable energy	MJ-eq.	5.418E-01	5.610E-01	2.063E-01	3.561E+00

Based on these values the intended scenarios are modelled and calculated and potential savings are added from substitution of virgin materials. Figure 2 presents the results on the assessed indicators for all scenarios for the waste treatment operations (positive values) and impact savings from substitution of primary materials (negative values).

The highest emissions from waste treatment are released at the landfilling scenario as no savings are accounted. The reuse scenario produces also considerable emissions as practically 44% of the waste (broken bricks and mortar) is sent to landfill. Scenarios 2a and 2b have insignificant release of emissions.

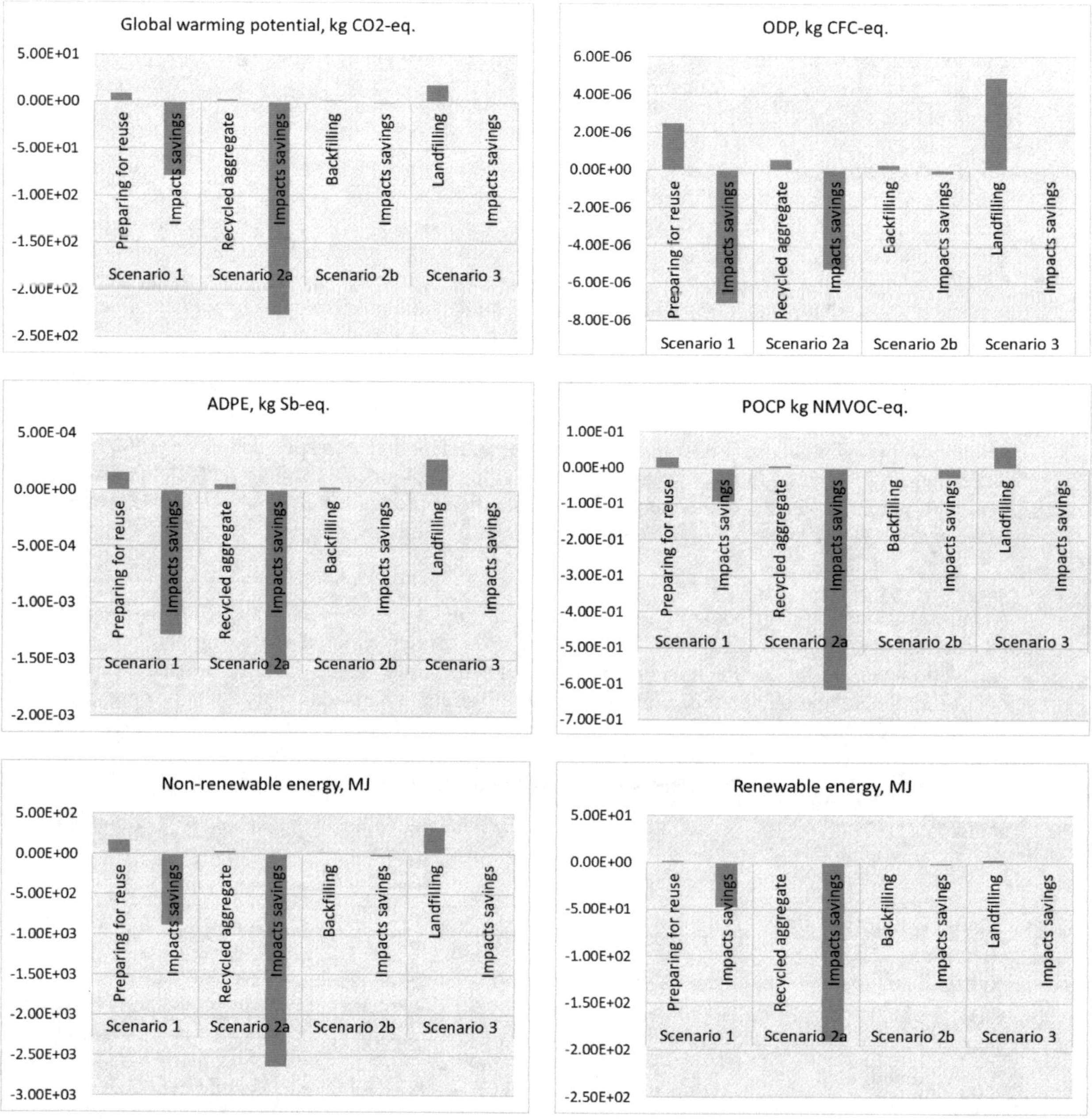

Figure 2. Impacts results for the scenarios under study (negative values are savings).

The largest savings are observed when bricks are recycled as lightweight concrete aggregates. This is due to the fact that the substituted materials (expanded clay materials) are manufactured by firing the clay to around 1100-1300 °C in predominantly oil-fuelled furnaces [37], so the impacts of the primary production are high. Reusing bricks can also significantly reduce the environmental impacts throughout the life cycle of ceramic bricks. The manufacturing of bricks uses significant amounts of natural gas for the firing of the bricks to temperatures up to 1100°C. ODP is the indicator that is sensitive to natural gas use, so the largest savings are achieved for the reuse scenario.

Scenarios 1 and 2a have high saving potentials for the depletion of mineral resources (ADPE), i.e. from resource efficiency standpoints these scenarios are most beneficial.

Additionally, three eutrophication potentials (freshwater, marine and terrestrial) and abiotic depletion potential of fossil resources were assessed and the trends are identical to most of the indicators (excluding ODP). The highest impacts are found to be releases for Scenario 3 (landfill) and the least emissions are formed, if bricks are recycled. The highest savings can be achieved with Scenario 2a (recycling as crushed stone).

4. Conclusion

Ceramic bricks form a considerable waste flow within the CDW that is expected to increase as there are many old buildings made with brick as primary bearing and non-bearing material. If the bricks have been protected from water absorption and freeze-thaw cycles, they are expected to preserve most of the initial technical properties. Bricks are one of the few building materials that can be directly reused after small interventions and conformity assessment and this potential is worth exploring given the large amounts of waste generated.

Proper demolition procedures (selective demolition, careful handling, storage and processing of bricks) can be planned to ensure that a great deal of the brick waste would be diverted from landfill and send for reuse, recycling or another type of recovery. There are diverse options for recycling of WCM but the most attractive, in terms of preserving the value of the waste, is to recycle the bricks as a lightweight aggregate for concrete, while the most used is the recycling into all-in fraction for backfilling.

Four end-of-life scenarios were assessed to compare the environmental impacts from waste treatment and to predict possible savings from utilisation of brick waste: reuse, recycling as concrete aggregate, recovery as backfilling material and landfill. The lowest emissions from waste treatment operations are released when the bricks are landfilled or reused. In terms of environmental savings, scenario 1 (reuse) and scenario 2a (recycling as lightweight concrete aggregates) are distinctively the most beneficial, because both prevent mining of natural resources and emissions from factory processes (clay firing). The highest savings are observed for the scenario 2a where the recycled brick waste substitutes expanded clay aggregates (ceramsite, agrloporite). The high temperature required for the expanding process consume a lot of energy resources (fossil fuels), so, avoiding this process brings significant emission reductions. The bricks reuse is the most beneficial also in terms of preserving the value of the construction products.

The study shows that proper treatment of CDW at the end-of-life stage helps for a more efficient resource use and can contribute to closing the loop and makes a step towards a circular economy in construction sector.

Acknowledgments

The research is performed within the Project BG05M2OP001-1.002-0019:„Clean technologies for sustainable environment – water, waste, energy for circular economy“ (Clean&Circle), for development of a Centre of Competence, financed by the Operational programme "Science and Education for Smart Growth" 2014-2020, co-funded by the EU through the European structural and investment funds.

References

[1] European Commission Construction and demolition waste. https://ec.europa.eu/environment/topics/waste-and-recycling/construction-and-demolition-waste_en
[2] National statistical institute Waste from economic activity https://nsi.bg/en/content/2556/waste-economic-activity (in Bulgarian)
[3] National Waste Management Plan 2021-2028 (in Bulgarian)
[4] European Commission Directive (EU) 2018/851 of the European Parliament and of the Council of 30 May 2018 amending Directive 2008/98/EC on waste
[5] Oluleye B I, Chan D W M, Saka A B and Olawumi T O 2022 Circular economy research on building construction and demolition waste: A review of current trends and future research directions *Journal of Cleaner Production* vol. 357 p. 131927
[6] Zegardło B 2022 Heat-resistant concretes containing waste carbon fibers from the sailing industry and recycled ceramic aggregates *Case Studies in Construction Materials* vol. 16 p. e01084
[7] Garzón E, Martínez S, Pérez-Villarrejo L and Sánchez-Soto P J 2022 Assessment of construction and demolition wastes (CDWs) as raw materials for the manufacture of low-strength concrete and bases and sub-bases of roads *Materials Letters* vol. 320 p. 132343
[8] Suchithra S, Sowmiya M and Pavithran T 2022 Effect of ceramic tile waste on strength parameters of concrete-a review *Materials Today: Proceedings* vol. 65 part 2 pp. 975-982
[9] Jain P, Gupta R and Chaudhary S 2022 A literature review on the effect of using ceramic waste as supplementary cementitious material in cement composites on workability and compressive strength *Materials Today: Proceedings* vol. 65 part 2 pp. 871-876
[10] Zhu L and Zhu Z 2020 Reuse of Clay Brick Waste in Mortar and Concrete *Advances in Materials Science and Engineering* vol. 2020 pp. 1–11
[11] D'Angelo G et al. 2021 Crushed Bricks: Demolition Waste as a Sustainable Raw Material for Geopolymers *Sustainability* vol. 13 no. 14 p. 7572
[12] Antunes A et al. 2021 Environmental Impacts and Benefits of the End-of-Life of Building Materials: Database to Support Decision Making and Contribute to Circularity *Sustainability* vol. 13 no. 22 p. 12659
[13] Butera S, Christensen T H and Astrup T F 2015 Life cycle assessment of construction and demolition waste management *Waste Management* vol. 44 pp. 196–205
[14] RAZ Consult EOOD 2018 Methodology for preparing for reuse of clay masonry products from construction and demolition waste and criteria for their submission for construction works (in Bulgarian)
[15] Zanni S, Simion I M, Gavrilescu M and Bonoli A 2018 Life Cycle Assessment Applied to Circular Designed Construction Materials *Procedia CIRP* vol. 69 pp. 154–159
[16] Ramos Huarachi D A, Gonçalves G, de Francisco A C, Canteri M H G and Piekarski C M 2020 Life cycle assessment of traditional and alternative bricks: A review *Environmental Impact Assessment Review* vol. 80 p. 106335
[17] Condeixa K, Haddad A and Boer D 2014 Life Cycle Impact Assessment of masonry system as inner walls: A case study in Brazil *Construction and Building Materials* vol. 70 pp. 141–147
[18] Lozano-Miralles J, Hermoso-Orzáez M, Martínez-García C and Rojas-Sola J 2018 Comparative Study on the Environmental Impact of Traditional Clay Bricks Mixed with Organic Waste Using Life Cycle Analysis *Sustainability* vol. 10 no. 8 p. 2917
[19] Maia de Souza D et al. 2016 Comparative life cycle assessment of ceramic brick, concrete brick and cast-in-place reinforced concrete exterior walls *Journal of Cleaner Production* vol. 137 pp. 70–82
[20] Murmu A L and Patel A 2018 Towards sustainable bricks production: An overview *Construction and Building Materials* vol. 165 pp. 112–125

[21] Brick Development Association 2020 Clay Brick: End of Life Cycle www.brick.org.uk
[22] Correia S L, Curto K A S, Hotza D and Segadães A M 2005 Clays from Southern Brazil: Physical, Chemical and Mineralogical Characterization *Materials Science Forum* vol. 498–499 pp. 447–452
[23] Galan-Marin C, Rivera-Gomez C and Garcia-Martinez A 2016 Use of Natural-Fiber Bio-Composites in Construction versus Traditional Solutions: Operational and Embodied Energy Assessment *Materials* vol. 9 no. 6 p. 465
[24] Berge B 2007 *Ecology of Building Materials* (Routledge Taylor & Francis Group)
[25] European Committee for Standardization (CEN) EN 771-1:2011 Specification for masonry units - Part 1: Clay masonry units
[26] Ordinance on construction waste management and on the use of recycled construction materials (In Bulgarian: Наредба за управление на строителните отпадъци и за влагане на рециклирани строителни материали (НУСОВРСМ)), last amendment: State Gazette issue 98 2017 (in Bulgarian).
[27] Ministry of Environment and Water 2015 Practical guidelines for application of the legislation related to the management of construction waste and recycled construction waste products in the Republic of Bulgaria (in Bulgarian)
[28] European Commission Commission Decision 2011/753/EU of 18 November 2011 establishing rules and calculation methods for verifying compliance with the targets set in Article 11(2) of Directive 2008/98/EC of the European Parliament and of the Council https://eur-lex.europa.eu/legal-content/EN/TXT/PDF/?uri=CELEX:32011D0753&from=EN
[29] European Committee for Standardization (CEN) EN 15804+A2:2019 - Sustainability of construction works - Environmental product declarations - Core rules for the product category of construction products
[30] International Organization for Standardization ISO 14044:2006 Environmental Management-Life Cycle Assessment-Requirements and Guidelines
[31] van der Harst E, Potting J and Kroeze C 2016 Comparison of different methods to include recycling in LCAs of aluminium cans and disposable polystyrene cups *Waste Management* vol. 48 pp. 565–583
[32] Allacker K et al. 2014 Allocation solutions for secondary material production and end of life recovery: Proposals for product policy initiatives *Resources, Conservation and Recycling* vol. 88 pp. 1–12
[33] European Committee for Standardization (CEN) EN 15978:2011 Sustainability of construction works - Assessment of environmental performance of buildings - Calculation method
[34] Malabi Eberhardt L C, van Stijn A, Nygaard Rasmussen F, Birkved M and Birgisdottir H 2020 Development of a Life Cycle Assessment Allocation Approach for Circular Economy in the Built Environment *Sustainability* vol. 12 no. 22 p. 9579
[35] GreenDelta GmbH https://www.greendelta.com
[36] Ecoinvent Association Ecoinvent v.3.8 database
[37] Szczygielski W 2021 Clay raw materials for lightweight aggregate production: General information and occurrence http://geoportal.pgi.gov.pl/surowce/skalne/ilaste_kruszywo

Impact of recycled aggregate properties on concrete performance

R Zaharieva[1] and G Dimitrov[1]

[1]University of Architecture, Civil Engineering and Geodesy, Building Materials Department, Sofia, Bulgaria

zaharieva_fce@uacg.bg

Abstract. One of the ways for construction sector to contribute to sustainability is to reuse and to recycle the huge amount of construction and demolition waste (CDW) it generates. Recycled concrete aggregate (RCA) is recognized by EN 206 standard as a legitimate concrete ingredient, but their full potential is far to be used, because of several reasons: availability of natural resources, heterogeneity of CDW, technology and costs issues and mistrust in performances. In Bulgaria, in addition, the national amendment to EN 206 is very restrictive because of lack of national know-how. This study aims to fill the gap and assess the impact of RCA on concrete performance. RCA from three different CDW sources is tested. RCA differs in porosity, water absorption and mechanical characteristics. The performance of recycled aggregate concrete with 30% and 50% replacement of coarse aggregate by RCA is compared to that of natural aggregate concrete. Mix design with two cement quantities and two water/cement ratios are investigated. The higher water absorption of RCA is taken into consideration while producing fresh concrete. The results demonstrate that the quality of primary concrete has certain effect on strength characteristics, E-modulus and water permeability. However, RCA up to 30% does not modify significantly the behaviour of both fresh and hardened concrete. RCA of 50% does not worsen the properties of the concrete, either, but some reduction of performance can be observed. In conclusion, RCA presents a solution for both natural resources preservation and appropriate CDW management.

1. Introduction

The main challenge for recycled concrete aggregate (RCA) to become a trustworthy replacement for natural concrete aggregate (NA) is the quality assurance of the aggregate bearing in mind their peculiarities in heterogeneity, presence of old cement paste, porosity, water absorption and their mechanical characteristics. The construction and demolition (C&D) concrete debris may contain different materials – concrete, bitumen, masonry, timber, etc. which might be present in RCA as impurities and may lower the mechanical properties of recycled aggregate concrete (RAC) [1]. Thus, a thorough attention must be paid to the selective demolition, cleansing and sorting the C&D waste subjected to recycling. The lack of practice and the observation of the long-term effects on the structures furthermore contributes to the prevention of usage of RCA in many countries, including Bulgaria although RCA and their use in concrete have been regulated by BDS EN 12620:2002+A1:2008 [2] and BDS EN 206:2013+A2:2021 [3]. Initially, RAC exhibited lower strengths, higher creep and shrinkage and lower durability in comparison to natural aggregate concrete

(NAC) [4]. Further investigations had shown that controlling the primary concrete quality, thorough recycling and cleansing of the aggregate and proper mix design and curing the RAC have a beneficial impact on the mechanical properties of concrete [5]. However, some durability issues, such as higher carbonization rate continue to persist [6], [7]. RAC's higher porosity also determines its lower frost resistance and thus there are some concrete uses, e.g.in structures exposed to aggressive environment, where the total replacement of NA by RCA is not possible [8].

The present study aims to quantify the impact of primary concrete quality on the properties of RCA and performances of RAC when the provisions of [3] are respected. Three sources of RCA are characterized and compared to conventional NA. Two replacement rates of NA by RCA are applied – 30% and 50%. The main properties of RAC are evaluated: compression strength, E-modulus and split resistance. Water permeability is investigated as an integral parameter representative for the concrete microstructure and thus for the durability of concrete [9].

2. Materials and methods

2.1. Coarse aggregate

For the present study, RCA fraction 4/22.4 mm, originating from three different parent sources has been used: a) from old Bulgarian railway concrete sleepers, b) from old concrete from Studena dam near the town of Pernik, and c) from a garage ramp in Sofia. The category of the three types of RCA as per BDS EN 933-11 [10] is Rc,90, because they do not contain impurities.

For comparing the properties of ordinary concrete and RAC, natural coarse aggregate from crushed dolomite (NA1) and limestone (NA2) fractions 4/11,2 and 11,2/22,4 were also used. The physical properties of the aggregate are summarised in table 1 (for RCA) and table 2 (for NA) below. The characteristics of the aggregates are assessed in compliance with [2].

Table 1. Properties of RCA.

Parameter	Method	Unit	RCA1	RCA2	RCA3
			Sleepers	Studena dam	Garage
Bulk density	BDS EN 1097-3 [11]	kg/m³	1290	1080	1240
Oven dried particle density	BDS EN 1097-6:2013 [12]	kg/m³	2830	2270	2380
Abrasion resistance	BDS EN 1097-1 [13]	LA coeff.	LA_{30}	LA_{35}	LA_{40}
Resistance to fragmentation	BDS EN 1097-2:2020 [14]	%	9.7	16.9	19.1
Water absorption, 24 h	BDS EN 1097-6:2022 [15]	%	3.40	4.20	5.30

The results in the present study confirm the findings of other authors. Verian [16] concludes that there are significant differences between RCA and NA, due to old mortar content. Generally, RCA has a lower density in comparison with NA again due to the presence of low-density old mortar attached to the aggregate [17].

Table 2. Properties of NA.

Parameter	Method	Unit	NA1		NA2	
			4/11,2	11,2/22,4	4/11,2	11,2/22,4
Density	BDS EN 1097-3 [11]	kg/m³	1490	1440	1500	1380
Oven dried particle density	BDS EN 1097-6:2013 [12]	kg/m³	2840	2840	2720	2720
Abrasion resistance	BDS EN 1097-1 [13]	LA coeff.	LA_{25}	LA_{30}	LA_{30}	LA_{25}
Resistance to fragmentation	BDS EN 1097-2:2020 [14]	%	7.6	8.5	7.8	8.9
Water absorption, 24h	BDS EN 1097-6:2022 [15]	%	0.4	0.5	0.4	0.5

However, when the RCA originates from a high-quality primary concrete (e.g., having remaining compression strength of about 50 to 60 MPa as in our case of dismantled railway sleepers, RCA1) its oven dried particle density is similar or even superior to that of natural aggregate (NA1 and NA2, respectively). The old Interfacial Transition Zone (ITZ) creates additional porosity and furtherly weakens the RCA – figure 1.

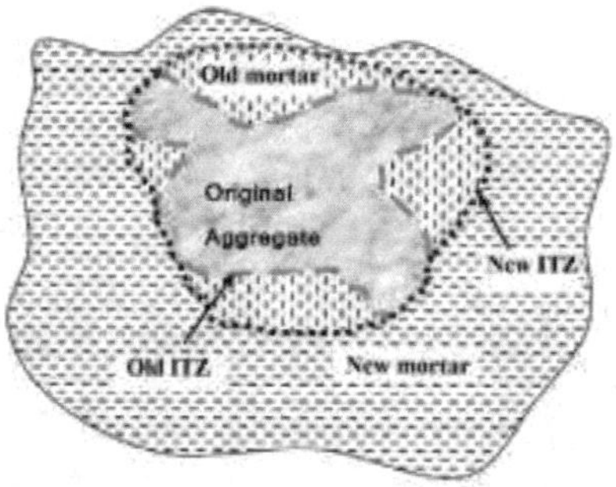

Figure 1. Scheme of old and new Interfacial Transition Zones (ITZs) [17].

The presence of the old mortar eventually increases the absorption capacity of RCA, but the pre-wetting of RCA helps the fresh concrete to maintain the workability and concrete to harden properly, without the aggregate draining the excess water during the cement hydration and thus forming capillary pores. The role of internal curing capacity has been proven by several authors [17], [18], [19]. It has been found that the water absorption of studied RCA varies between 3,4% and 5,3% and thus is about ten times higher than that of NA (of order of 0,4% to 0,5% only). However, the absorption of RCA is not uniform during the immersion process. The most intensive is the absorption during the first two hours because of filling of the coarser pores in RCA. This amount of water namely was used for the pre-wetting the RCA: 1,3% for RCA1, 3% for RCA2 and 4% for RCA3.
The results for resistance to fragmentation and LA coefficient show that all three types of RCA have an inferior performance in comparison with NA, as per [20]. However, the resistance to fragmentation of RCA1 is of the same order as that of NA. Since it is below 12%, according to BDS EN 206:2013+A2:2021/NA:2021 [3], there are no limitations for their use. The resistance to fragmentation of RCA2 and RCA3 is much lower and since it is above 16%, those aggregate would not be allowed to be applied in concrete, unless the RAC is proven as meeting the requirements for the class of concrete. As expected, the LA coefficient of RAC2 (LA_{35}) and RAC3 (LA_{35}) is also higher than that of NA (LA_{25} and LA_{30}) and of RCA1 (LA_{30}), but still acceptable for recycled aggregate, according to [3],

because its lower than the maximum allowed (LA_{50}). Despite of lower mechanical performances, the presence of rough and broken RCA surface contributes to a stronger mechanical bonding with the new cement paste and thus the new ITZ might be of a better quality than that with dense and less crushed NA [20].
The overall evaluation of RAC allows its classification as type A according to BDS EN 206:2013+A2:2021 [21] and can be used even for concrete classes above C30/37 and in some aggressive environment. To achieve the same granular skeleton of RAC and NAC, NA of two fractions 4/11.2 mm and 11.2/22.4 mm have been combined in a proportion 40% and 60% respectively. Despite of some differences, the gradings of NA and RCA are quite similar and all easily enter the standard zone for fraction 4/22.4 mm, as shown in figure 2.

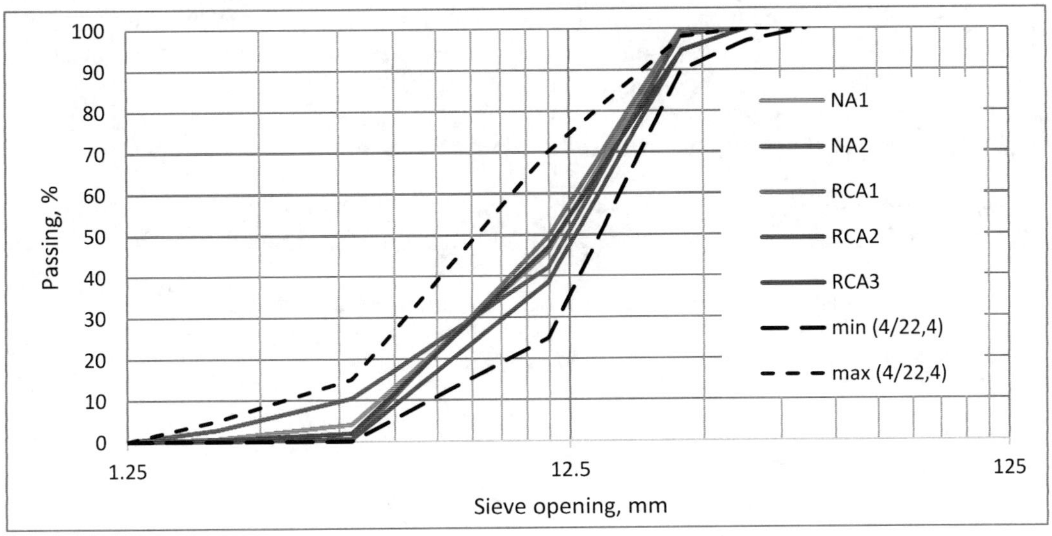

Figure 2. Grading curves of coarse aggregate.

2.2. *Fine aggregate*
River sand fraction 0/4 mm, having bulk density of 1470 kg/m³ and true density of 2530 kg/m³ is used as fine aggregate. Its fineness modulus is 3.0.

2.3. *Binder*
Portland cement CEM-I 52,5R in compliance with BDS EN 196-1 [22] is used in two different quantities – 350 kg/m³ and 400 kg/m³. According to [23], an increase of up to 30% in the quantity of cement is suggested to compensate the RAC compressive strength reduction due to the incorporation of RCA.

2.4. *Superplasticizer (SP)*
Superplasticizer (SP) for concrete, based on polycarboxylates, is used in the amount of 0.5% of the cement mass for NAC and 0.6% for RAC, because the slump test usually shows that RAC has a slightly lower workability than NAC, probably due to the presence of old mortar in the RCA [24].

2.5. *Concrete mix design*
The target class of concrete compressive strength is C30/37. The aimed class of consistency is S2. Two replacement rates of NA by RCA are applied – 30% and 50% respectively. The proportion between fine and coarse aggregate is kept constant, sand being 43% of the total aggregate. Recycled aggregate are pre-wetted by the quantity of water, corresponding to their water absorption after 2 hours of immersion, as explained in 2.1. Effective water/cement ratio (w/c) is fixed to 0.5, because

[23] reported that the variation of w/c does not significantly affect the difference between NAC and RAC. However, two mixes – one NAC and one RAC with 30% of RCA have w/c=0.55.

2.6. Test specimens

The production and testing of the samples took place in the Laboratory to Building Materials and Insulations department, University of Architecture, Civil Engineering and Geodesy, Sofia, Bulgaria. From each concrete batch 6 cubes with dimensions 15cm x 15cm x 15cm, 6 cylinders with 15cm x 15 cm and 3 prisms - 10 cm x 10 cm x 40 cm are produced. The samples are demolded after 24 hours and stored in water until the 28th day. The testing of compressive strength as per BDS EN12390-3 [24] is tested at 7 and 28 days. The prisms are produced to measure the modulus of elasticity in accordance with BDS EN12390-13 [25] and the three cylinders are used for testing the splitting tensile resistance (BDS EN12390-6 [26]). Three more cylinders are used for the water impermeability determination as per BDS EN 12390-8 [27]. The E-modulus, splitting tensile resistance and water permeability are tested at the age of 28 days. Average values are reported.

3. Properties of recycled aggregate concrete (RAC)

3.1. Fresh concrete properties

Table 3 below summarizes the differences between the batches and the properties of fresh concrete.

Table 3. Concrete mix design varieties and fresh concrete properties.

Mix design designation	Type of NA	Replacement rate by RCA, %	Type of RCA	Cement content, kg/m^3	Density of fresh concrete, kg/m^3	Slump, cm
NAC-1	NA1	0	RCA1	400	2420	6
RAC30-1	NA1	30	RCA1	400	2340	6
RAC50-1	NA1	50	RCA1	400	2280	8
NAC-2	NA1	0	RCA1	350	2420	1
RAC50-2	NA1	50	RCA1	350	2330	1
NAC-3	NA1	0	RCA1	350	2410	4
RAC30-2	NA1	30	RCA1	350	2360	5
RAC50-3	NA1	50	RCA1	350	2310	3
NAC-4	NA2	0	RCA1	350	2350	13
RAC-30-3	NA2	30	RCA1	350	2270	8
NAC-5	NA2	0	RCA2	350	2410	2
RAC30-4	NA2	30	RCA2	350	2270	6
RAC30-5	NA2	30	RCA2	350	2280	9
RAC50-4	NA2	50	RCA2	350	2270	12
NAC-6	NA2	0	RCA2	350	2370	4
RAC50-5	NA2	50	RCA2	350	2280	3
NAC-7	NA2	0	RCA3	350	2380	8
RAC30-6	NA2	30	RCA3	350	2360	6
RAC50-6	NA2	50	RCA3	350	2310	10

According to [28], the slump is barely higher in NAC than in RAC. The results in table 3 shows that the consistency of mixes with RCA is difficult to be controlled – there is a huge variation of slump

values. However, it seems the pre-wetting of RCA and the increased amount of SP are able to compensate the slump reduction due to the incorporation of RCA. In mixtures with higher quantity of RCA, namely RAC50, the higher values for the slump test may be granted to the extra water used for pre-wetting of the aggregate, especially when RCA2 and RCA3 are used, because a bigger amount of water for their pre-wetting is applied, while the mixing time is kept constant and probably some of pre-wetting water is acting as free water. It is known that the excess free water improves the RAC mixture's consistency [29]. According to [30] the density of fresh concrete is also affected by the porosity, shape, texture and size of the aggregate.
The study confirms that the density of fresh concrete is lower when the quantity of the RAC is increased. The calculations for theoretical density show difference of 1% to 5% with measured density for all mixtures, but no correlation related to RCA content or slump value can be established.

3.2. Hardened concrete properties
The partial or fully replacement of RCA in lieu of NA affects the compressive strength, tensile strength and elastic modulus with the value of the decrease depending on percentage replaced [31]. Numerous studies concluded that the increase of the quantity of recycled aggregate decreases the overall performances of the mix and hardened specimen [32], [33]. Replacement of NA by 30% of RCA of good quality (RCA1), practically does not change the compressive strength of concrete, taking into consideration the standard deviation of measured values – figure 3 and figure 4. No significant difference in strength gain of NAC and RAC-30 is observed either.

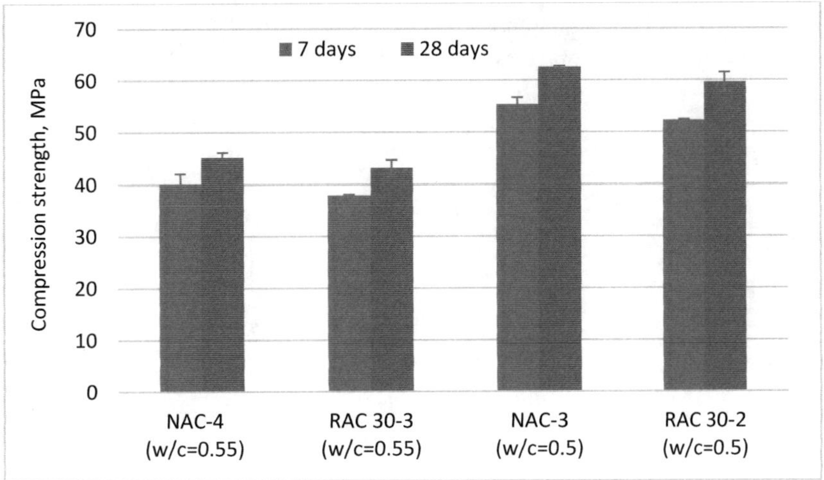

Figure 3. Effect of water-cement ratio on compressive strength of concrete.

Just as for NAC, the impact of w/c ratio on compressive strength of RAC is much more pronounced - the higher w/c ratio (w/c = 0.55) resulted in lower compressive strength by approximately 26.7% and 26.68%, respectively, compared to that of mixtures with w/c = 0.50 (figure 3). As expected, the increase of cement content from 350 kg to 400 kg per 1m^3 leads to an increase of the compressive strength at 28 days. However, despite the replacement of 50% of NA by RCA is associated with a certain strength reduction (of order of 12.3% for series 1 and 8.85% for series 2) – figure 4, for achieving the class C30/37 there is still enough margin in strength with 350 kg cement per 1m^3 and no need to increase the cement quantity.

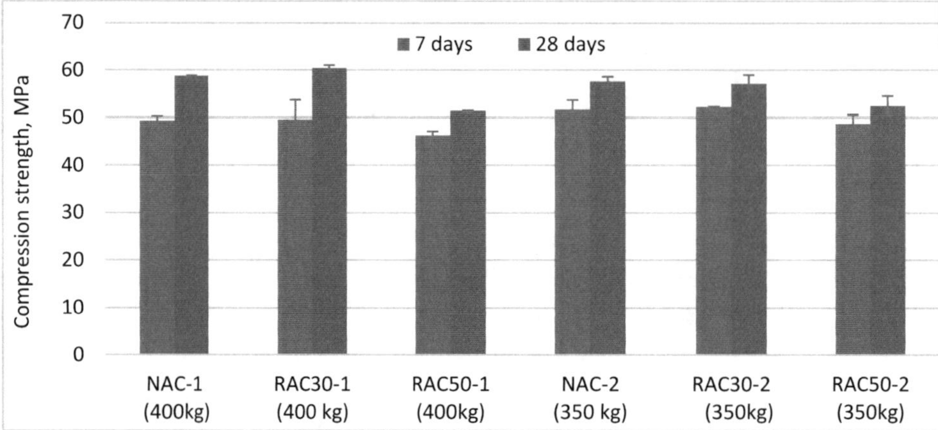

Figure 4. Effect of cement content on compressive strength of concrete.

Figure 5 illustrates the impact of the different aggregate on the concrete performance. The differences in compressive strength at age of 28 days ($f_{c,28}$) of concrete, produced with RCA1 and RCA3 and NAC are mainly due to the replacement rate - RAC with 30% of RCA has identical or even superior (up to 8.5%) $f_{c,28}$ values, while replacement of 50% leads to a decrease of RAC of 6% (with RCA3) to 9% (with RCA1), compared to that of NAC-7 and NAC-2, respectively. These results are surprising, taking into consideration that the resistance to fragmentation of RCA1 is much bigger than that of RCA3 and the same is for the abrasion resistance (table 1).

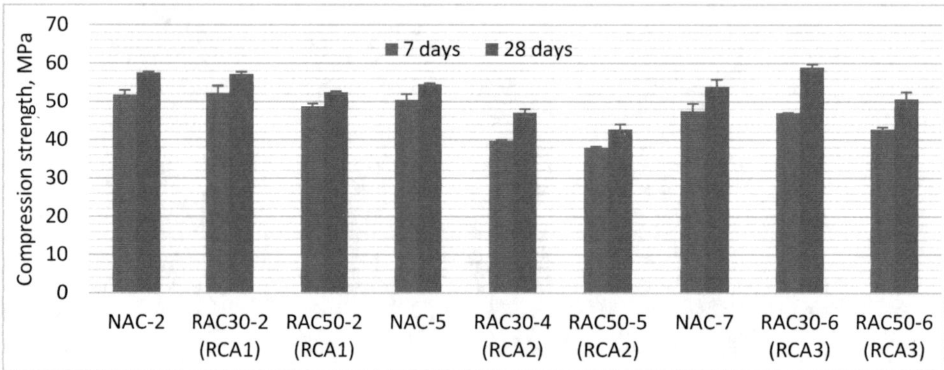

Figure 5. Effect of recycled aggregates from different sources.

The incorporation of RCA2 leads to more significant strength reduction, compared to that of the corresponding NAC (NAC-5) – 14% for RAC30-4 and 22% for RAC50-5 for $f_{c,28}$. It shall be mentioned, that RCA2 does not have the worst resistance to fragmentation and has intermediate LA value between RCA1 and RCA2 (table 1), but at the same time RCA2 has the lowest oven dried particle density. Thus, a correlation between the mechanical properties of RCA and compression strength of RAC could not be established. The possible explanation might be that the lower density, abrasion and fracture resistance of RCA are due to several factors, among which are the quality of old cement paste, the presence of old ITZ, but also the nature of aggregate in the parent concrete. There are no peculiarities in the compression strength gain process – the values at age of 7 days follow the same trends as those at 28 days, for all types of concrete – figure 3, figure 4 and figure 5.
The tensile splitting strength of RAC is of order of 3.0 to 3.5 MPa. The replacement of NA by 30% does not impact the split resistance average value but leads to an increase of the standard deviation of results. This effect is even more expressed for 50% content of RCA. With the higher replacement rate,

when RCA2 and RCA3, having lower density and mechanical performance, a certain reduction of about 15% of tensile splitting strength is also recorded.
The modulus of elasticity of RAC with high quality RCA (RCA1) is even less influenced by the replacement of natural by recycled aggregate – it remains practically the same for NAC, RAC30 and RAC50 and is of order of 35 GPa, being marginally lower for concrete with 50% of RCA. Almost the same, but slightly inferior, are the values for E-modulus when RCA2 is used, although RCA2 is less dense and mechanically weaker than RCA2 and RCA1. The trend that the E-modulus decreases, even to a small extent, with the decrease of resistance to abrasion and to fragmentation of aggregate is confirmed by the results for concrete with RCA3 - its E-modulus is lower by 10% to 14% compared to concrete with RCA1 and RCA2. For concrete with RCA3 the most significant absolute reduction of E-modulus is also noticed – up to 25%, compared to that of NAC-7.

Table 4. Tracking the splitting tensile strength, modulus of elasticity and water permeability.

Mix design designation	Type of aggregate	Tensile splitting strength, 28th day [MPa]	Modulus of elasticity [GPa]	Water permeability at 0.6 MPa [mm]
NAC-1	NA1	3.7 ± 0.07	35 ± 0.6	24 ± 5
RAC30-1	RCA1	3.8 ± 0.14	38 ± 2.4	25 ± 3
RAC50-1	RCA1	3.4 ± 0.21	36 ± 0.8	25 ± 8
NAC-4	NA2	3.4 ± 0.08	35 ± 0.6	20 ± 2
RAC30-3	RCA2	3.5 ± 0.12	36 ± 0.6	22 ± 4
RAC50-4	RCA2	2.9 ± 0.34	33± 0.7	18 ± 4
NAC-7	NA2	3.3 ± 0.09	40 ± 0.8	25 ± 6
RAC30-6	RCA3	3.3 ± 0.13	31 ± 0.7	14 ± 5
RAC50-6	RCA3	2.8 ± 0.18	30 ± 1.5	14 ± 6

The water permeability of concrete is considered as a convenient parameter to characterize the flow properties of concrete, RAC included [34]. The average results for the water penetration depth at a water pressure of 0,6 MPa of the three types of concrete (NAC, RAC-30 and RAC-50) are quite the same when RCA1 and RCA2 are used for the replacement of NA. It can be concluded that the RAC susceptibility to carbonation and penetration from chlorides would most likely be similar to that of NAC.

4. Conclusions

Recycling concrete CDW as aggregate is a sustainable path towards circular economy in construction. Despite recycled concrete aggregate (RCA) is recognized by EN 206 standard as a legitimate concrete ingredient and many successful examples can be found, in many countries, including Bulgaria, recycled aggregate are not practically implemented and several challenges are to be overcome.
This study has proven some reasons for mistrust and reluctance to implement RCA were not justified such as significantly inferior performance of recycled aggregate concrete (RAC), or need of bigger amount of cement and thus increased costs. At the same time, it has been found that some relationships are not easily applicable to recycled aggregate concrete.
Actually, the incorporation of RCA leads or not to a modification of RAC performances in function of the replacement rate: 30% of RCA do not practically change the compression strength gain and values, neither the tensile splitting strength, E-modulus and water permeability; RCA of 50% does not significantly worsen the properties of the concrete, either, but some reduction of performance can be observed – up to 10% for the compressive strength, up to 15% for the tensile splitting strength, up to 25% for the E-modulus. The water impermeability, measured as per BDS EN12390-8 at a pressure of 0.6 MPa is however increased, due to the complex porous structure of RCA [9], [18]. The impact of

replacement rate is much more pronounced when the properties of RCA differ noticeably from those of natural aggregate. However, it is not always possible to establish a correlation between the mechanical properties of RCA (such as resistance to fragmentation and resistance to abrasion) and mechanical properties of RAC, because of many factors – overall particle porosity, quality of old cement paste, presence of old ITZ, nature of natural aggregate in the parent concrete, might play important role. It seems that the oven dried particle density is more reliable parameter for assessing the quality of RCA. The desired behavior of fresh RAC can be achieved by applying a combination of pre-wetting of RCA and an increase in the superplasticizer quantity. The amount of water for the pre-wetting shall be adjusted according to the water absorption process by RCA.
Finally, the peculiarity of RCA shall be taken into consideration when using them in RAC, but there are various ways to design and control the required concrete performances.

Acknowledgment

The research presented in the paper was carried out in the framework of the Clean & Circle Project BG05M2OP001-1.002-0019: "Clean technologies for sustainable environment — waters, waste, energy for circular economy", funded by the Operational Programme "Science and education for smart growth" 2014–2020, co-financed by the EU through the European structural and investment funds.

References

[1] Domone P and Illston J, 4th edition published in 2010, *Consturction Materials, Their Nature and Behaviour*, page 203.
[2] BDS EN 12620:2002+A1:2008, Aggregates for concrete - https://bds-bg.org/en/project/show/bds:proj:75461
[3] BDS EN 206:2013+A2:2021/NA:2021, Concrete - Specification, performance, production and conformity. National Annex - https://bds-bg.org/bg/project/show/bds:proj:119283
[4] De Brito J and Saikia N, 2013, *Recycled Aggregate in Concrete. Use of Industrial, Construction and Demolition Waste, Springer*, ISBN 978-1-4471-4539-4.
[5] Taveh B, AlSaffar D, Alyousef R, July–August 2020, *The Utilization of Recycled Aggregate in High Performance Concrete: A Review*, Elsevier, Journal of Materials Research and Technology, Volume 9, Issue 4, Pages 8469-8481.
[6] Guo H et al, May 2018, *Durability of recycled aggregate concrete – A review*, Elsevier, Cement and Concrete Composites, Volume 89, Pages 251-259.
[7] Silva R V, Neves R, De Brito J, Dhir R K, Accepted 29 April 2015, *Carbonation behaviour of recycled aggregate concrete*, Cement & Concrete Composites, Elsevier Ltd.
[8] Zaharieva R, Buyle-Bodin F, Wirqin E, Accepted 25 Februrary 2004, *Frost resistance of recycled aggregate concrete*, Cement and Concrete Research, Elsevier Ltd.
[9] Buyle-Bodin F, Zaharieva R, *Influence of industrially produced recycled aggregates on flow properties of concrete*, September 2002, Materials and Structures 35(8):504-509 , DOI: 10.1007/BF02483138
[10] BDS EN 933-1:2012, Tests for geometrical properties of aggregates – Part 1: Determination of particle size distribution – Sieving method - https://bds-bg.org/bg/project/show/bds:proj:84399
[11] BDS EN 1097-3:2000, Tests for mechanical and physical properties of aggregates – Part 3: Determination of loose bulk density and voids - https://bds-bg.org/bg/project/show/bds:proj:3476
[12] BDS EN 1097-6:2013, Tests for mechanical and physical properties of aggregates – Part 6: Determination of particle density and water absorption - https://bds-bg.org/bg/project/show/bds:proj:82392
[13] BDS EN 1097-1:2021, Tests for mechanical and physical properties of aggregates – Part 1: Determination of resistance to wear (micro-Deval) https://bds-bg.org/bg/project/show/bds:proj:82424
[14] BDS EN 1097-2:2020, Tests for mechanical and physical properties of aggregates – Part 2:

Methods for the determination of resistance to fragmentation https://bds-bg.org/bg/project/show/bds:proj:105694
[15] BDS EN 1097-6:2022, Tests for mechanical and physical properties of aggregates – Part 6: Determination of particle density and water absorption, https://bds-bg.org/bg/project/show/bds:proj:111372
[16] Verian K P, Ashraf W, Cao Y, Accepted 6 February 2018, *Properties of recycled concrete aggregate and their influence in new concrete production*, Resources, Conserving and Recycling, Elsevier Ltd.
[17] Kisku N, Joshi H, Ansari M, Panda S K, Nayak S, Dutta S, Accepted 7 November 2016, *A critical review and assesment of usage of recycled aggregate as sustainable construction material*, Construction and Building materials, Elsevier Ltd.
[18] Zaharieva R, Buyle-Bodin F, Skoczylas F, Wirqin E, 2002, *Assessment on the surface permeation properties of recycled aggregate concrete*, Cement & Concrete Composites, Elsevier ltd.
[19] Ozbakkaglu T, Gholampour A, Xie T, November 2017, *Mechanical and Durability Properties of Recycled Aggregate Concrete: Effect of Recycled Aggregate Properties and Content*, Journal of Materials in Civil Engineering, ResearchGate.
[20] Rao A G, Prasad R, February 2022, *Influence of the roughness of Aggregate Surface on the Interface Bond Strength*, Cement and Concrete Research, 32(2):253-257, DOI 10:1016/S0008-8846(01)00668-8.
[21] BDS EN 206:2013+A2:2021, Concrete – Specification, performance, production and concrete, https://bds-bg.org/bg/project/show/bds:proj:114927
[22] BDS EN 196-1:2016, Methods of testing cement – Part1: Determination of strength, https://bds-bg.org/bg/project/show/bds:proj:97859
[23] Hasaba S, Kawamura M, Toorii K, Takemoto K, 1981, *Drying shrinkage and durability of concrete made with recycled concrete aggregates*, Trans. Jpn. Concr. Inst. 3, 55-60.
[24] BDS EN 12390-1:2021, Testing hardened concrete – Part1: Shape, dimensions and other requirements for specimens and moulds, https://bds-bg.org/bg/project/show/bds:proj:111362
[25] BDS EN 12390-13:2013, Testing hardened concrete – Part 13: Determination of secant modulus of elasticity in compression, https://bds-bg.org/bg/project/show/bds:proj:86721
[26] BDS EN 12390-6:2009, Testing hardened concrete – Part 6: Tensile splitting strength of test specimens, https://bds-bg.org/bg/project/show/bds:proj:79096
[27] BDS EN 12390-8:2019, Testing hardened concrete – Part 8: Depth of penetration of water under pressure, https://bds-bg.org/bg/project/show/bds:proj:103634
[28] Kabir A, January 2015, *Effect of Cement Proportion on Strength of Recycled Aggregate Concrete*, International Journal of Structural and Civil Engineering Research Vol. 4, No. 4, Researchgate.
[29] De Brito J, Lavado J, Bogas J, Hawreen A, Accepted 17 October 2019, *Fresh properties of recycled aggregate concrete*, Construction and Building Materials, Elsevier Ltd.
[30] Poon C S, Shui Z H, Lam L, Fok H, Kou S C, 2004, *Influence of moisture states of natural and recycled aggregates on the slump and compressive strength of concrete*, Cem. Concr. Res
[31] Mehta P K, Monteiro P J M, 1994, *Concrete: Structure, properties, materials (in Portuguese).*
[32] Garcia-Gonzalez J, Rodriguez-Robles D, Juan-Valdes A, Moran J M – del Pozo and M.I. Guerra-Romero, September 2014, *Pre-Saturation Technique of the Recycled Aggregates: Solution to the Water Absroption Drawback in the Recycled Concrete Manufacture*, Materials (Basel).
[33] Exteberia M, Vasuez E, Mari A, Barra M, Accepted 7 February 2007, *Influence of amount of recycled coarse aggregates and production process on properties of recycled aggregate concrete*, Cement and Concrete research, Elsevier Ltd.
[34] Picconali A, Diotti A, Plizzari G, Sorlini S, Published 22 Feb 2022, *Impact of Recycled Aggregate on the Mechanical and Environmental Properties of Concrete: A Review*, Materials (Basel).

Effect of coarse aggregate characteristics on the permeability properties of pervious concrete

M Daukšys[1], A Dunauskas[1], S Juočiūnas[1]

[1]Faculty of Civil Engineering and Architecture, Kaunas University of Technology, Studentų str. 48, LT-51367 Kaunas, Lithuania

mindaugas.dauksys@ktu.lt

Abstract. This study analyses the effect of the geometric characteristics of coarse aggregate particles on the physical-mechanical properties and durability of hardened permeable concrete prepared by vibropressing compaction method. Four types of coarse aggregate were used to prepare the concrete mixture: gravel and crushed gravel stone of particle size fraction 4/16 and 8/16. The particle size and shape (rounded and rectangular) were considered. Four different concrete mix compositions were tested. The following properties of hardened permeable concrete were analysed in comparison: total porosity, water permeability, density, compressive strength, frost resistance, and abrasion resistance. The findings of this study showed that particle size and shape influence the physical-mechanical properties of permeable concrete. The total porosity and water permeability values were higher in the test specimens where crushed gravel stone of fraction 4/8 mm and 8/16 mm was used as coarse aggregate. The resistance of permeable concrete to volumetric freezing and thawing depends on the frost resistance of the coarse aggregate. It was also found that abrasion resistance of permeable concrete is affected by ravelling, i.e. disintegration of angular coarse aggregate particles from the concrete matrix.

Keywords: permeable concrete, physical-mechanical properties, durability

1. Introduction

Permeable concrete (PC) is a material that comprises cementitious binders, coarse aggregates, fine aggregate, water, admixtures and additives. PC is also called porous concrete, pervious concrete, no-fines concrete, gap-graded concrete, and enhanced-porosity concrete [1]. The main application of permeable concrete is in pavements as it rapidly drains and removes rainwater from the pavement surface, filters pollutants and preserves the natural hydrologic functions. It also can be used for parking lots, sidewalks, pathways, tennis courts, patios, slope stabilization, swimming pool decks, greenhouse floors, zoo areas, road shoulders, drains, noise barriers, friction courses for highway pavements, permeable bases under a standard concrete paving, and low-volume roads.

PC is relatively porous with most of the volume filled with coarse aggregate, which creates more pores in the structure, leading to higher water infiltration [2]. The infiltration is the main parameter that describes the functionality of PC [3]. The increase in the porosity of PC leads to an increase in permeability. The grading of aggregate for PC influences the permeability coefficient [4]. A higher permeability can be obtained by leaving out fine aggregate, designing single grain graded mixes or double size grain graded mixes with the gap in the grading.

The porous structure may compromise the structural strength of PC [5] because it is determined by the properties of cement paste (cement content, water to cement ratio, compaction level), coarse aggregate and the interface between the coarse aggregate and the cement paste [5-7]. The stiff consistency of PC mixture requires considerable compaction efforts to achieve the desired mechanical performance, without neglecting permeability properties, which strongly depend on the open porosity of PC [8]. Resistance of PC to freeze-thaw cycles and abrasion challenge its durability properties. The pores in PC products can provide frost resistance if water passes through the pores freely without being trapped inside the pore structure before freezing [9]. Sand, latex and air-entraining agents added to PC mixture improve the resistance of PC to freeze-thaw cycles as they preventing de-bonding between the aggregates and the cement paste. Air entrainment in a cement paste can also improve frost resistance of PC [1]. PC specimens made with smaller-size aggregates and tested by different methods produced better abrasion resistance values than the specimens made with larger-size aggregates [10]. The addition of latex can improve the abrasion resistance of PC, while the use of polypropylene fibres did not show any significant effect on abrasion.

This study aims to determine the effect of the size and shape of the local coarse aggregate particles on the physical-mechanical properties and durability of hardened permeable concrete.

2. Materials and methodology

Portland cement CEM I 42.5 R from Heidelberg Cement Group (Sweden) satisfying the requirements of standard EN 197-1 was used for the tests. The cement has the following physical and mechanical properties: compressive strength after 2 days ≥20.0 MPa, after 28 days ≥42.5≤62.5 MPa, initial setting time 180 min, volume stability 1.0 mm and specific surface area (by Blaine method) 575 m^2/kg.

Two types of coarse aggregates from the local quarries in Lithuania were used: gravel and crushed gravel stone. Particle size fractions 4/16 mm and 8/16 mm of both aggregates were tested. Gravel particles had a round shape and smooth surface. Crushed stone particles had a rectangular shape and rough surface. The granulometric properties of the coarse aggregates were determined according to the standard EN 12620 and are presented in Table 1.

Table 1. Sieve analysis data of the coarse aggregates

Sieve opening size, mm	Passing, %	
	Gravel fraction 4/16 mm	Crushed stone fraction 4/16 mm
32.0	100.0	100.0
22.0	100.0	100.0
16.0	96.6	95.5
11.0	73.9	71.7
8.0	48.4	44.0
5.6	26.0	15.7
4.0	8.2	6.8
2.0	1.6	1.8
1.0	1.0	1.0
0.5	0.8	0.7
0.125	0.6	0.6
0.063	0.5	0.5
On the sieve bottom	0.3	0.3

Gravel of particle size fraction 8/16 mm was obtained from the gravel of particle size fraction 4/16 mm by sieving with the sieves of appropriate opening sizes. Crushed gravel stone of particle size fraction 8/16 mm was prepared in the same way from crushed gravel stones of particle size fraction 4/16. The coarse aggregates used in the research were not washed and contained some dust. The gravel of particle size fraction 4/16 mm had the following physical properties: dry bulk density 1547 kg/m^3, specific gravity 2655 kg/m^3, water absorption 1.69%, particle shape index (*SI*) 5%, particle flakiness index (*FI*) 4%, resistance to fragmentation according to Los Angeles test (EN 1097-2) 25% and strength of coarse aggregate according to fragmentation index M_{fi} (LST 1476.7) 8.4 MPa. The crushed gravel stone of particle size fraction 4/16 mm had the following physical properties: dry bulk density 1451 kg/m^3, specific gravity 2667 kg/m^3, water absorption 1.58%, particle shape index (*SI*) 13%, particle flakiness index (*FI*) 11%, resistance to fragmentation according to Los Angeles test (EN 1097-2) 26% and strength of coarse aggregate according to fragmentation index M_{fi} (LST 1476.7) 9.0 MPa. The shape index of the coarse aggregates was calculated according to EN 933-4 as the ratio of the mass of the test portion to the mass of the non-cubical particles. The length (*L*) to thickness (*E*) ratio of non-cubical particles was > 3. Particle flakiness index was determined according to EN 933-3 as the percentage (by mass) of aggregate particles having an average least dimension of less than 0.6 times their average dimension.

Four different concrete mix compositions were analysed in the experimental part of the research basing on literature analysis. The water to cement ratio in all mixes was 0.34, coarse aggregate to cement ratio was 5.8 and no admixtures were added. The concrete mix composition differed by the type of coarse aggregate used. The clump shaping method was used to check if the chosen PC mix proportions were adequate [1]. A small amount of mix was extracted by hand and a clump was shaped and observed with palm open. The concrete mix proportions were satisfactory because the clump of fresh concrete preserved its shape and no cement paste bleeding was observed. And vice versa, If the clump of fresh concrete not preserved its shape the concrete mix proportions were not satisfactory. However, a small amount of cement paste segregated at the bottom of the mould in test specimens PC-3 and PC-4 while preparing the specimens and applying vibration and additional load simultaneously. It means that the content of the cement paste should be reduced when bigger-size aggregates (8/16 mm) are used. The concrete mix designs with the amounts of materials (kg per m^3) used are presented in Table 2.

Table 2. Compositions of tested mixture series

Materials	Unit	Specimen designation			
		PC-1	PC-2	PC-3	PC-4
Portland cement	kg	350	350	350	350
Water	l	119	119	119	119
Gravel fraction 4/16	kg	2030	-	-	-
Crushed gravel stone fraction 4/16	kg	-	2030	-	-
Gravel fraction 8/16	kg	-	-	2030	-
Crushed stone fraction 8/16	kg	-	-	-	2030
W/C	–	0.34	0.34	0.34	0.34

The concrete was mixed from dry materials in a rotating pan mixer Zyklos ZZ 50 HE from Pemat (Germany) under laboratory conditions. The mixing was done in two stages. First, coarse aggregates and 2/3 of the water were poured into the moistened mixer and mixed for about one minute. After the first cycle the mixture was left to stand for two minutes. Second, Portland cement and 1/3 of water were added and the mixture was mixed for about 2 minutes.

Slump test of fresh concrete was conducted according to EN 12350-2 when the mix was prepared.

The test specimens were cast in moulds in the laboratory and were compacted using a vibration table with an oscillation frequency of 50 Hz and amplitude of 0.5 mm. Concrete test specimens of 71×71×71 mm dimensions and 100×100×100 mm dimensions were cast in metal moulds. To imitate the vibropressing compaction method a load of 11.63 kg (with the size of 100×295×50 mm) was used to compact two test specimens with the size of 100×100×100 mm in each mould (Figure 1). Vibration time during the compaction of the fresh mixture was 15 ± 1 s. The test specimens with the size of 71×71×71 mm were used to determine the wear resistance of concrete, while the test specimens with the size of 100×100×100 mm were used to determine the total porosity, density, compressive strength and frost resistance of concrete. Concrete test cylinders with the diameter of 110 mm and height of 320 mm were cast in plastic moulds. These test specimens were used to determine water permeability of concrete. After the casting and compacting of concrete mixtures, all test specimens were kept in moulds in the laboratory at the temperature of 20±5°C for approximately 20 hours. Then the test specimens cast in metal moulds were demoulded and cured in the environmental chamber for 28 days at the temperature of 20±2°C and humidity of 90%. The test cylinders with the diameter of 110 mm and height of 320 mm were not demoulded and kept in plastic moulds in the same environmental chamber for the curing period of 28 days.

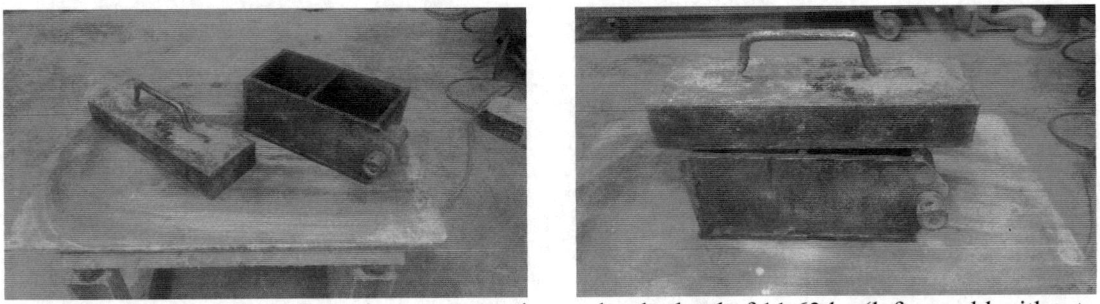

Figure 1. View of the test specimen compaction under the load of 11.63 kg (left: mould without load, right: mould with load).

The total porosity of permeable concrete specimen with the size of 100×100×100 mm was calculated from the equation (1):

$$P = \left[1 - \left(\frac{W_1 - W_2}{q_w \cdot V}\right)\right] \times 100, \% \quad (1)$$

here: P is total porosity of permeable concrete, %; W_1 is mass of the permeable concrete specimen air-dried for 24 hours, kg; W_2 is mass of the permeable concrete specimen submerged in water, kg; V is volume of the permeable concrete specimen, m^3; q_w is density of water, kg/m^3.

Water permeability of permeable concrete was defined by the permeability coefficient k, which was calculated from the equation (2):

$$k = \frac{Q \cdot L}{A \cdot h \cdot t}, mm/s \quad (2)$$

here: k is permeability coefficient, mm/s; Q is total volume of water, mm^3; L is length of a permeable concrete specimen, mm; A is cross-sectional area of a permeable concrete specimen, mm^2; h is water head, mm; t is time of water passing through a permeable concrete specimen, s.

The density of hardened concrete specimens was determined according to EN 12390-7, the compressive strength according to EN 12390-3, the wear resistance of concrete according LST 1428-15, and the frost resistance according to LST 1428-17.

3. Results and discussions

The slump test of fresh concrete was conducted after the concrete mixture preparation procedure. The slump values of the tested concrete mixtures with different particle size fractions of the coarse aggregate were about 0 mm. The tested mixtures were stiff and had low consistency because of the special composition of mix. It means that considerable compaction effort and adequate methods are required to achieve the desired mechanical performance of concrete [8]. To this end, an additional load of 11.63 kg was used during the compaction of fresh concrete.

The tests of physical and mechanical properties and durability of hardened PC were carried out after 28 days of curing. The comparison of the average values of total porosity and permeability coefficient *k* of PC specimens made with different types of coarse aggregate is presented in Figure 2. Statistical analysis of the total porosity of PC specimens calculated from the equation (1) is presented in Table 3.

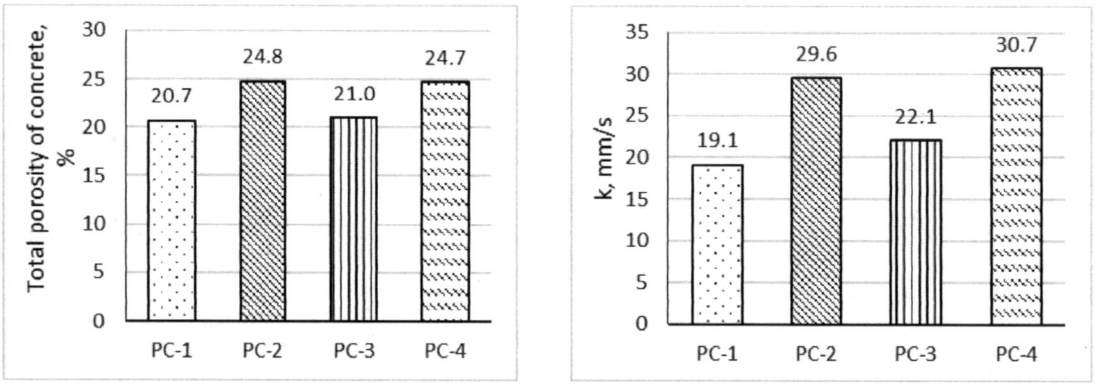

a) b)

Figure 2. The results of the average values of total porosity (a) and water permeability (b) of PC (PC-1, PC-2, PC-3 and PC-4 - compositions of tested mixture series).

Fig. 2 illustrates that the total porosity (volume of pores) of PC depends on the geometrical properties of the coarse aggregate. PC specimens with gravel of particle size fraction 4/16 mm and 8/16 mm had a similar total porosity, i.e. 20.7% and 21.0% respectively, and a similar coefficient *k*, i.e. 19.1 mm/s and 22.1 mm/s respectively. The total porosity of PC increased from about 21.0% to 25.0% and coefficient *k* increased from about 20 mm/s to 30 mm/s when crushed gravel stone of particle size fraction 4/16 mm and 8/16 mm was used. As expected, a correlation between the total porosity and permeability coefficient *k* of PC was determined. PC with lower total porosity values had a lower permeability coefficient *k* and vice versa. The velocity of water flow through concrete structure increases with the increase of the porosity of PC.

Table 3. Statistical analysis of total porosity data

Description	PC-1	PC-2	PC-3	PC-4
Number of values	17	18	18	20
Min., %	16.3	20.6	16.4	21.4
Max., %	25.0	28.0	27.4	29.1
Arithmetic mean, %	20.7	24.8	21.0	24.7
Median, %	20.8	25.1	20.9	24.7
Variance, %	5.602	5.190	7.610	5.547
Standard deviation, %	2.367	2.278	2.758	2.355
Coefficient of variation	0.114	0.091	0.131	0.096

Table 3 illustrates that the total porosity of PC-1 series specimens varied in the range from 16.3% to 25.0%, while the total porosity of PC-2 series specimens varied in the range from 20.6% to 28.0%. The total porosity of PC-3 series specimens varied in the range from 16.4% to 27.4% and the total porosity of PC-4 series specimens varied in the range from 21.4% to 29.1%. Crushed gravel stone has angular and jagged edges formed during the crushing process and natural gravel has round shapes. Angularity and the jagged edges of crushed gravel stone create higher open porosity of PC-2 and PC-4 series specimens because of improper placement of coarse aggregate particles in the volume.

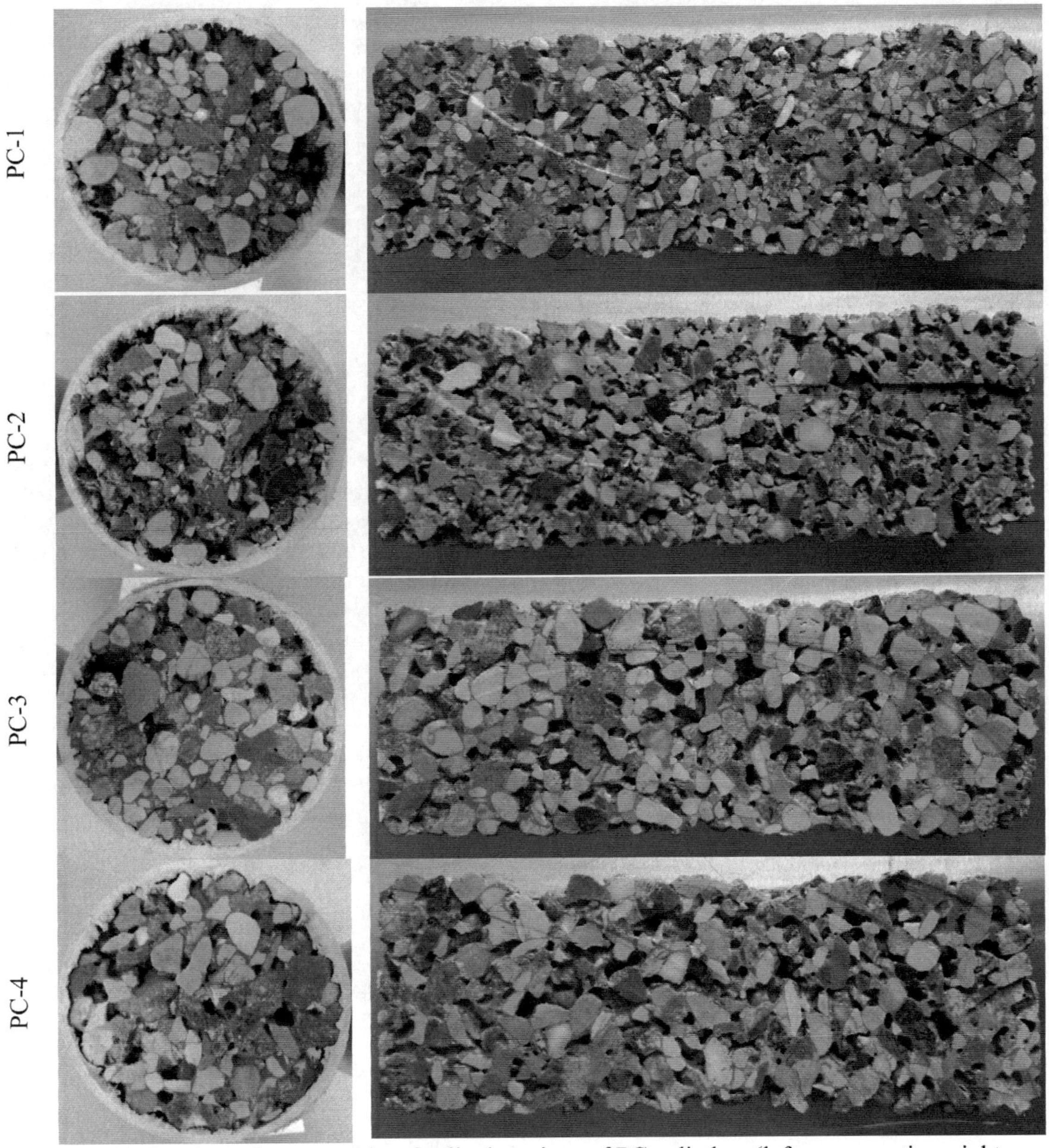

Figure 3. The view of cross and longitudinal sections of PC cylinders (left: cross section, right: longitudinal section).

Gravel of 4/16 fraction had the following geometric properties: particle shape index 5%, particle flakiness index 4%. The shape index and particle flakiness index of crushed gravel stone of 4/16 fraction were 13% and 11% respectively. The increased total porosity of PC-2 and PC-4 series specimens is seen in the pictures (Figure 3) of cross and longitudinal sections of those test specimens.

Figure 4 illustrates the comparison of the average values of PC density and compressive strength, when different types of coarse aggregate were used.

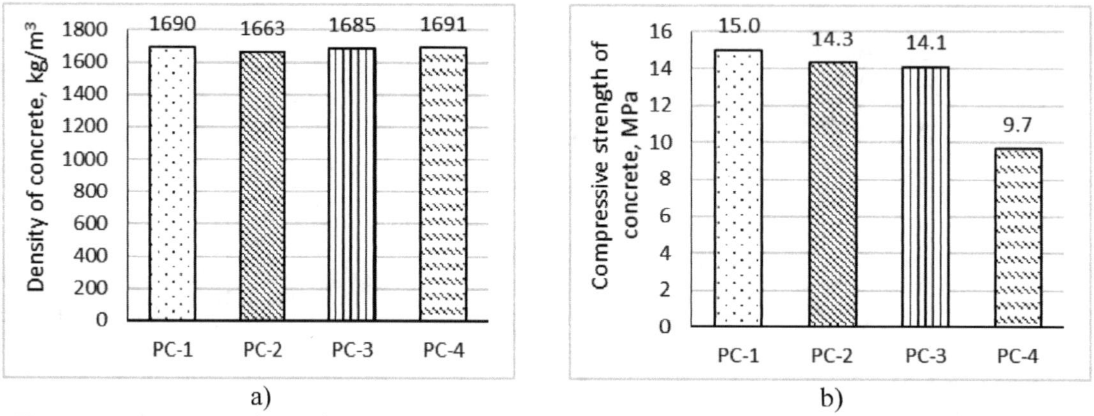

Figure 4. Average values of PC density (a) and compressive strength (b) (PC-1, PC-2, PC-3 and PC-4 - compositions of tested mixture series).

The results given in Figure 4 show that the average values of the density of PC series specimens varied from 1663 kg/m^3 to 1691 kg/m^3, while the compressive strength varied from 9.7 MPa to 15.0 MPa. The PC test specimens with bigger-size aggregates (8/16 mm) had lower compressive strength than the test specimens when smaller-size aggregates (4/16 mm). A three-phase model of the porous concrete was adopted: aggregates, cement paste and interfacial transition zone. However, the results of compressive strength showed that not all coarse aggregate particles were properly covered with cement paste, i.e., there was no interfacial transition zone between the coarse aggregates and the cement paste in some places observed. PC made with smaller-size aggregates and angular aggregate grains (crushed gravel stone) generally requires more cement paste to produce the specified workability than PC made with round aggregate grains (gravel). The load of 11.63 kg used in vibropressing compaction process may have been too high and was the reason for a very thin layer of cement paste that bonds coarse aggregate particles. The interface between the aggregate and the paste needs to be improved to improve the strength of PC. Authors [6] noted that porous concrete fails at the binder aggregate interface because the layer of the cement paste in PC is very thin to bond coarse aggregate particles. Therefore, the compressive strength is low.

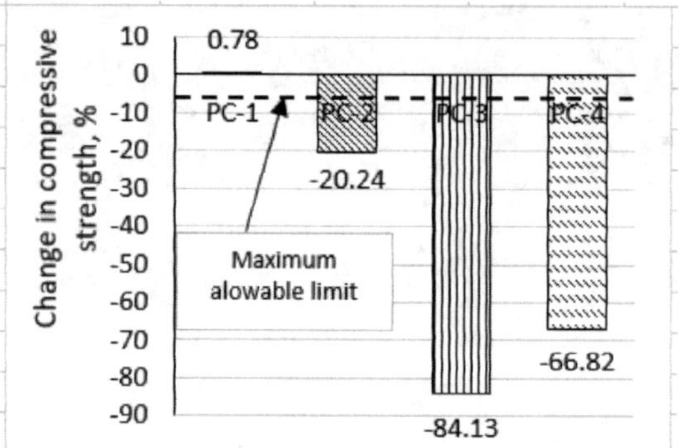

Figure 5. The results of resistance to volumetric freezing and thawing test of PC (PC-1, PC-2, PC-3 and PC-4 - compositions of tested mixture series).

Fig. 5 illustrates the results of resistance to volumetric freezing and thawing test of PC series specimens. The appearance of PC specimens before and after a different number of freeze-thaw cycles is presented in Figure 6. The first defects were observed in the PC-4 series specimens (Figure 6) after 64 freezing and thawing cycles. Later, after 108 freezing and thawing cycles, the first defects were also observed PC-1, PC-2 and PC-3 series specimens (Figure 6). The observed defects were related to the cracks in the test specimens between the hardened cement paste and the aggregate, also damaged gravel particles were observed in some places. The addition of air-entraining admixtures created entrained air bubbles in the cement paste and improved the freeze-thaw durability [1]. To avoid the damage of gravel grains used it should be replaced by crushed granite stone, which is frost resistant.

According to LST 1428-17, the change in compressive strength of PC specimens compared to control specimens must not decrease more than 5%. The results given in Figure 5 show that the compressive strength of PC-1 series specimens did not decrease more than 5% when the number of freezing and thawing cycles was 108. It means that the frost resistance mark for the test specimens of PC-1 series is F100 according to LST 1428-17. The PC-2, PC-3 and PC-3 series specimens did not pass the frost resistance test because the change in compressive strength of PC specimens compared to control specimens decreased more than 5%.

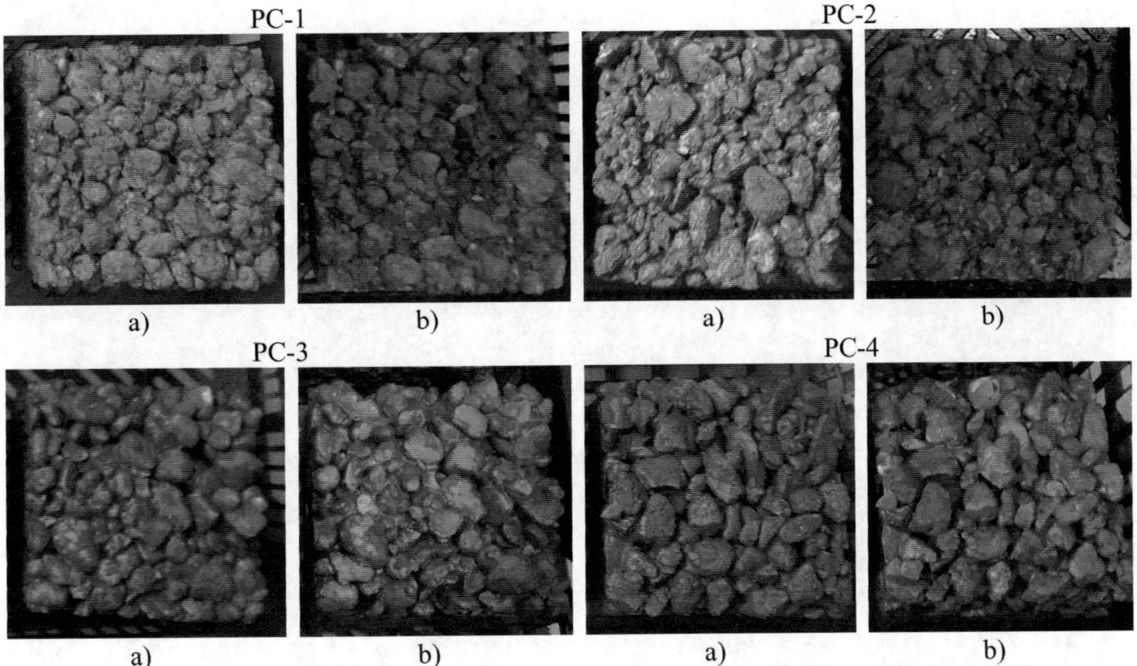

Figure 6. The view of PC series specimens before (a) and after a different number of freeze-thaw cycles (b) (PC-1, PC-2, PC-3 and PC-4 - compositions of tested mixture series).

The results of the average wear by mass of PC series specimens are given in Figure 7, while the appearance of PC series specimens before and after the wear resistance test is presented in Figure 8. According to LST 1428-15, the wear resistance of concrete is determined by the difference in the test specimen mass before and after the abrasion resistance test performed on a 600-meter abrasion road. Figure 7 shows that the specimens of PC-1 series had the lowest average wear by mass after the abrasion resistance test on 600-meter abrasion road whereas the specimens of PC-2 series had the highest values of wear by mass. It should be noted that the highest wear of PC-2 series specimens was caused by the ravelling of the angular particles of the coarse aggregate from the concrete matrix. It means, that not all angular particles of the coarse aggregate were properly covered by the cement paste and therefore ravelled from the concrete matrix during the abrasion resistance test.

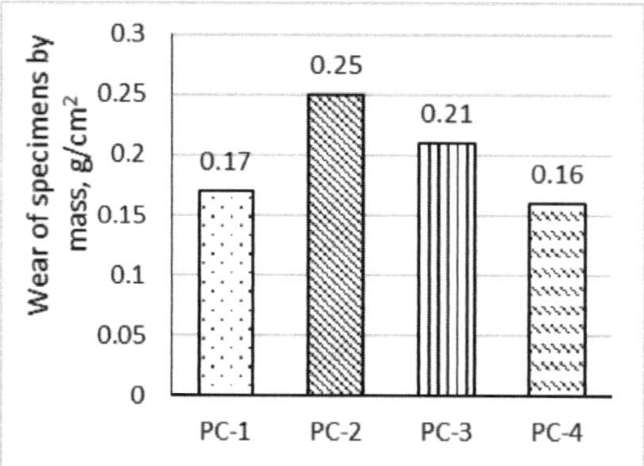

Figure 7. The results of abrasion resistance of PC (PC-1, PC-2, PC-3 and PC-4 - compositions of tested mixture series).

Figure 8. The view of PC specimens series before a test (a) and after a wear resistance test (b) (PC-1, PC-2, PC-3 and PC-4 - compositions of tested mixture series).

The application and final performance of permeable concrete, i.e., the rate of permeability, strength, resistance to freeze-thaw cycles, and resistance to wear should be considered while designing the permeable concrete mix. Further experiments need to be carried out to improve the properties of permeable concrete. Especially, the interface between the aggregate and the paste needs to be improved.

4. Conclusions

The type of coarse aggregates used for permeable concrete influences the permeability coefficient, compressive strength, freeze-thaw resistance and abrasion resistance. The specimens containing crushed gravel stone of fraction 4/16 mm and 8/16 mm had a higher permeability coefficient and the total porosity than the specimens where gravel of fraction 4/16 mm and 8/16 mm was used as the coarse aggregate. The total porosity of PC increased from about 21.0% to 25.0% and permeability coefficient *k* increased from about 20 mm/s to 30 mm/s when crushed gravel stone of fraction 4/16 mm and 8/16 mm was used as the coarse aggregate. Angularity and jagged edges of crushed gravel stone influenced higher open porosity of permeable concrete. Higher compressive strength (15.0 MPa) was observed in the specimens containing gravel with the particle size fraction of 4/16 mm. The test specimens of permeable concrete with bigger-sized aggregates (8/16 mm) had lower compressive strength compared to the test specimens with smaller-sized aggregate (4/16 mm). The results of compressive strength test revealed that not all coarse aggregate particles were properly covered by the cement paste, i.e., in some places there was no interfacial transition zone between the coarse aggregates and the cement paste observed. The load of 11.63 kg using in the vibropressing compaction procedure may have been too high. The permeable concrete with 4/16 fraction gravel was frost resistance after 108 freeze-thaw cycles. The lowest wear resistance of permeable concrete was determined when gravel with a particle size fraction of 4/16 mm was used. Consistency tests by slump method showed that the tested concrete mixes were stiff and had low consistency. Further studies of the properties of previous concrete should focus on improving the interface between the aggregate and the cement paste.

References

[1] Tennis PD, Leming ML, Akers DJ 2004 Pervious Concrete Pavements. *Portland Cement Association* Maryland USA 36 p.
[2] Meininger C 1988 No-fines pervious concrete for paving *Concr. Int.* **10** 20-27.
[3] Gola L, Vaclavik V, Valicek J, Harnicarova M, Kusnerova M, Dvorsky T 2015 Drainage concrete based on cement composite and industrial waste *Mech. Mater. Eng. Mod. Struct. Comp. Des.* **70** 155-165.
[4] Mrakovčić S, Čeh N, Jugovac V 2014 Effect of aggregate grading on pervious concrete properties *Građevinar* **66** 107-113.
[5] Larrard F, Belloc A 1997 Influence of aggregate on the compressive strength of normal and high-strength concrete *ACI Mater. J.* **94** 26-33.
[6] Yang J, Jiang G 2003 Experimental study on properties of pervious concrete pavement materials *Cem. Concr. Res.* **33** 381-387.
[7] Lian C, Zhuge Y 2010 Optimum mix design of enhanced permeable concrete – An experimental investigation *Constr. Buil. Mater.* **24** 2664-2671.
[8] Sonebia M, Bassuoni M, Yahia A 2016 Pervious Concrete: Mix Design, Properties and Applications *RILEM Technical Letters* **1** 109-115.
[9] Kerven JT, Schaefer VR 2013 Mixture proportioning considerations for improved freeze-thaw durability of pervious concrete *J. Mater. Civ. Eng.* **25** 886-892.
[10] Wu H, Huang B, Shu X, Dong Q 2011 Laboratory evaluation of abrasion resistance of Portland cement pervious concrete (PCPC) *J. Mat. Civil. Eng.* **23** 697-702.

Thermal investigation on high temperature treatment of cement mortars with high content of marble powder

V Stoyanov*[1,2], V Petkova[3], B Kostova[4], T Kaljuvee[5]

[1] University of Structural Engineering and Architecture „Lyuben Karavelov“, 175 Suhodolska Str., 1373 Sofia, Bulgaria
[2] Academy of the Ministry of Interior, Faculty of Fire Safety and Civil Protection, 171 Pirotska Str., 1309 Sofia, Bulgaria
[3] Institute of Mineralogy and Crystallography, Bulgarian Academy of Sciences, Acad. G. Bonchev Str., bldg.107, 1113 Sofia, Bulgaria
[4] New Bulgarian University, Department of Natural Sciences, 21 Montevideo Blvd., 1618 Sofia, Bulgaria
[5] Tallinn University of Technology, Ehitajate tee 5, 19086, Tallinn, Estonia

* E-mail: vensy.stoyanov@vsu.bg

Abstract. The object of the study is cement compositions based on white Portland cement, characterized by a high content of marble powder and a low water-to-cement ratio. The samples were prepared with different concentrations of the components, affecting their densities and strengths determined at 28 days or 120 days of curing. The IR spectroscopy and scanning electron microscopy supplement the thermal analysis (heating rate of 10°C/min in the temperature range of 20-1200°C). This work aimed to characterize cement mortars by powder X-ray diffraction (PXRD) and thermal (TG-DTG-DSC) analysis up to 1200°C. The study obtained data on phase composition, binder-to-aggregate and water-to-cement ratio, and spectroscopic and thermal properties of cement mortars. As a result of the study, an increase in the decomposition temperatures of composites with marble powder was found. Substitution of a part of the Portland cement clinker with a marble powder leads to the formation of carbo-sulpho-aluminates.

1. Introduction

White or colored Portland cement and/or aggregates are used to achieve an aesthetic and decorative surface for cement mortars and concretes [1, 2, 3]. The microstructure of these cement composites determined their properties [4]. They are composed of calcium-silicon, calcium-aluminum, etc. minerals formed in the cement minerals hydration process [5, 6]. Such hydrated phases have different amounts of crystallization and structurally bound water. Their type and composition depend on active and inert mineral additives included in the bonding system. Most often, the production of cement mortars and concretes was by man-made waste additives, reducing the obtained composites price while maintaining or improving their durability to environmental impact [7, 8]. This work deals with the influence of inert additive white marble powder on prepared white cement mortars.

Our previous research presents the results of mechanical, physical, chemical, and spectroscopic investigations of the properties of cement mortars based on white cement, quartz sand, and marble powder [9, 10, 11]. The obtained results prove the formation of calcium-silicon hydrosilicates and establish the difficulties in identifying the formed hydrate phases, which provoke continuing the research in these systems based on cement binder.

The present work aims to study the influence of i) marble powder (mineral additive) amount and ii) water-cement factor on the hydration and phase formation processes of hydro-sulphate and hydro-sulpho-carbonate calcium-silicate phases occurring in white cement mortars. The study was accomplished by thermal analysis (TG-DTG-DSC) and powder X-ray diffraction (PXRD).

2. Samples and methods

2.1. Samples preparation

The white Portland cement - CEM I 52.5 N (BDS EN 197-1), produced by Devnya Cement (Bulgaria) with chemical composition SiO_2 – 24.3; Al_2O_3 – 2.1; Fe_2O_3 – 0.2; CaO – 68.3; MgO – 0.3; Na_2O – 0.13; K_2O – 0.02; Free lime – 1.9 (weight %). The following mineral composition calculated by Bogue method were defined - C_3S – 72.13; C_2S – 15.28; C_3A – 5.23; C_4AF – 0.61 (weight %). The mortar samples were prepared of two types of aggregates – river sand and marble powder (Table 1). The Q-samples were prepared with river sand with SiO2 content of over 85%. The M-samples are with marble powder. The marble powder, produced by the company White Marble Products AIAS S.A. (Greece), was with the following chemical composition $CO_2 + H_2O$ – 45.7; SiO_2 – 0.12; Al_2O_3 – 0.38; Fe_2O_3 – 0.14; CaO – 32.9; MgO – 20.0; Na_2O – 0.05; K_2O – 0.19; MnO – 0.01 (weight %), which define the mineral composition of dolomite $CaCO_3.MgCO_3$ and calcite $CaCO_3$.

Table 1. Sample composition.

Samples	Aggregate	Ratio		
		cement-aggregate	water- cement	water-fines***
Q028* and Q120** (referent samples)	river sand	1:3	0.50	0.500
M028* and M120**	marble powder	1:2	0.60	0.353

*28 days stored underwater; ** 120 days stored underwater; ***all particles with sizes below 125 μm.

The samples were prepared with distiller water. The ratios of cement- aggregate, water-cement, and water-fines are shown in Table 1. The referent Q-samples were prepared with composition satisfying the EN 196-1 (section 6) requirements [12, 13] for the determination of the mortar compressive strength. The M-samples were prepared with composition satisfying the EN 196-1 with a maximum of water where the sedimentation or segregation borders were not observed 10 min after mixing [12, 14].

These samples were prepared by mechanical mixing, and compacting in 40×40×160 mm moulds, by the standard, for 1 day in a moist atmosphere (> 95% RH and 20°C). Then the unmolded samples were stored underwater (20°C) until strength testing (28 and 120 days).

2.2. Methods

Powder X-ray diffraction (PXRD) measurements were collected by D2 Phaser BrukerAXS, CuKα radiation (λ = 0.15418 nm) (operating at 30 kV, 10 mA) from 5 to 70 °2θ with a step of 0.05°, 1 s/step (ground sample weight – 1.0±0.1 mg and particle size below 0.075 mm). The PDF (Powder Diffraction File (PDF) 2001, copyright by International Centre for Diffraction Data (ICDD) database was used for determining the phases and minerals in the samples.

Thermal analysis - TG/DTG-DSC was carried out on a Setsys Evolution 2400, SETARAM, France, combined with an OmniStar mass-spectrometer, in the temperature range room temperature (RT) – 1100°C; in gas mediums – static dry air atmosphere, with a heating rate of 10°C.min^{-1} and sample mass of 19.8±20.0 mg.

3. Results and discussion

3.1. PXRD

Figures 1, 2, and Table 2 show the results of the PXRD analysis.

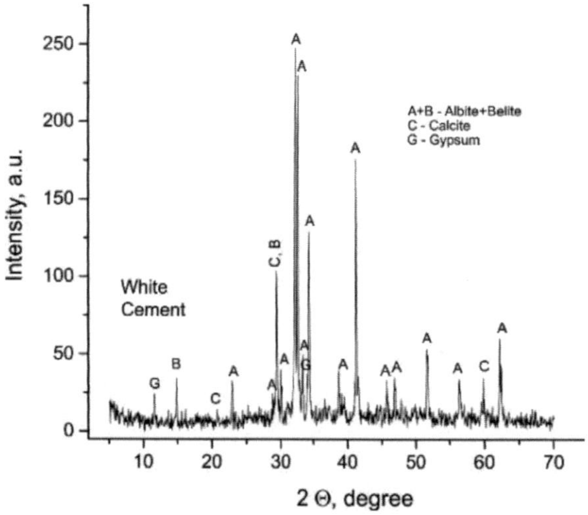

Figure 1a. PXRD pattern of white Portland cement.

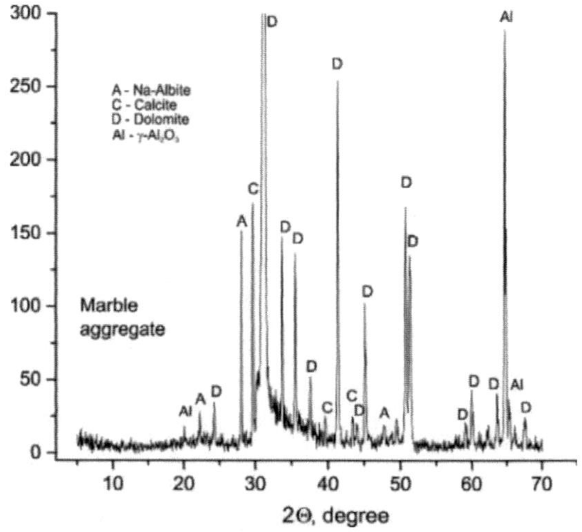

Figure 1b. PXRD pattern of marble powder.

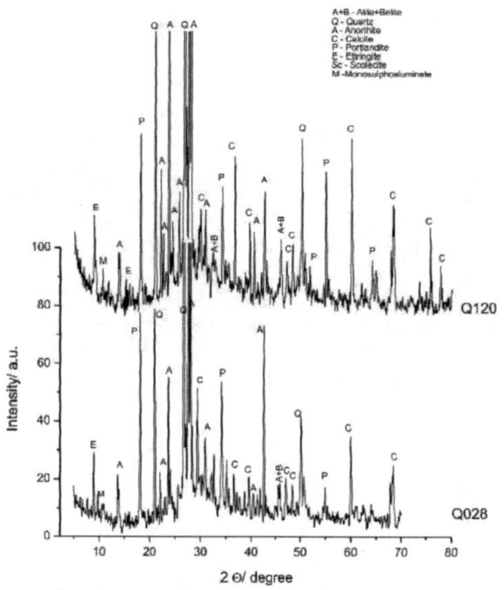

Figure 2a. PXRD patterns of Q028 and Q120.

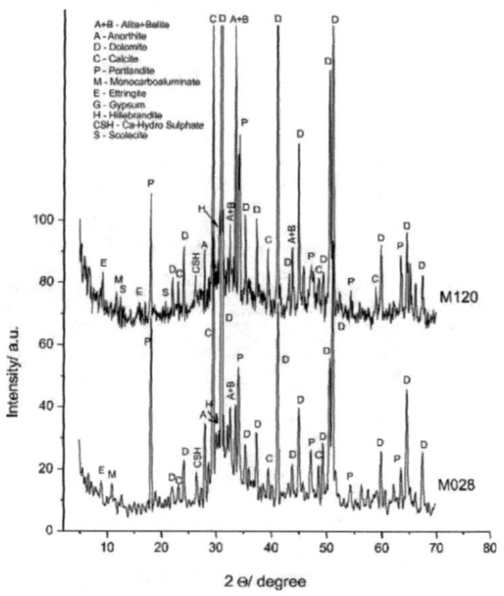

Figure 2b. PXRD patterns of M028 and M120.

The relict phases from the raw materials were detected - belite, alite, anorthite, and Mg-calcite for all samples, quartz for Q-samples, and dolomite for M-samples (Table 2). The relict raw phases were diagnosed because of uncomplete hydration of binder phases due to insufficient water.

The newly formed hydrated phases, with lower water content, were detected. The $Ca(OH)_2$ (portlandite), formed at the beginning of the hydration processes (Q028 and M028), is also present in Q120 and M120.

Carbonate and sulphate ions also involved in the formation of hydrate phases. Such are detected by PXRD - monocarboaluminate $3CaO.Al_2O_3.CaCO_3Ca(OH)_2.5H_2O$ (for M120), hemicarboaluminate $3CaO.Al_2O_3.0.5CaCO_3.0.5Ca(OH)_2.5.5H_2O$ (for M028), ettringite $6Ca(OH)_2.Al_2(SO_4)_3.26H_2O$ (samples Q028, Q120, M028, and M120), monosulphoaluminate $Ca_4Al_2(OH)_{12}(SO_4).6H_2O$ (Q028, Q120) and calcium hydrogensulphate $Ca(HSO_4)_2$ (Q028, Q120, M028, M120).

The crystal hydrates that form the C-S-H hydrosilicate gel are hillebrandite and scolecite. In M-samples, the hydrated calcium-aluminum silicates and aluminates have a greater variety – monocarboaluminate, hemicarboaluminate and ettringite. The comparison of the PXRD patterns shows that in Q-samples (white Portland cement, quartz sand, and water) ettringite is formed (Fig. 2a), while in M-samples (white Portland cement, marble powder, and water) - ettringite and carboaluminates (Fig. 2b). Despite the lower water-cement ratio, portlandite and calcium hydrosilicates were formed in the mortar samples.

Table 2. Results from PXRD analysis of raw materials (white Portland cement, marble powder, and sand) and Q028, Q120, M028, and M120.

No	Description	Sample	Identified phases (name, ICDD card number; formula)
1.	Minerals of raw materials and their relicts in samples		
		white Portland cement, Q028, Q120; M028, M120	Belite (C_2S), #49-1673 – Ca_2SiO_4 Alite (C_3S), #11-0593 – Ca_3SiO_5 Calcite, #47-1743 – $CaCO_3$
		white Portland cement	Tricalcium aluminate (C_3A), #38-1429 – $Ca_3Al_2O_6$ Gypsum, #33-0311 – $CaSO_4.2H_2O$
		marble powder, M028, M120	Dolomite, #36-0426 – $CaMg(CO_3)_2$ Mg-rich Calcite, #43-0697 – $CaCO_3$
		Q028, Q120	Quartz, #46-1045 – SiO_2
2.	Newly formed phases		
2.1.	hydrate phase	Q028, Q120 M028, M120	Portlandite (CH), #44-1481 – $Ca(OH)_2$
2.2.	C-S-H gel phases	Q028, Q120 M028, M120	Hillebrandite, #29-0373, #42-0538 – $Ca_2SiO_3(OH)_2$
		Q028, Q120	Scolecite, #41-1355 – $CaAl_2Si_3O_{10}.3H_2O$
2.3.	OH^- and HCO_3^- / CO_3^{2-} phases	M028, M120	Monocarboaluminate, #41-0219 – $Ca_4Al_2(OH)_{12}(CO_3).5H_2O$
		M028, M120	Hemicarboaluminate, #41-0221 – $Ca_4Al_2(OH)_{12}(OH)(CO_3)_{0.5}.4H_2O$
2.4.	OH^- and HSO_4^- / SO_4^{2-} phases	Q028, Q120 M028, M120	Ettringite, #41-1451 – $6Ca(OH)_2.Al_2(SO_4)_3.26H_2O$
		M028, M120	Calcium hydrogensulphate, #85-1271 - $Ca(HSO_4)_2$
		Q028, Q120	Monosulphoaluminate, #50-1607 – $Ca_4Al_2(OH)_{12}(SO_4).6H_2O$

3.2. Thermal analysis

Figures 3-6 and Table 3 show the result of thermal analysis. The measured mass loss (ML) are distributed in four ranges: I range with ML between 1.00-1.53%, II range - 0.46-0.70%, III range – 2.60-

4.36%, and IV range – 3.30-30.10%. The total ML shows an increase from 11.44% (Q028) to 40.70% (M120), due to the increase of the ML in the I range (dehydration) and in the IV range (decarbonation).

The ML distribution is also associated with the different temperature ranges in which dehydration processes take place at temperatures of 40-200°C (I range), dehydroxylation at 420-470°C (II range), dehydroxylation and decarbonation at 520-700 C (III range), and decarbonation at 700-830°C (IV range) [6, 9, 15-21]. The defined four temperature ranges are typical for raw materials (Figs. 3, 4) and for mortar samples (Figs. 5, 6). Differences are registered in the ML in the III and IV ranges, produced by different amounts of vaporized crystallization- and structurally-bound water and diverse CO_2 realisation during decarbonation. These differences are due to the raw materials used, their ratios in prepared samples, and the type of microstructure formed.

Ist temperature range: dehydration processes - 40-200°C

In the low-temperature range, the dehydration of the crystallisation water from the newly-formed hydrate phases takes place. Data from PXRD analysis prove the existence of phases that form gypsum and C-S-H gel. The C-S-H gel is composed of hydrosilicates with a fine crystalline structure formed during the cement phases hydration, often X-ray amorphous [9, 10, 22]. The registered ML from dehydration is of C-S-H gel phases, and the low value is due to their low content in the mortar samples. For the present study, changes in higher temperature ranges are of importance.

IInd temperature range –dehydroxylation process - 420-470°C

The thermal effects registered in this narrow temperature range are weak but well manifested and constant for all mortar samples. The temperature dependences and ML in this range are due to dehydroxylation of portlandite [6, 9-11, 15-21]. The obtained lower values of ML in Q- and M-samples are via the smaller water-cement ratio and type of water bonding in hydrosilicates, forming C-S-H gel.

The hydration processes are established with the registered decrease of portlandite amount (bound water in portlandite) and increase of water content of the C-S-H phase despite the measured low ML (Table 3).

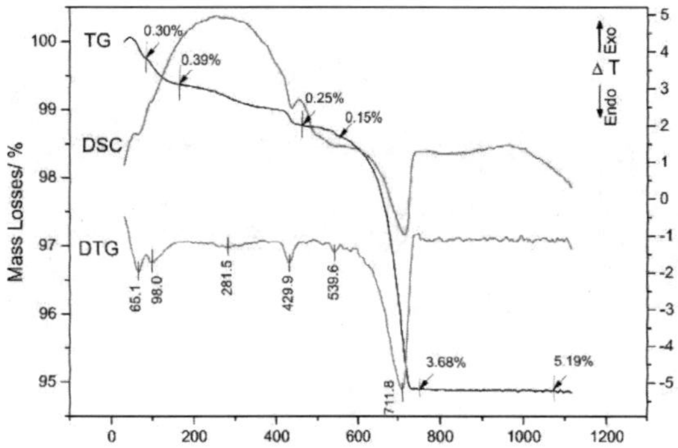

Figure 3. TG-DTG-DSC curves of white Portland cement.

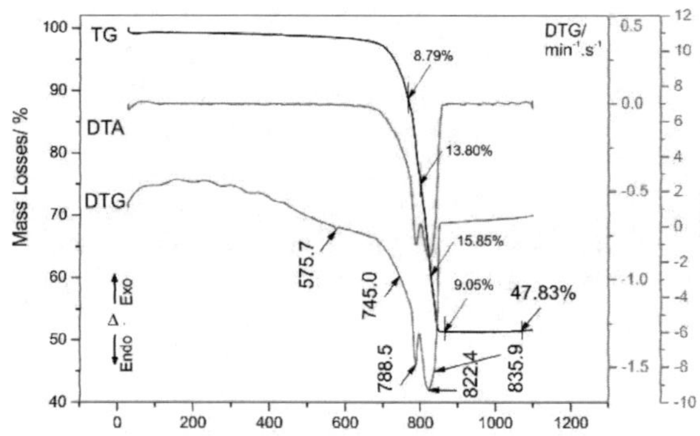

Figure 4. TG-DTG-DSC curves of marble powder.

Table 3. Data on temperature intervals, inflection point temperatures ($T_{infl.}$) and mass losses in thermal decomposition of series *Q* and *M* samples

Process / Sample	I range 40-200°C Dehydratation		II range 420-470°C Dehydroxylation of $Ca(OH)_2$		III range 520-700°C Dehydratation and dehydroxylation of C-S-H gel		IV range 700-830°C Decarbonation of $MgCa(CO_3)_2$		Total ML / %
	$T_{infl.}$* / °C	ML / %	$T_{infl.}$ / °C	ML / %	$T_{infl.}$ / °C	ML / %	$T_{infl.}$ / °C	ML / %	
White Portland cement	65.1	0.30	429.9	0.25	539.6	0.15	711.8	3.68	5.19
	98.0	0.39							
Marble powder	-	-	-	-	575.7	7.70	745.0	13.33	47.85
							788.5	15.85	
							822.4	9.05	
Q028	108.5	1.01	465.3	0.70	696.7	1.46	742.5	3.33	11.44
	177.1	0.52			625.1	1.90			
	-	-			682.8	1.00			
Q120	100.4	0.68	479.2	0.56	562.9	2.46	743.3	1.87	10.48
	130.2	0.36			-	-	764.6	0.59	
	169.5	0.44			675.7	1.82			
M028	85.2	0.50	469.0	0.91	529.5	4.15	794.3	11.91	42.80
	116.2	0.82			685.5	1.68	814.4	8.69	
					734.7	4.49	827.3	5.95	
M120	89.1	1.10	472.0	0.62	560.0	4.23	775.1	9.87	41.85
					703.2	3.00	806.9	12.49	
					754.0	3.24	824.5	4.92	

$T_{infl.}$ – Temperature of the point of inflection

IIIrd temperature range –dehydroxylation and decarbonation processes – 520-700°C

The ML in the temperature range 520-700° C do not exceed 6.4% (M120), but their distribution shows changes in the amount of structurally bound water. The thermal analysis was supplemented by measuring evolving gas composition to clarify and correctly interpret the reaction mechanism.

The analysis of thermal decomposition results of the cement composites in this temperature range shows the preservation of the ML for the Q samples (a total of 4.36% for Q028 and 4.28% for Q120) (Table 3, Fig. 5). The M samples show a slight increase in ML up to 10.32% was measured (M028) and 10.47% - M120 (Fig. 6). The more than twice higher ML values are explained by bigger amounts of hydroxyl, carbonate, and/or sulfate ions in the structure of the M-samples.

In this temperature range, the dehydroxylation of the structurally bound water in the C-S-H gel phases occurs [15-18, 20]. The decomposition processes are complex – the thermal reactions overlapping occurs in which hydroxyl and carbonate ions are released. The mass spectrometric analysis of the evolving gases established parallel dehydroxylation and decarbonation. From the analysis of the results obtained for the studied mortar samples, it can be assumed that in the Q samples carbonate ions are released together with the hydroxyl one (identified as OH^- and CO_2 -Fig. 5) [23, 24]. The Q samples do not contain sulphur oxides. This fact proves that during the hydration process, the formation of hydroxyl- and carbonate-containing phases such as ettringite and mono/hemi-carboaluminates together with portlandite $Ca(OH)_2$ and gypsum occur. The simultaneous evolution of carbonate and hydroxyl ions implies both their independent emission from the solid phase and the joint emission in the form of hydroxycarbonate ions.

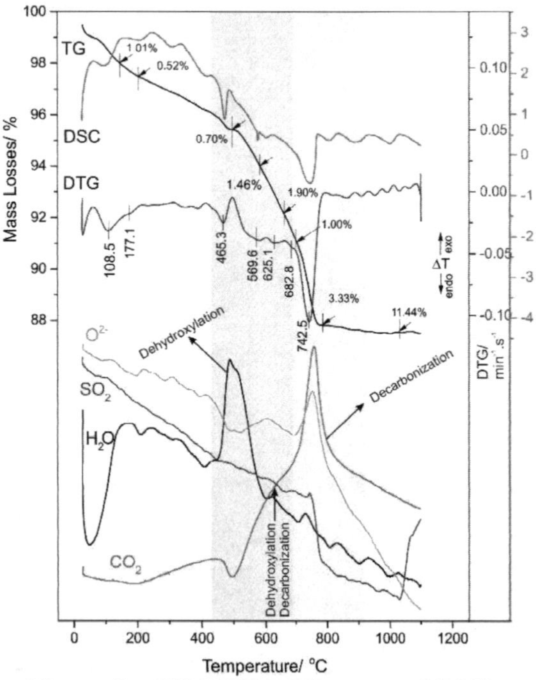

Figure 5a. TG-DTG-DSC curves of Q028.

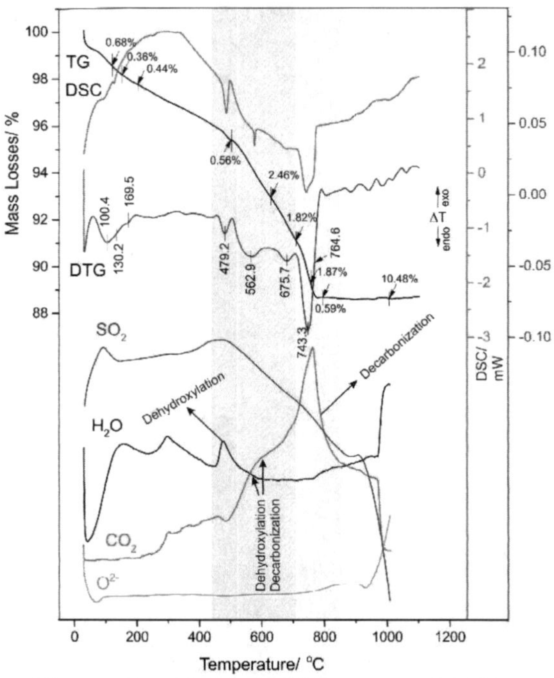

Figure 5b. TG-DTG-DSC curves of Q120.

In the M samples, in addition to the carbonate and hydroxyl ions, also sulphur oxides are registered (Fig. 6). It supposes the existence of hydroxyl, carbonate/hydroxycarbonate, and sulphate phases. During the IV temperature range, the sulphate (gypsum, hemihydrate, and anhydrite) and carbonate (dolomite, calcite) phases are thermally stable. The presence of OH^-, SO_x, CO_2, and O^{2-} could be due to release from phases, with such ions at different crystallographic positions.

Thermal reactions in the temperature range 520-700°C could be referred to as dehydration-decarbonation with the release of carbonate ions from hydrogen-carbonate phases.

IVth temperature range – decarbonation process - 700-830°C

The decomposition of the carbonate phases was at higher temperatures compared to the carbonate phases in the raw white cement and marble powder (Table 3, Fig. 3-6). In the decarbonation range, significant differences were registered between the due to different amount of carbonates: 3.68% in the white cement, 38.23% in marble powder, 27.28% in M120. It is observed the maintenance with minor deviations of decomposition temperature ranges. For marble powder, it is characteristic that decarbonation occurs with several parallel reactions at close decomposition temperatures leading to DTG/DSC peaks splitting, observed both in raw marble powder thermal analysis (Fig. 4) and in the M samples (Fig. 6). The registered ML is of dolomite and calcite decomposition.

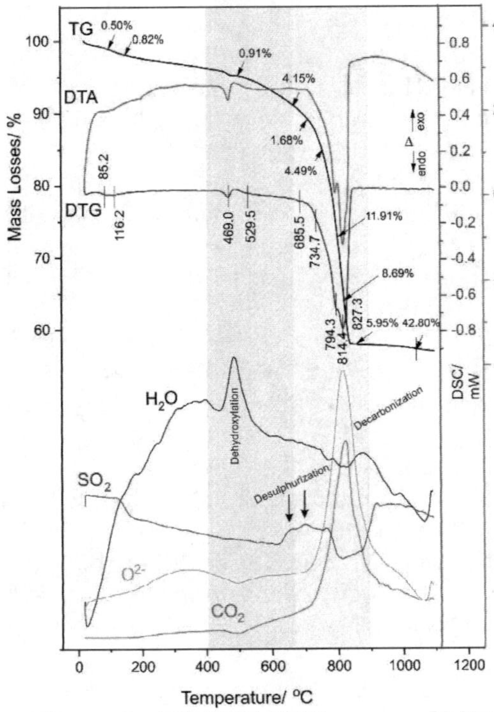

Figure 6a. TG-DTG-DSC curves of M028.

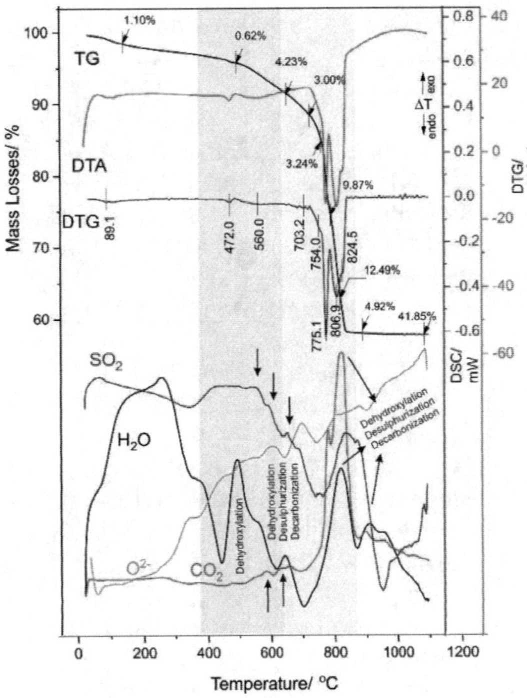

Figure 6b. TG-DTG-DSC curves of M120.

4. Conclusion

The analysis of thermal investigations outlines the following conclusions:

i. Marble powder is considered an inert additive that does not participate in the hydration process. Nevertheless, its use as a partial substitute for Portland cement clinker results in hardened mortars with a more variety of hydrated phases reduced amount of portlandite, and increased content of structurally bound water. The mortar structure is denser (M samples in comparison to Q samples), which affects the physical and mechanical properties.

ii. Water-cement ratio is essential for the proper hydration of cement minerals. Its independent research and significance have long been known, and it has an impact in combination with other parameters controlling the hydration process - mineral and chemical additives, etc. The present study proved that the added water is insufficient for the complete phase's hydration, and the measured small amounts of ML during dehydration range, evidence it. The effect of the water-cement ratio can be better assessed by varying it in combination with other additives and fillers.

Acknowledgments

This work was realized with financial support by the Operational Program "Science and Education for Intelligent Growth", co-financed by the European Union through the European Structural and Investment Funds under grant BG05M2OP001-1.001-0008 of National Centre for Mechatronics and Clean Technology (V. P.) and Bilateral project between Tallinn University of Technology, Laboratory of Inorganic Materials of Estonian Academy Sciences and "I. Kostov" Institute of Mineralogy and Crystallography of Bulgarian Academy of Sciences (2022-2023) (V.S., V.P., B. K., T. K.).

References

[1] *Lea's Chemistry of Cement and Concrete* 2004 (Ed. P.C. Hewlett). 4th Ed., Butterworth-Heinemann, 1092 p.

[2] Hamad B S 1995 *Advanced Cement Based Materials* **2** 161–167 https://doi.org/10.1016/1065-7355(95)90017-9

[3] Ling T-C and Poon C-S 2011 *Materials & Design,* **32**, 2675–2684.
[4] Taylor H F W 1997 *Cement chemistry* (London: Thomas Telford Publishing) p 459
[5] Bensted J 1999 *Cement and Concrete Composites*, 117-121
[6] Boris R, Antonovič V, Kerienė J and Stonys R, 2016 J Therm Anal Calorim, **125**, 1061-1070
[7] Aïtcin P-C S 2007, *Binders for Durable and Sustainable Concrete*, Spon Press; 528 p.
[8] Aïtcin P-C S 2011 *Mindess, Sustainability of Concrete*, Spon Press 328 p.
[9] Stoyanov V, Kostova B, Petkova V and Pelovski Y 2012 *J Therm Anal Calorim*, **110** (1), 405-412 https://doi.org/10.1007/s10973-012-2433-0
[10] Petkova V, Stoyanov V, Pelovski Y 2012 *J Therm Anal Calorim* **109** (2), 797-806. https://doi.org/10.1007/s10973-012-2447-7
[11] Kostova B, Petkova V, Stoyanov V, Serafimova E 2016 In Proc.: *XV Int. Conf. on Thermal Analysis and Calorimetry in Russia* (RTAC-2016), 16–23 September, 2016, Saint-Petersburg, Russia, ISBN 978-5-7422-5447-8 (vol. I), Peter the Great St. Petersburg Polytechnic University, 1, 435-439
[12] EN 196-1:2006, Methods of testing cement - Part 1: Determination of strength
[13] Stoynova I, Handruleva A and Dimitrova A 2021 Modeling of rectangular plates by the elastic scheme method, *IOP Conf. Ser.: Mater. Sci. Eng.* **1141**, 012039
[14] Stoyanov V 2010 Investigation of properties of decorative mortars besed on design of experiments. In: Proc.: *Int. Conf. on Civil Engineering Design and Construction. (Eurocodes – Science and Practice*) DCB 2010, September 09–11, 2010, Varna, Bulgaria, 463-470 (in Bulgarian)
[15] Tavangarian F and Emadi R 2011 *Materials Letters* **65** 2354-2356 https://doi.org/10.1016/j.matlet.2011.04.108
[16] Pacewska B, Wilinska I and Nowacka M 2011 *J Therm Anal Calorim*, **106**, 859-868 https://doi.org/10.1007/s10973-011-1570-1
[17] Olszak-Humienik M and Jablonski M 2015 *J Therm Anal Calorim* **119** 2239-2248 https://doi.org/10.1007/s10973-014-4301-6
[18] Huang Y, Wang S, Hou P, Chen Y, Gong C and Lu L 2015, *J Therm Anal Calorim*, **119**(3), 1731–1737 https://doi.org/10.1007/s10973-014-4340-z
[19] Jo D, Leonardo R S, Cartledge F K, Mendoza Reales O A and Toledo Filho R D 2016 *J Therm Anal Calorim*, **123**, 1053-1062 https://doi.org/10.1007/s10973-015-5078-y
[20] Frankeová D and Slížková Z 2016 *J Therm Anal Calorim*, **125**, 1115-1123 https://doi.org/10.1007/s10973-016-5360-7
[21] Hassen S, Platret G and Vincent J 2016 *J Therm Anal Calorim*, **126**, 1023-1043 https://doi.org/10.1007/s10973-016-5639-8
[22] Barnett S J, D Adam, A R W Jackson and Hywel-Evans P D 1999 *Cement and Concrete Composites*, **21**, 123-128 https://doi.org/10.1016/S0958-9465(98)00019-5
[23] Carmona-Quiroga P M and Blanco-Varela M T 2013 *Cement and Concrete Research*, **52**, 140–148 https://doi.org/10.1016/j.cemconres.2013.05.021
[24] Tu Z, Guo M-z, Poon C S and Shi C, 2016 *Cement and Concrete Composites*, **72** 9-16 https://doi.org/10.1016/j.cemconcomp.2016.05.019

Influence of the concrete strength class and the span/depth ratio on the sustainability of reinforced concrete beam slabs

T Chardakova[1] **and M Traykova**[1]

[1] Department of Reinforced Concrete Structures, University of Architecture, Civil Engineering and Geodesy, 1 Hristo Smirnenski Blvd, Sofia, Bulgaria

E-mail: tanya.chardakova@gmail.com

Abstract. Slabs form a significant part of the mass of reinforced concrete structures in buildings. Therefore, their sustainable design is crucial for the overall sustainability of the structure. A critical factor for the sustainability of slabs is their depth. The initial choice is usually based on conservative span/depth ratios or span/effective depth ratios, found in literature. However, thinner slabs may often be designed at the expense of greater reinforcement ratios and/or greater concrete strength class. This paper seeks to find the most effective choice of span/depth ratio of beam slabs for different concrete strength classes in terms of their sustainability. Quantitative assessment of the slab's embodied energy per m^2 and embodied CO_2 per m^2 is made for slabs with different static schemes and with different variable loads. Only the cradle-to-gate embodied energy and CO_2 of the concrete and the reinforcement steel of the slab itself are considered. Other aspects of the sustainability of slabs, such as their influence on other structural elements, insulation properties, etc., are shortly discussed. Based on the analysis of the obtained results, some general conclusions are made and recommendations for the optimal choice in terms of sustainability of concrete strength class and span/depth ratio are given.

1. Introduction

The construction sector is a major contributor for carbon dioxide (CO_2) emissions, accounting for a significant part of all energy-related greenhouse gas emissions. The issue of sustainable development and the impact of construction activities on the environment are often discussed. Compared with other sectors, these impacts are rather significant, especially in terms of consumption of raw material resources and in terms of greenhouse gas production. It is very important to minimize these environmental impacts by suitable design and material selection.

In the same time, it can be specified, that one of the most important building material in the world is the concrete and considering the volume of concrete produced and the environmental impacts associated, the design of optimized reinforced concrete structures is a very important goal to the sustainable development.

The slabs and their behavior are closely related to the performance of the whole structure. The potential of minimising embodied energy and embodied carbon dioxide in reinforced concrete slabs by parametrically varying the strength class of concrete and effective span to effective depth ratio is one good opportunity to contribute to the sustainable design of building structures.

The studied parameters are mainly related to the serviceability limit states. The design slab thickness could be limited by fire safety, compression failure (instead of under-reinforced ductile failure), shear failure and deflection control. Deflection of slabs is a principal criterion in design and governs the final choice of the thickness. Deflection is usually controlled by limiting span/depth ratio. Relaxing the deflection limit and reducing slab depths can reduce the negative impact but can lead to problems during the service life.

2. Scope of the study

2.1. Objectives of the study
The study is focused on the environmental impact of reinforced concrete beam slabs (slabs, supported by linear supports) in buildings. The environmental impact is assessed quantitatively by the embodied energy for square meter of the slab and the carbon dioxide emissions for square meter of the slab. Only the cradle-to-gate part of the life cycle is considered.

The main objective of the study is to find the most appropriate (in terms of their environmental impact) effective span to effective depth ratios of different types of beam slabs, depending on the applied imposed loads. Another objective is to fine the optimal choice of concrete strength class in this same context.

2.2. Limitations of the study
The quantitative part of the study does not consider:
- The social or economical terms of the sustainability of the slab.
- The environmental impact, associated with water and other resources consumption.
- The environmental impact, associated with green house gasses emissions, other than carbon dioxide emissions.
- The environmental impact, associated with actual construction or demolition process, since they are very dependent on the specifics of each individual construction site.
- The environmental impact, associated with the lifetime operation of the building, since they are dependent on the behaviour of the occupants. In any case, this impact is mostly associated to the insulation thermal properties of the building, which are impossible to be assessed only based on the thickness of the slab.
- The environmental impact, associated with construction waste management, both during construction and after demolition.
- The environmental impact of the supporting beams or other structural or non-structural elements, besides the slab itself.

2.3. Parameters of the study
It should be noted, that, while the parameters of the study are chosen in order to cover some of the most commonly met practical cases, it is impossible to include the endless variety of possible (and probable) cases, that the structural design practice can present. However, the results, presented in this paper, can be extrapolated for other cases, with some engineering judgment.

Specifically, the following cases are considered:
- Simply supported one-way slabs with effective spans 3.0m, 4.5m and 6.0m, respectively.
- Fixed at both supports one-way slabs with effective spans 3.0m, 4.5m and 6.0m, respectively.
- Simply supported at all four sides two-way slabs with effective spans 4.0m/4.0m, 6.0m/6.0m, 8.0m/8.0m, 4.0m/6.0m and 6.0m/9.0m respectively.
- Fixed at all four sides two-way slabs with effective spans 4.0m/4.0m, 6.0m/6.0m, 8.0m/8.0m, 4.0m/6.0m and 6.0m/9.0m respectively.

The considered concrete classes are C20/25, C25/30 and C30/37, as per [1]. The steel class is always B500B, as per [1].

The slabs are loaded with permanent distributed load with characteristic value, equal to 1,5kN/m^2 plus the self weight of the slab (calculated for volume weight of the reinforced concrete 25kN/m^3). Each slab is solved for three different variations of the imposed loads:

- imposed load, corresponding to Category of use A (areas for domestic and residential activities), as per [2], with characteristic value q_k=2kN/m^2, and quasi-permanent value as per [3] $\psi_2.q_k$=0,3.2=0,6kN/m^2; the results can also be considered relevant to Category of use B (office areas);
- imposed load, corresponding to Category of use C (areas, where people may congregate), as per [2], with characteristic value q_k=5kN/m^2, and quasi-permanent value as per [3] $\psi_2.q_k$=0,6.5=3,0kN/m^2; the results can also be considered relevant to Category of use D (shopping areas);
- imposed load, corresponding to Category of use E (areas of storage and industrial use), as per [2], with characteristic value q_k=7,5kN/m^2, and quasi-permanent value as per [3] $\psi_2.q_k$=0,8.7,5=6,0kN/m^2.

3. Description of the study

For each considered case a wide range of possible slab thicknesses h is considered. Slabs, thinner than 10cm, are not considered for technology and fire protection reasons. Each thickness is multiple by 2cm (e.g. 10cm, 12cm, 14cm, etc.). The range starts with the thinnest slab, which can still be designed to meet ultimate and survivability limit state criteria, under the following assumptions:

- the slabs are designed only with tensile reinforcement (no reinforcement in compression zone is included in the calculations);
- ductile failure is required – the depth of the compression zone is limited to 0,45.d, where d is the corresponding effective depth;
- the reinforcement, required by calculation for bending moment, can be increased by no more than 50% to fulfil shear force and deflection verifications;

If any of the ultimate or serviceability limit states is not complied under these assumptions, this thickness is excluded from the analysis. All calculations are carried out as per [1].

For any given slab, the loads and action effects (bending moment and shear force) are first calculated. The one-way slabs are solved as elastic beams with 1 meter width. The two-way slabs are solved, using Marcus' Method. The nominal concrete cover is assumed 20mm, when the imposed load is 2kN/m^2 (and bars with diameter no larger than 10mm are expected), and 25mm, when the imposed load is higher (in this case the diameter of the bars is limited to 14mm). The minimal bar diameter used is 6mm.

The shear force is verified as for an element, without shear reinforcement. Shear failure is not relevant for any of the cases. The deflection is controlled indirectly by the limit value of effective span/effective depth ratio, given in [1]. Cracking is controlled indirectly by limiting the bar diameter and by introducing minimal reinforcement.

The spacing of the bars is taken between 10cm and the maximum spacing, allowed by [1]. Since the spacing is rounded down to 1cm, there is some overdesign of the reinforcement. Wherever possible, this is kept under 5%.

After the design is finished, the total quantity of the reinforcement is calculated, using the following detailing rules:

- all bars are anchored to n times the diameter of the bar, where n=47 for concrete class C20/25, n=40 for C25/30 and n=36 for C30/37 (calculated as per [1]);
- all bar lengths are rounded up to 5cm;
- the quantity of the reinforcement of one-way slabs is calculated for a strip with width 1 meter; reinforcement over the short supports is not considered;
- the bottom reinforcement has length, equal to the span plus half the anchorage length at each support (the reduction is possible, due to the decrease of steel stresses near the supports);
- the top reinforcement above fixed supports has length, equal to 30% of the span plus anchoring;

- simply supported slabs have top reinforcement above each support, with area no less than the minimal and at least 25% of the bottom one; this reinforcement have length, equal to 20% of the span plus anchoring;
- the distribution reinforcement in one way slabs is taken no less than 20% of the calculated bottom reinforcement;
- in slabs with thickness, greater than 20cm, there is at least minimal top reinforcement at both directions along the entire slab;
- where there is top reinforcement only in one direction, assembly reinforcement is introduced at a distance of 35cm;
- where there is any top reinforcement, 3 spacers per square meter are calculated; the length of a spacer is 50cm plus two times the distance from the formwork to the lowest top bar.

This total mass of the reinforcement is then divided to the area of the slab, to receive the steel mass per m^2 W_s. The volume weight of the steel is adopted 7850kg/m^3. The mass per m^2 of the concrete W_c, also found in kg/m^2, is calculated for its net volume (the volume of the reinforcement is subtracted), which is multiplied by a volume weight of 2400kg/m^3. The embodied energy per m^2 of the slab *ee* and the embodied carbon dioxide per m^2 of the slab *ecd* are calculated, based on the inventory data, found in [4] (see table 1), using equations (1) and (2):

$$ee = W_c.I_{ee,c} + W_s.I_{ee,s}, \text{ [MJ/m}^2\text{]} \quad (1)$$

$$ecd = W_c.I_{ecd,c} + W_s.I_{ecd,s}, \text{ [kgCO}_2\text{/m}^2\text{]} \quad (2)$$

Table 1. Inventory data of the materials (source [4]).

	Concrete class C20/25	Concrete class C25/30	Concrete class C30/37	Steel class B500
embodied energy of material *j*, $I_{ee,j}$ [MJ/kg]	0,74	0,78	0,85	17,4
embodied CO_2 of material *j*, $I_{ecd,j}$ [kgCO2/kg]	0,100	0,106	0,1175	1,31

4. Analysis of the results

The results are plotted at graphs, shown in figure 1, in which the abscise is the effective span to effective depth ratio, and the ordinate – the embodied energy per m2 of the slab or the embodied CO_2 of the slab, respectively. For the two-way slab, the minimal effective span is taken, and its respective effective depth, which is always the larger one (first row of reinforcement).

In each graph is given the information for one type of slab (e.g. one-way simply supported) at a given value of the imposed load q_k. The results for the different spans and concrete classes are presented with differently styled lines. For the two way slab, the results for slabs with spans 4,0m/6,0m and 6,0m/9,0m are given at a separate graph for clarity of presentation.

It can be seen at the graphs, that the general tendency is that thinner slabs (with larger effective span to effective depth ratio) have smaller ecological footprint (especially if the slab thickness is under 20cm, in which case there is no minimal top reinforcement for the entire slab), but this is not always the case. In some cases the thinnest slabs, which can still be designed under the assumptions, given in section 3, have significantly higher ecological footprint, especially in terms of embodied energy, compared to the second, or sometimes even the third chosen depth. This is a direct result of the increase of the reinforcement for deflection control – the provided reinforcement is more than the calculated for bending. This effect is most pronounced in the case of simply supported slabs, which are the most sensitive to deflection, but can also be seen in the other slabs, in the cases with large spans and/or low concrete class – the graphs, where there is an upward trend at its right end (largest effective span to effective depth ratio). In this respect, the thinnest slab, for which the deflection is verified without increase of the reinforcement, can be considered the most sustainable.

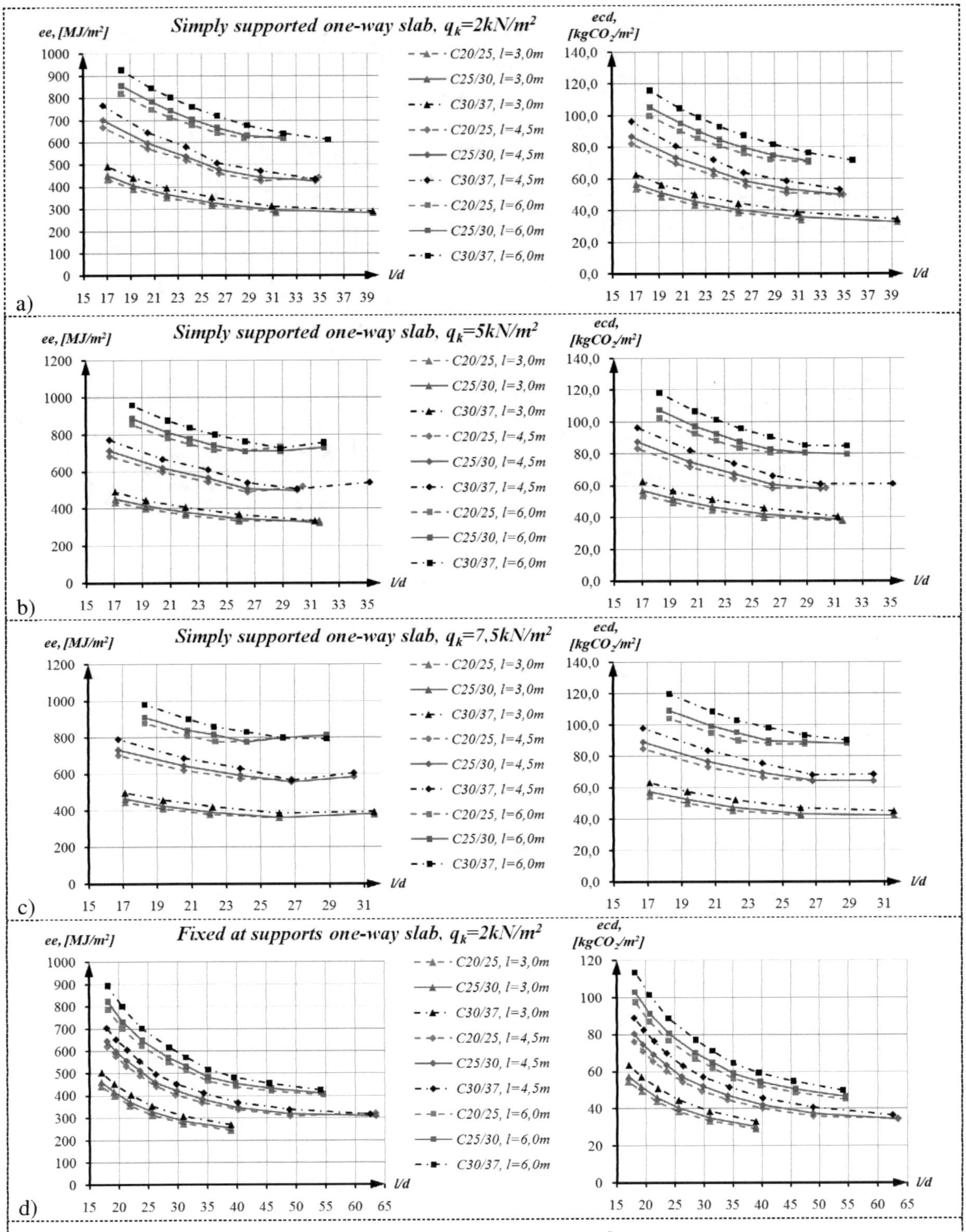

Figure 1. Embodied energy (ee) and embodied CO_2 (ecd) per m^2 of the slab to effective span/effective depth ratio for: a) simply supported one-way slab, q_k=2kN/m^2; b) simply supported one-way slab, q_k=5kN/m^2; c) simply supported one-way slab, q_k=7,5kN/m^2; d) fixed at supports one-way slab, q_k=2kN/m^2.

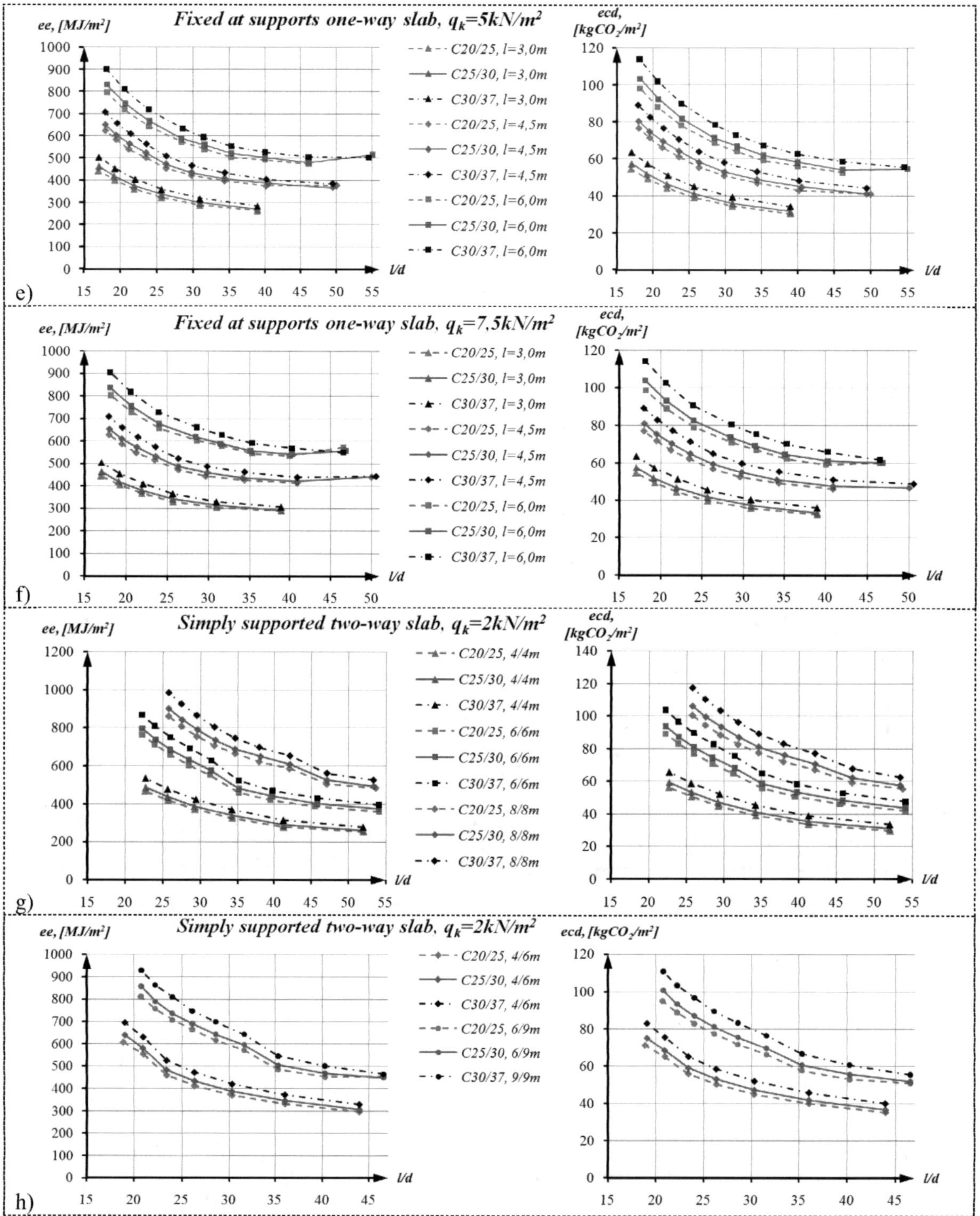

Figure 1 – continuation. Embodied energy (ee) and embodied CO_2 (ecd) per m^2 of the slab to effective span/effective depth ratio for: e) fixed at supports one-way slab, q_k=5kN/m^2; f) fixed at supports one-way slab, q_k=7,5kN/m^2; g) simply supported two-way square slab, q_k=2kN/m^2; h) simply supported two-way rectangular slab, q_k=2kN/m^2.

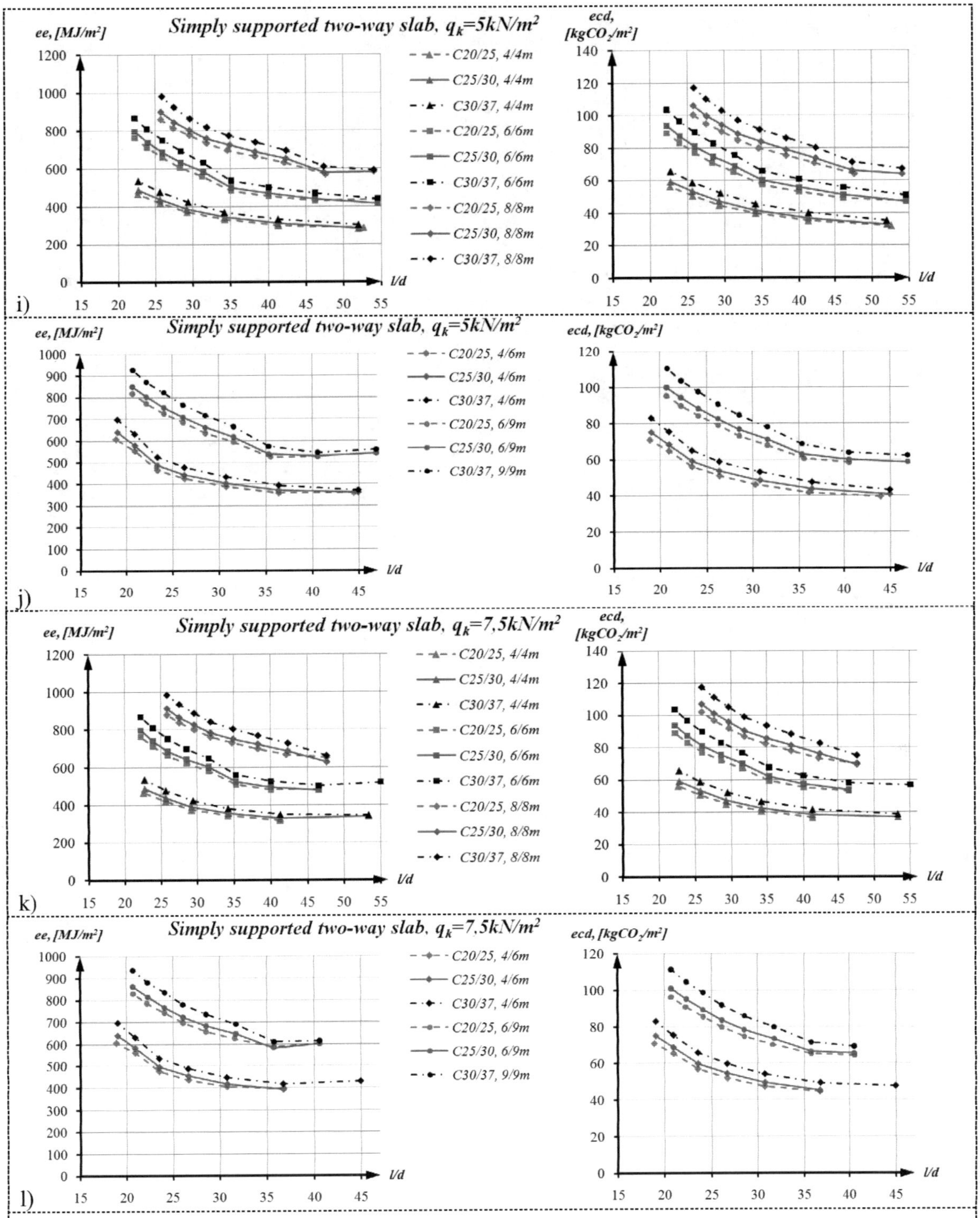

Figure 1 – continuation. Embodied energy (ee) and embodied CO_2 (ecd) per m^2 of the slab to effective span/effective depth ratio for: i) simply supported two-way square slab, q_k=5kN/m^2; j) simply supported two-way rectangular slab, q_k=5kN/m^2; k) simply supported two-way square slab, q_k=7,5kN/m^2; l) simply supported two-way rectangular slab, q_k=7,5kN/m^2.

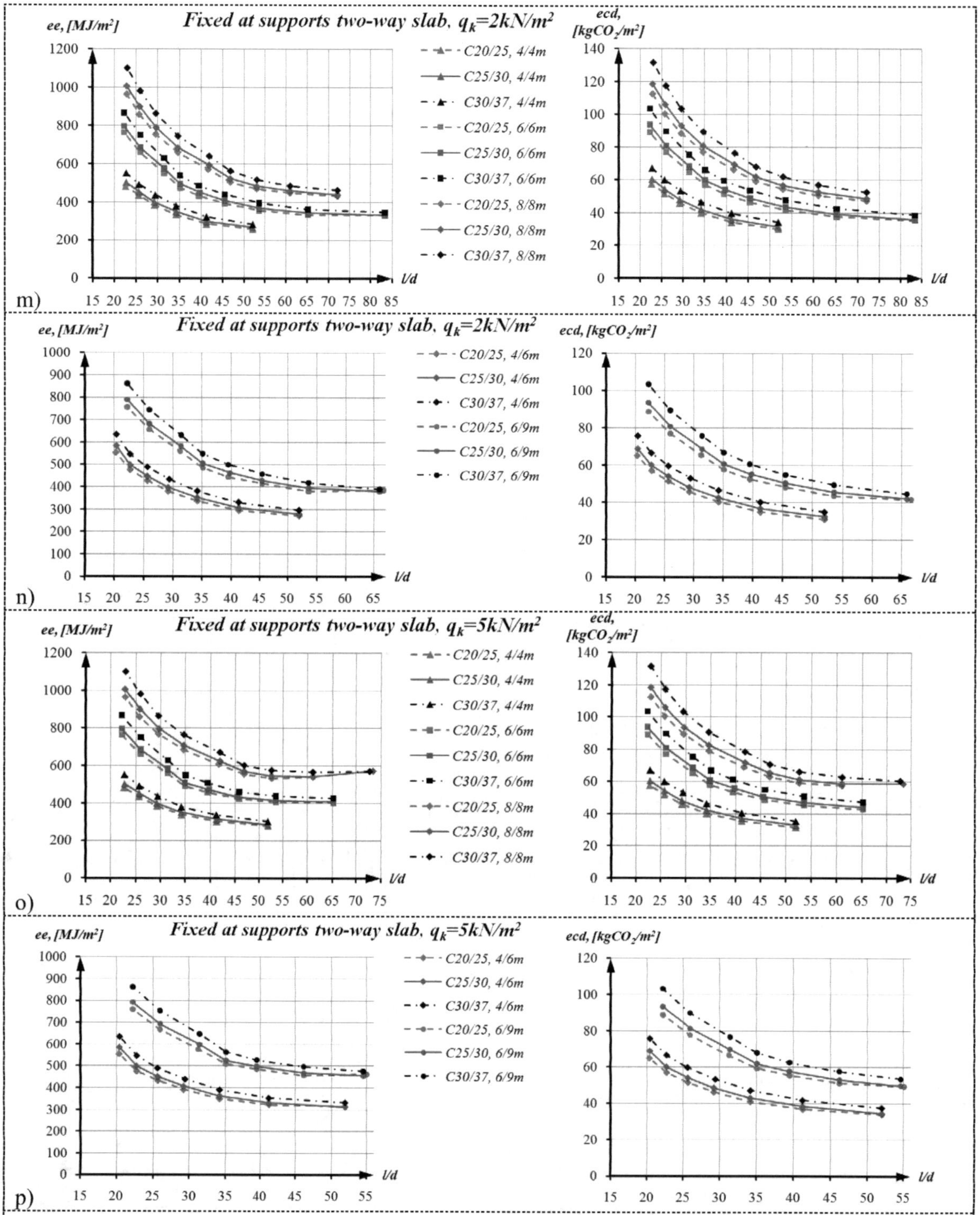

Figure 1 – continuation. Embodied energy (ee) and embodied CO_2 (ecd) per m^2 of the slab to effective span/effective depth ratio for: m) fixed at supports two-way square slab, q_k=2kN/m^2; n) fixed at supports two-way rectangular slab, q_k=2kN/m^2; o) fixed at supports two-way square slab, q_k=5kN/m^2; p) fixed at supports two-way rectangular slab, q_k=5kN/m^2.

IOP Conf. Series: Materials Science and Engineering 1276 (2023) 012006 doi:10.1088/1757-899X/1276/1/012006

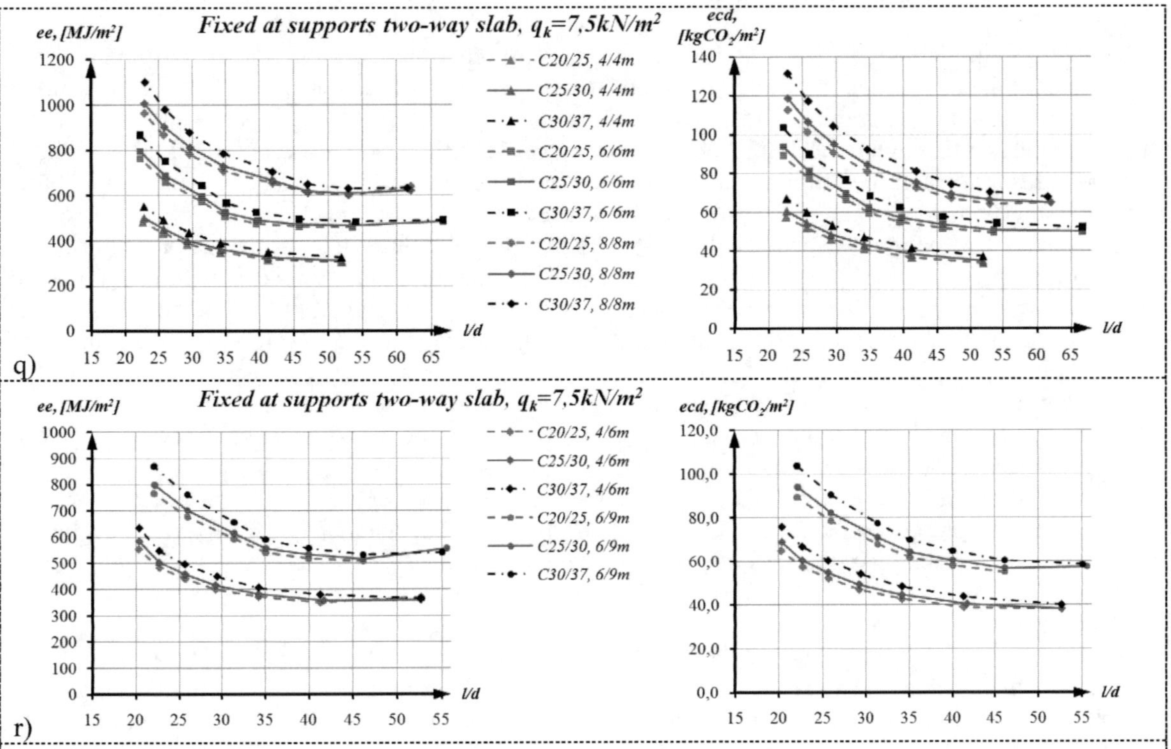

Figure 1 – continuation. Embodied energy (ee) and embodied CO_2 (ecd) per m^2 of the slab to effective span/effective depth ratio for q) fixed at supports two-way square slab, q_k=7,5kN/m^2; r) fixed at supports two-way rectangular slab, q_k=7,5kN/m^2.

Another general observation that can be done is that for the same effective span/effective depth ratio, the slabs with higher concrete class have larger ecological footprint. This is due to the fact, that for slabs the compressive strength of the concrete does not have significant effect on the total quantity of the reinforcement. However, it can be seen from the results, that the higher concrete class allows the use of larger effective span/effective depth ratio, and, ultimately, to more sustainable slabs.

The recommended effective span/effective depth ratio ranges of different types of slabs are shown in table 2. The recommendations for the most appropriate choice from the range are given in the notes under the table. In any case, the deflection verification should dictate the final choice of slab thickness.

Table 2. Optimal span/effective depth ratio range of beam slabs for different categories of use [a].

type of slab	Residential and office areas	Congregation and shopping areas	Storage areas
one-way – simply supported	28-32 (30)	24-28 (26)	22-26 (24)
one-way – fixed at both supports	50-65 (55)	40-50 (45)	35-45 (40)
two-way – simply supported	40-55 (45)	35-50 (40)	30-45 (35)
two-way – fixed at both supports	55-75 (65)	45-65 (55)	40-60 (50)

[a] Values in brackets are generally recommended. The higher end of the range (thinner slabs) is suitable for smaller spans and sufficient concrete class, and, in the case of two-way slabs – low maximum to minimum span ratio (close to 1). The lower end of the range (thicker slabs) should be preferred if the span is large or the concrete class is low, or, in the case of two-way slabs – high maximum to minimum span ratio (larger than 1,5).

5. Influence of the depth of the slab and the concrete class on other aspects of the sustainability of the building

The depth of the slab has critical role for the design of all other elements in a building structure, such as beams, columns, structural walls and foundations, because the self weight of the slab forms a significant part of the gravity loads. Additionally, in seismic regions, the increased mass of the building, related to a greater slab thickness, increases the seismic action effects proportionally. This is another reason to seek the thinnest possible slab, which is structurally adequate. As said above, this can be achieved with the choice of sufficient concrete class.

However, an important aspect of the sustainability of slabs, are their insulation properties. The thermal insulation relates to the ecological footprint of the building during its occupation, and also to the comfort of habitation (which is related to the social aspect of sustainability). The thickness of the slab is crucial for the sound insulation, and therefore, again, for the comfort of habitation. Fire protection, which is related not only to the safety of the inhabitance (social aspect) but also to the durability of the building (ecological aspect), is also depended on the thickness of the slab. All this governs the conclusion, that thicker slabs may be considered more sustainable, considering the social aspects of sustainability.

Lastly, the durability of the building has a great importance for its overall sustainability. With reinforced concrete structure, longer life is ensured by providing greater concrete cover and higher concrete class, especially in the case of unfavorable environment.

6. Conclusion

Based on the results of the study and their analysis, the following conclusion can be drawn:

1) Optimizing slab thickness and concrete strength class based on different criteria (deformations, insulation properties, durability, etc.) of reinforced concrete slabs can lead to significant improvement of the sustainability of the entire building.

2) Generally, thinner slabs (with larger effective span/effective depth ratio) tend to have smaller ecological footprint in terms of embodied energy and embodied carbon dioxide. However, if the deflection check is not fulfilled, it is more sustainable to increase the depth of the slab and recalculate it, rather than increase the provided reinforcement. The recommended effective span to effective depth ratio ranges, given in this paper, can be used for initial choice of slab thickness. In any case, the thickness of the slab should be sufficient to ensure adequate structural behavior and insulation properties.

3) While the choice of lower concrete class generally leads to smaller ecological footprint, it should be high enough to allow the use of thinner slabs, since this ultimately leads to more sustainable results. In addition, structures with higher concrete class are more durable and have longer operational life, which is crucial factor for their sustainability. The most suitable concrete class in the context of this study is C25/30.

References

[1] European Committee for Standardization 2004 *Eurocode 2: Design of concrete structures - Part 1-1: General rules and rules for buildings*
[2] European Committee for Standardization 2002 *Eurocode 1: Actions on structures - Part 1-1: General actions - Densities, self-weight, imposed loads for buildings*
[3] European Committee for Standardization 2002 *Eurocode 0: Basis of structural design*
[4] Hammond G and Jones C 2011 *The Inventory of Carbon and Energy (ICE)* ed. F Lowrie and P Tse (BSRIA) chapter 3 pp 13-20

Dynamic analysis of structures using MATLAB

J Melcer[1] and V Valašková[1]

[1] University of Zilina, Faculty of Civil Engineering, Depatrment of Structural Mechanics and Applied Mathematics

veronika.valaskova@uniza.sk

Abstract. MATLAB is a very important program tool for engineers in practice. For its effective use, it is necessary to prepare students already during their studies at the university. Structural dynamics is an important engineering discipline. The presented paper deals with some applications of the MATLAB in the teaching of Structural Dynamics at the University of Zilina. It follows the legacy of Prof. Koloušek, the author of the slope deflection method in dynamics. It focuses on solving the eigenvalue problems of structures in teaching dynamics. In the appendix it presents lists of programs in MATLAB usable for calculation of natural values of beam structures and vehicle models in teaching process and engineering practice.

1. Introduction

The University of Žilina is the successor organization of the University of Transport Technology. At the Department of Structural Mechanics of the University of Transport Technology a world format scientist worked, Prof. Ing. Vladimír Koloušek, DrSc., Fig. 1, [1].

Figure 1. V. Koloušek author of Slope-Deflection Method in dynamics.

He focused on the dynamics of structures. His benefit lies in the fact that he introduced a slope-deflection method into dynamics. Not only did he create the complete mathematical apparatus needed to solve dynamic problems by the slope-deflection method, but he also processed tables of numerical values of frequency functions, which he defined himself and which are used in the method. This was an amazing thing, because at a time when there was no computational technique, these tables allowed engineers to do the calculation manually. His book Dynamics in Engineering Structures [2] has been translated into all world languages. His name is known at all world universities. The University of Žilina is proud of prof. Koloušek and wants to stay true to his legacy. Therefore, due attention is paid to the teaching of the subject Dynamics of Structures. One of the authors of the article is a student of Prof. Koloušek. When teaching dynamics, we work with modern computational methods such as the Finite Element Method and the Component Element Method. We use effective tools such as MATLAB. Our effort is to combine theoretical knowledge with an effective numerical solution when teaching dynamics and at the same time teach students to program. Although student programs do not always reach the professional level, this activity has the advantage of forcing the student to know in detail the solution process. This is also the direction of our submitted contribution. The authors presented two examples from real practice, where they show how to process the theoretical apparatus in the form of MATLAB procedures. The authors offer much wider application options to teachers and students within the textbook Application of the Program System MATLAB in Solving Problems of Structural Dynamics [3].

2. Calculation of natural frequencies and natural modes

2.1. Models of beam structures

Discrete computational models with multi degrees of freedom can be solved in the spirit of the stiffness constant method or the compliance constant method.

The stiffness constants method, in the older literature called the method of elastic constants, is based on the conditions of dynamic equilibrium of forces acting on mass points, which are expressed through stiffness constants. Due to the fact that there is a certain analogy with the slope-deflection method, some authors call it the slope-deflection method.

The system of n equations of motion can be briefly written in matrix equation

$$[\mathbf{m}]_{\mathbf{D}} \cdot \{\ddot{\mathbf{v}}(t)\} + [\mathbf{k}] \cdot \{\mathbf{v}(t)\} = \{\mathbf{0}\}. \tag{1}$$

Particular solutions have a form

$$\{\mathbf{v}(t)\} = \{\mathbf{v}^{\mathbf{0}}_{(\mathbf{j})}\} \cdot \sin(\omega_{(j)} \cdot t) \text{ and } \{\mathbf{v}(t)\} = \{\mathbf{v}^{\mathbf{0}}_{(\mathbf{j})}\} \cdot \cos(\omega_{(j)} \cdot t). \tag{2}$$

We get the frequency equations in the form

$$([\mathbf{k}] - \omega^2_{(j)} \cdot [\mathbf{m}]_{\mathbf{D}}) \cdot \{\mathbf{v}^{\mathbf{0}}_{(\mathbf{j})}\} = \{\mathbf{0}\}, \tag{3}$$

$$[\mathbf{A}(\omega_{(j)})] \cdot \{\mathbf{v}^{\mathbf{0}}_{(\mathbf{j})}\} = \{\mathbf{0}\}, \tag{4}$$

where

$$[\mathbf{A}(\omega_{(j)})] = [\mathbf{k}] - \omega^2_{(j)} \cdot [\mathbf{m}]_{\mathbf{D}} \tag{5}$$

is the dynamic stiffness matrix of the system. The secular equation has the form

$$\det[\mathbf{A}(\omega_{(j)})] = \mathbf{0}. \tag{6}$$

Multiply equation (8.27) from the left by the inverse matrix $[\mathbf{m}]^{-1}_{\mathbf{D}}$

$$([\mathbf{m}]^{-1}_{\mathbf{D}} \cdot [\mathbf{k}] - \omega^2_{(\mathbf{j})} \cdot [\mathbf{m}]^{-1}_{\mathbf{D}} \cdot [\mathbf{m}]_{\mathbf{D}}) \cdot \{\mathbf{v}^{\mathbf{0}}_{(\mathbf{j})}\} = \{\mathbf{0}\}, \tag{7}$$

$$([\mathbf{D}]^{-1} - \omega^2_{(j)} \cdot [\mathbf{E}]_{\mathbf{D}}) \cdot \{\mathbf{v}^{\mathbf{0}}_{(\mathbf{j})}\} = \{\mathbf{0}\}. \tag{8}$$

From the mathematical point of view, this matrix equation represents a special problem of eigenvalues, where the numerical values $\lambda_{(j)} = \omega^2_{(j)}$ are calculated as eigenvalues of the matrix $[\mathbf{D}]^{-1}$ and vectors $\{\mathbf{v}^{\mathbf{0}}_{(\mathbf{j})}\}$ describing the natural modes of vibrations are calculated as eigenvectors of the matrix $[\mathbf{D}]^{-1}$. Matrix

$$[\mathbf{D}]^{-1} = [\mathbf{m}]^{-1}_{\mathbf{D}} \cdot [\mathbf{k}] = [\mathbf{k_r}] \tag{9}$$

is called the inverse dynamic matrix or the reduced stiffness matrix.

To deflections $v^0_{i(j)}$ (i = 1, ..., n), (j = 1, ..., n), defining the natural modes of vibrations, we can also come in the following way. The relative ratios of the deflections of the individual mass points for each vibration mode are equal to the mutual ratios of the algebraic complements of the subdeterminants of the matrix $[\mathbf{A}(\omega_{(j)})]$ calculated with respect to the elements of any row (index i) or column (index k) of this matrix.

$$v^0_{1(j)} : v^0_{2(j)} : v^0_{i(j)} : ... : v^0_{n(j)} = A_{i1(j)} : A_{i2(j)} : A_{ii(j)} : ... : A_{in(j)} = \\ .... = A_{1k(j)} : A_{2k(j)} : A_{kk(j)} : ... : A_{nk(j)} \,. \tag{10}$$

The values of the respective algebraic complements or their *normalized values* (values obtained by dividing them by any constant) can be considered as deflections $v^0_{i(j)}$. All natural modes of vibration, vectors

$$\{\mathbf{v^0_{(j)}}\} = [v^0_{1(j)}, v^0_{2(j)}, v^0_{i(j)}, ..., v^0_{n(j)}]^{\mathbf{T}} , \tag{11}$$

can be arranged into so-called a *modal matrix* in which the elements of each column j define the jth mode of vibration

$$[\mathbf{v^0}] = [\{\mathbf{v^0_{(1)}}\}, \{\mathbf{v^0_{(2)}}\}, ..., \{\mathbf{v^0_{(j)}}\}, ..., \{\mathbf{v^0_{(n)}}\}] \,. \tag{12}$$

The general integral of the system of motion equations (1), *free vibration*, is described in matrix form by the equations

$$\{\mathbf{v_0(t)}\} = \sum_{j=1}^{n} \left(\mathrm{A}^0_{(j)} \cdot \sin(\omega_{(j)} \cdot \mathrm{t}) + \mathrm{B}^0_{(j)} \cdot \cos(\omega_{(j)} \cdot \mathrm{t}) \right) \{\mathbf{v^0_{(j)}}\} \tag{13}$$

or

$$\{\mathbf{v_0(t)}\} = \sum_{j=1}^{n} q^0_{(j)} \cdot \{\mathbf{v^0_{(j)}}\} \cdot \sin(\omega_{(j)} t + \varphi^0_{(j)}) = \sum_{j=1}^{n} q_{(j)}(t) \cdot \{\mathbf{v^0_{(j)}}\} \tag{14}$$

The constant $q^0_{(j)}$ expresses the proportion of each natural mode in the resulting vibration, and the phase angles $\varphi^0_{(j)}$ express the phase shift of each natural mode in time. The following matrix equations apply to velocity and acceleration

$$\{\dot{\mathbf{v}}_\mathbf{0}(\mathbf{t})\} = \sum_{j=1}^{n} \omega_{(j)} \left(A^0_{(j)} \cdot \cos(\omega_{(j)} t) - B^0_{(j)} \cdot \sin(\omega_{(j)} t) \right) \cdot \{\mathbf{v^0_{(j)}}\} , \tag{15}$$

$$\{\ddot{\mathbf{v}}_\mathbf{0}(\mathbf{t})\} = -\sum_{j=1}^{n} \omega^2_{(j)} \left(A^0_{(j)} \cdot \sin(\omega_{(j)} t) + B^0_{(j)} \cdot \cos(\omega_{(j)} t) \right) \cdot \{\mathbf{v^0_{(j)}}\} . \tag{16}$$

The *compliance constant method* or also the *method of influences* is a method analogous to the force method. The dynamic deflections at the individual points of the system are calculated here and the coordinates of the influence lines of deflections are used to express them. The equation of motion has the form

$$\{\mathbf{v}(t)\} = -[\boldsymbol{\delta}] \cdot [\mathbf{m}]_\mathbf{D} \cdot \{\ddot{\mathbf{v}}(t)\} \quad \text{or} \quad \{\mathbf{v}(t)\} = -[\mathbf{D}] \cdot \{\ddot{\mathbf{v}}(t)\} , \tag{17}$$

where

$$[\mathbf{D}] = [\boldsymbol{\delta}] \cdot [\mathbf{m}]_\mathbf{D} = [\boldsymbol{\delta}_\mathbf{r}] \tag{18}$$

is the *dynamic matrix* or the *reduced compliance matrix*.

Particular solutions have a form

$$\{\mathbf{v}(t)\} = \{\mathbf{v^0_{(j)}}\} \cdot \sin(\omega_{(j)} \cdot t) \text{ and} \{\mathbf{v}(t)\} = \{\mathbf{v^0_{(j)}}\} \cdot \cos(\omega_{(j)} \cdot t) \,. \tag{19}$$

Substituting one of the particular solutions into (17) we get the *frequency equations* in the form

$$([\boldsymbol{\delta}]\cdot[\mathbf{m}]_{\mathbf{D}}-\frac{1}{\omega_{(j)}^{2}}\cdot[\mathbf{E}]_{\mathbf{D}})\cdot\{\mathbf{v}_{(\mathbf{j})}^{\mathbf{0}}\}=\{\mathbf{0}\}, \tag{20}$$

$$([\mathbf{D}]-\frac{1}{\omega_{(j)}^{2}}\cdot[\mathbf{E}]_{\mathbf{D}})\cdot\{\mathbf{v}_{(\mathbf{j})}^{\mathbf{0}}\}=\{\mathbf{0}\} \tag{21}$$

or

$$[\mathbf{B}(\omega_{(j)})]\cdot\{\mathbf{v}_{(\mathbf{j})}^{\mathbf{0}}\}=\{\mathbf{0}\}\ \text{where}\ [\mathbf{B}(\omega_{(j)})]=[\mathbf{D}]-\frac{1}{\omega_{(j)}^{2}}\cdot[\mathbf{E}]_{\mathbf{D}}. \tag{22}$$

From the mathematical point of view, the matrix equation (21) represents a special problem of eigenvalues, where numerical values $1/\lambda_{(j)}=1/\omega_{(j)}^{2}$ are calculated as eigenvalues of the matrix $[\mathbf{D}]$ and vectors $\{\mathbf{v}_{(\mathbf{j})}^{\mathbf{0}}\}$ describing the natural modes of vibrations are calculated as eigenvectors of the matrix $[\mathbf{D}]$. The secular equation has the form

$$\mathbf{det}[\mathbf{B}(\omega_{(j)})]=0. \tag{23}$$

The relative ratios of the deflections of the individual mass points for each mode of vibration are equal to the mutual ratios of the algebraic complements of the subdeterminants of the matrix $[\mathbf{B}(\omega_{(j)})]$ calculated with respect to the elements of any row (index i) or column (index k) of this matrix.

$$v_{1(j)}^{0}:v_{2(j)}^{0}:v_{i(j)}^{0}:\ldots:v_{n(j)}^{0}=B_{i1(j)}:B_{i2(j)}:B_{ii(j)}:\ldots:B_{in(j)}=$$
$$\ldots\ldots\ldots\ldots=B_{1k(j)}:B_{2k(j)}:B_{kk(j)}:\ldots:B_{nk(j)}. \tag{24}$$

Furthermore, equations (11) to (16) apply as for the stiffness constants method.

It is velid for orthonormed natural modes

$$\{\mathbf{V}_{(\mathbf{j})}\}^{\mathbf{T}}\cdot[\mathbf{m}]_{\mathbf{D}}\cdot\{\mathbf{V}_{(\mathbf{j})}\}=1, \tag{25}$$

then the vectors $\{\mathbf{V}_{(\mathbf{j})}\}$ describe normed natural modes and call them ortho-normed. We calculate them according to the relationship

$$\{\mathbf{V}_{(\mathbf{j})}\}=\frac{\{\mathbf{v}_{(\mathbf{j})}^{\mathbf{0}}\}}{\sqrt{\{\mathbf{v}_{(\mathbf{j})}^{\mathbf{0}}\}^{\mathbf{T}}\cdot[\mathbf{m}]_{\mathbf{D}}\cdot\{\mathbf{v}_{(\mathbf{j})}^{\mathbf{0}}\}}}. \tag{26}$$

Relationships (25) and (26) written in scalar form are

$$\sum_{i=1}^{n}m_{i}\cdot V_{i(j)}^{2}=1, \tag{27}$$

$$V_{i(j)}=\frac{v_{i(j)}}{\sqrt{\sum_{i=1}^{\mathrm{n}}\mathrm{m_i}\cdot {v_{i(j)}^{0}}^{2}}}. \tag{28}$$

The ortho-normed modal matrix $[\mathbf{V}]$, composed of ortho-normed vectors $\{\mathbf{V}_{(\mathbf{j})}\}$, representing its individual columns, is given by

$$[\mathbf{V}]=\left|\{\mathbf{V}_{(1)}\},\{\mathbf{V}_{(2)}\},\{\mathbf{V}_{(\mathbf{j})}\},\ldots,\{\mathbf{V}_{(\mathbf{n})}\}\right|. \tag{29}$$

The orthogonality conditions written by the ortho-normed modal matrix $[\mathbf{V}]$ express the equations

$$[\mathbf{V}]^{\mathbf{T}}\cdot[\mathbf{m}]_{\mathbf{D}}\cdot[\mathbf{V}]=[\mathbf{E}]_{\mathbf{D}}, \tag{30}$$

$$[\mathbf{V}]^{\mathbf{T}}\cdot[\mathbf{k}]\cdot[\mathbf{V}]=[\boldsymbol{\omega}^{2}]_{\mathbf{D}}. \tag{31}$$

2.2. Models of vehicles

Discrete computational model of a vehicle (lumped mass model with n degrees of freedom) can be created in the sense of Component Element Method (CEM) [4]. CEM defines the properties of mass objects and joining members. Their combination creates a computational model of the structure. If we denote by $\{\mathbf{r}\}$ the vector of the displacement components of mass objects corresponding to the degrees of freedom, the calculation of natural frequencies and natural modes leads to the special eigenvalue problem

$$([\mathbf{k_r}]-\omega_{(j)}^2[\mathbf{E}]_\mathbf{D})\cdot\{\mathbf{r_{(j)}}\}=\{\mathbf{0}\}, \tag{32}$$

where

$$[\mathbf{k_r}]=[\mathbf{m}]_\mathbf{D}^{-1}\cdot[\mathbf{k}]. \tag{33}$$

Matrix $[\mathbf{m}]_\mathbf{D}^{-1}$ is inverse matrix to diagonal mass matrix $[\mathbf{m}]_\mathbf{D}$ and $[\mathbf{k}]$ is stiffness matrix. Second powers of angular natural frequencies $\omega_{(j)}^2$ are calculated as eigenvalues $\lambda_{(j)}=\omega_{(j)}^2$ of the matrix $[\mathbf{k_r}]$ and natural modes $\{\mathbf{r_{(j)}}\}$ are calculated as eigenvectors of the matrix $[\mathbf{k_r}]$. The matrices $[\mathbf{m}]_\mathbf{D}$ and $[\mathbf{k}]$ are needed as entering values. The stiffness matrix $[\mathbf{k}]$can be calculated as:

$$[\mathbf{d_{ED}}]=[\mathbf{A}]^\mathbf{T}\cdot[\mathbf{E_D}]_\mathbf{D}, \tag{34}$$

$$[\mathbf{F_{K,ED}}]=[\mathbf{k_i}]_\mathbf{D}\cdot[\mathbf{d_{ED}}], \tag{35}$$

$$[\mathbf{k}]=[\mathbf{A}]\cdot[\mathbf{F_{K,ED}}], \tag{36}$$

where$[\mathbf{A}]^\mathbf{T}$ is transposed static matrix of vehicle computational model. The elements of the stiffness matrix $[\mathbf{k}]$ are generalized forces, which cause the unit displacement of mass points corresponding to the degree of freedom. The matrix $[\mathbf{E_D}]_\mathbf{D}$ is the identity matrix of unit displacements of mass points corresponding to degrees of freedom. The matrix $[\mathbf{d_{ED}}]$represents the deformation of joining members corresponding to the unit displacements stored in an array$[\mathbf{E_D}]_\mathbf{D}$. The matrix $[\mathbf{k_i}]_\mathbf{D}$is the diagonal stiffness matrix of joining members. Forces in joining members arising due to their deformation are in the matrix$[\mathbf{F_{K,ED}}]$. Static equivalents of these forces acting on the degrees of freedom are actually elements of stiffness matrix$[\mathbf{k}]$.

2.3. Calculation of natural values using MATLAB

MATLAB is an effective tool for the implementation of numerical calculations in engineering practice, especially if the mathematical formulation of the problem is done in matrix form [5]. It is therefore necessary for future engineers to become familiar with such a tool during their studies. At the Department of Structural Mechanics and Applied Mathematics, University of Zilina, MATLAB has become an organic part of the teaching of all subjects taught by the Department. We try to ensure that students not only passively use what MATLAB offers, but that they learn to create their own programs. Mainly their homework is oriented in this direction. Although such programs do not always reach the professional level, it forces the student to understand the solution algorithm and to analyze the task in detail. In this paper we want to show an examples of working with MATLAB in the subject Dynamics of structures especially in connection with the calculation of natural values of structures. The textbook [3] presents several simple programs enabling the calculation of the natural values of beam structures (simply supported beam, beam fixed at both ends, one-side fixed beam, cantilever beam, continuous beam with two and three fields) and calculation models of vehicles (2D and 3D models of lorry Tatra). These programs can be used to solve specific tasks, or they can serve as aids in creating other programs.

As an example, in the Appendix 1 there is the m-file **NatVibSBndf.m**, which calculates and plots the natural frequencies and natural modes of a simply supported beam with n degrees of freedom. The program asks as input values: number of mass points, degrees of freedom n, beam span l, modulus of elasticity E, moment of inertia I and mass intensity mi. Based on this data, it calculates matrices $\mathbf{[k]}$, $\mathbf{[m]_D}$, $\mathbf{[k_r]}$, $\{\boldsymbol{\omega}_{(j)}\}$, $\{\mathbf{f}_{(j)}\}$, $\mathbf{[V]}$ and plots all natural modes $\{\mathbf{V}_{(j)}\}$. For illustration, the first three natural modes for the steel beam profile $I_{No.2000}$, $l = 6$ m, $n = 5$, are shown in Fig. 2. Natural frequencies are in Table 1.

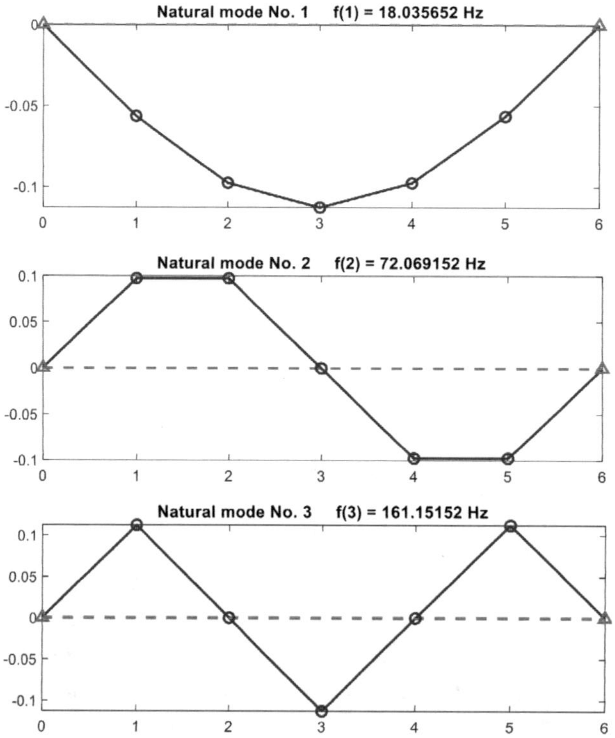

Figure 2. Natural modes of simply supported beam.

Table 1. Values of natural frequencies of beam model.

	Natural frequencies	
j	$\omega_{(j)}$ [rad/s]	$f_{(j)}$ [Hz]
1	113.321345	18.035652
2	452.823839	72.069152
3	1012.544887	161.151523
4	1753.779189	279.122627
5	2509.257159	399.360680

In the Appendix 2 there is the m-file **NatVibT148PM**, which calculates the natural frequencies and natural modes of planar model of Tatra T148 lorry. The vehicle model is shown in Fig. 3. It is a discrete model with 5 degrees of freedom. Damping is not considered when calculating natural values.

An important step is to create a transposed static matrix $\mathbf{[A]^T}$.

$$\mathbf{[A]}^{\mathbf{T}} = \begin{bmatrix} 1 & -a & -1 & 0 & 0 \\ 1 & b & 0 & -1 & 0 \\ 0 & 0 & 1 & 0 & 0 \\ 0 & 0 & 0 & 1 & -c \\ 0 & 0 & 0 & 1 & +c \end{bmatrix} \tag{37}$$

Its elements have the meaning of deformations of joining members due to unit displacements of mass points, corresponding to the degrees of freedom. The number of rows is equal to the number of joining members and the number of columns is equal to the number of degrees of freedom, (37).

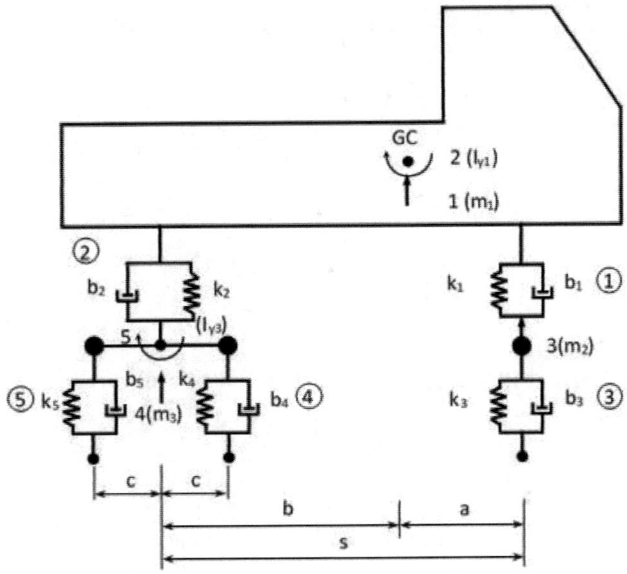

Figure 3. Planar model of Tatra T148 lorry.

Vehicle dimensions
$a = 3.399$ m; $b = 0.951$ m; $c = 0.660$ m;
Vehicle mass parameters
$m_1 = 11500.0$ kg, $I_{y1} = 33500.0$ kg.m2, $m_2 = 440.0$ kg,
$m_3 = 1060.0$ kg, $I_{y3} =$ 461.7 kg.m^2.
Stiffness of connecting members
$k_1 = 197965$ N/m, $k_2 = 1572881$ N/m; $k_3 = 1200000$ N/m, $k_4 = 2400000$ N/m, $k_5 = 2400000$ N/m.
All vehicle parameters are entered as default values in the program. The program calculates and prints the matrices $[\mathbf{k}_i]_D$, $[\mathbf{m}]_D$, $[\mathbf{A}]^T$, $[\mathbf{A}]$, $[\mathbf{k}]$, $[\mathbf{k}_r]$, $\{\boldsymbol{\omega}_{(j)}\}$, $\{\mathbf{f}_{(j)}\}$ and ortho-normed modal matrix **[R]**. At the end, the conditions of orthogonality of natural modes are checked. The calculated natural frequencies are in Table 2.

Table 2. Values of natural frequencies of vehicle model.

	Natural frequencies	
j	$\omega_{(j)}$ [rad/s]	$f_{(j)}$ [Hz]
1	8.532451	1.357981
2	11.587989	1.844285
3	56.476407	8.988499
4	67.295281	10.710376
5	77.829290	12.386916

The natural modes are shown in Fig. 4.

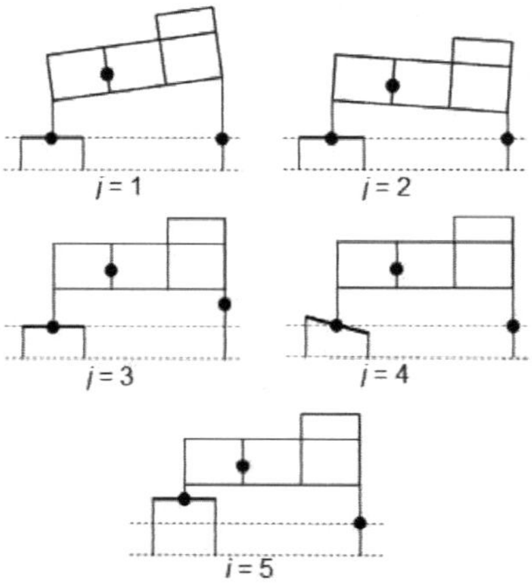

Figure 4. Natural modes of Tatra T148 lorry model.

3. Conclusion

The dynamics of structures is an important and demanding engineering discipline. To its teaching should be paid serious attention. Modern numerical methods such as FEM CEM and higher-level programming languages such as ATLAB contribute to the development of students' skills. MATLAB enables efficient implementation of numerical calculations, replaces manual work and, within programming, forces the student to penetrate deeper into the essence of the problem and analyze the solution process in detail. Universities should be prepared for this and provide students with the widest possible space to take advantage of these opportunities.

Acknowledgments

This work was supported by the Grant National Agency VEGA of the Slovak Republic, project No. 1/0006/20.

References

[1] Melcer J 2021 *50 years of Department of Structural Mechanics on RU in Prague and UTT, UTCT and ZU in Zilina*, in Slovak (EDIS, University of Žilina) p 163
[2] Koloušek V 1973 *Dynamics of Civil Engineering Structures* (London, Butterworth) p 345
[3] Melcer J and Lajčáková G 2011 *Application of the Program System MATLAB in Solving Problems of Structural Dynamics*, in Slovak (EDIS, University of Žilina) p 166
[4] Levy S and Wilkinson J P D 1976 *The Component Element Method in Dynamics with Application to Earthquake and Vehicle Engineering* (McGraw–Hill, New York) p 363
[5] MATLAB 7.0.4 The Language of Technical Computing, Version 7. The MathWorks, June 2005

Appendix 1

```
% NatVibSBndf.m
% Program for calculating natural frequencies and natural modes
% simply supported beam with n degrees of freedom, EI=constant.
text=char(' ', 'Program for calculating natural frequencies and natural modes',...
              'simply supported beam with n degrees of freedom, EI=constant',' '); disp(text)
n=input('Number of mass points, degrees of freedom n = '); l=input('Beam span l = ');
E=input('Modulus of elasticity E = '); I=input('Moment of inertia I = '); mi=input('Mass intensity mi = ');
for i=1:n
   pb=i/(n+1);
   ppb=1-pb;
   for j=i:n;
```

```
        xb=j/(n+1);
        xpb=1-xb;
        Delb(j,i)=pb*xpb*(1-pb*pb-xpb*xpb);
        Delb(i,j)=Delb(j,i);
    end
end
Delb=(1/6)*Delb; Suc=(l*l*l)/(E*I); Del=Suc*Delb; kb=inv(Delb); Suci=1/Suc;
text='Matrix of stiffness constants [k]'; disp(text)
k=Suci*kb

text='Diagonal mass matrix md'; disp(text)
d=l/(n+1);                % length of the element, the beam is divided into elements of the same length
m=mi*d;                   % the lumped mass at the point i
mr=m*ones(1,n);           % row-shaped mass matrix
md=diag(mr,0)                                % diagonal mass matrix md
mdinv=diag(ones(n,1)./diag(md,0),0);         % inverse matrix to mass matrix
text='Reduced stiffness matrix kr'; disp(text)
kr=mdinv*k

% Calculation of eigenvalues
[v,om2]=eig(kr);                   % eigenvectors and eigen numbers of the matrix kr
om2v=diag(om2,0);                  % vector of quadrats of natural angular frequencies
omv=sqrt(om2v);                    % vector of natural angular frequencies
fv= omv./(2*pi);                   % vector of natural frequencies
text='Frequencies and natural modes arranged in order'; disp(text)
omv=sort(omv)                      % arranges elements of omv vector by size
[fv,i]=sort(fv)                    % arranges elements of fv vector by sizeu
                                   % and stores the order of the indexes in the vector {i}
v=v(:,i);                          % swaps the order of the columns of the matrix in the order of their natural modes

% Calculation of orthonormal modal matrix
V=zeros(n);
for j=1:n
    vj=v(:,j);
    VJ=vj./sqrt(vj'*md*vj);
    V(:,j)=VJ;
end

text='Orthonormal modal matrix V'; disp(text)
V
text='Check of calculation of V matrix, Ekont=unit matrix'; disp(text)
Ekont=V'*md*V          % check of orthonormal natural modes

% Plotting of results
x=linspace(0,l,n+2);
xp=[0,l]; yp=[0,0]; xv=x(2:n+1);
for j=1:n
    VJ=[0;V(:,j);0];
    jr=int2str(j);
    fr=num2str(fv(j,1),8);
    out=['Natural mode No. ',jr,'    f(',jr,') = ',fr,' Hz'];
    plot(x,VJ,'b-',xp,yp,'r^--',xv,V(:,j),'bo','LineWidth',1.5)
    title(out)
    pause
end

text='End'; disp(text)
```

IOP Conf. Series: Materials Science and Engineering 1276 (2023) 012007 doi:10.1088/1757-899X/1276/1/012007

APPENDIX 2

```
% NatVibT148PM.m
% Program for calculating the natural frequencies and natural modes of a planar model of vehicle TATRA T148.
% Fully loaded vehicle.
text=char('','Calculation of natural frequencies and natural modes of a planar model',...
          'of vehicle TATRA T148. Fully loaded vehicle',' '); disp(text)
text='Input of data and matrices defining the properties of the vehicle'; disp(text)
text='Number of degrees of freedom of the vehicle n'; disp(text)
n=5                              % number of degrees of freedom of the vehicle
text='Stiffness matrix of joining members [ki]'; disp(text)
kiV=[197965 1572881 1200000 2400000 2400000]; % stiffness matrix of joining members in vector form
kiD=diag(kiV,0)                  % diagonal stiffness matrix of joining members
kiDin=inv(kiD);                  % calculation the inverse matrix to the matrix kiD

text='Vehicle mass matrix [m]'; disp(text)
mV=[11500 33500 440 1060 461.7]; % vehicle mass matrix in vector form vo vektorovom tvare
mD=diag(mV,0)                    % diagonal vehicle mass matrix mD

text='Transposed static matrix of the vehicle [AT]'; disp(text)
AT=[1 -3.399 -1 0 0; 1 0.951 0 -1 0; 0 0 1 0 0; 0 0 0 1 -0.66; 0 0 0 1 0.66]   % transposed static matrix of the vehicle AT
text='Static matrix of the vehicle [A]'; disp(text)
A=AT'                % static matrix of the vehicle A
Ain=inv(A);          % inverse matrix to matrix A
ATin=Ain';           % transposed matrix to matrix Ain
E=eye(n);            % it generates the identity matrix
ED=E;                % matrix of unit displacements of mass points

text='Matica tuhosti vozidla [k]'; disp(text)
dED=AT*ED;                       % deformations of joining members due to unit deformation states
FkED=kiD*dED;                    % forces in the springs of joining members due to unit deformation states
FED=A*FkED;                      % forces acting on masses due to unit deformation states
k=FED                            % stiffness matrix of vehicle

text=char('','Eigenvalues - stiffness constants method.', 'Reduced stiffness matrix'); disp(text)
kr=inv(mD)*k                     % reduced stiffness matrix of vehicle
[r,om2]=eig(kr);                 % eigen vectors and eigen numbers of matrix kr
om2v=diag(om2,0);                % vector of squares of natural angular frequencies
omv=sqrt(om2v);                  % vector of natural angular frequencies
fv= omv./(2*pi);                 % vector of natural frequencies

text='Frequencies and modes of vibration arranged in order'; disp(text)
omv=sort(omv)        % it sorts members of vector omv in ascending order
[fv,i]=sort(fv)      % it sorts members of vector {fv} in ascending order
                     % and stores the order of indexes in the vector {i}
r=r(:,i);            % it changes the order of the columns of matrix r according to the order of natural modes

% Calculation of orthonormal modal matrix
R=zeros(n);
for j=1:n
    rj=r(:,j);
    RJ=rj./sqrt(rj'*mD*rj);
    R(:,j)=RJ;
end

text='Orthonormal modal matrix [R]'; disp(text)
R
text='The check matrix Ekont should be the identity matrix'; disp(text)
Ekont=R'*mD*R        % control of orthonormal natural modes
text='End'; disp(text)
```

Assessment of the effects of strong earthquakes on the city of Plovdiv

D Stefanov, D Solakov and J Milkov

National Institute of Geophysic, Geodesy and Geography (NIGGG), Bulgarian Academy of Sciences, Acad.G.Bonchev str., Block 3, Sofia 1111, Bulgaria

E-mail: dstefanov@geophys.bas.bg

Abstract. The assessment of the effects of strong earthquakes is based on procedures for integrating the seismic hazard and the seismic vulnerability of the load-bearing structures of the building stock. The analyses are made for seismic excitation, determined by the analysis of seismic hazard - a deterministic earthquake scenario (expressed in macroseismic intensity). The scenario takes into account the effects of soil amplification, using the geotechnical information for the considered urban area. Based on the collected information about the building stock and the performed analysis, an assessment of the direct damages and destructions is obtained. The distribution of the buildings is graphically presented, respectively by number and by total built-up area, at different degrees of damage. Economic losses are calculated based on the results obtained for damage and destruction in three main groups of buildings. The main task of this study is to quantify the effects of strong earthquakes for the city of Plovdiv based on an accurate assessment of seismic hazard. This information will enable the municipal administration to make informed management decisions about planning of rescue equipment and people, hospital beds, shelter for the homeless people and financial resources to restore the damaged buildings.

1. Introduction

The increment in seismic losses over years is mainly due to the rapid growth of exposure and population in earthquake-prone areas worldwide. This is largely true for the city of Plovdiv, which has undergone rapid development in recent years. New industrial zones are arising with many new enterprises. The people who come to work in these factories need housing accordingly. This leads to a process of continuous growth of the city. The gathering of more people in a certain area increases the influence of the exposure factor, which is a valid reason for more precise assessments of the seismic risk in connection with the planning of the relevant post-earthquake activities.

The city of Plovdiv is located in the central part of Southern Bulgaria, with a population of about 340 000 people. Plovdiv is situated along the two banks of the river Maritsa and also on a number of picturesque hills (called "tepe") that are part of the town's charm and beauty. Plovdiv is known as the "City of Seven Hills". These natural features have been appreciated by people since ancient times. Archeological findings from the Neolithic age, about 6000 years old, have been discovered. Philip II of Macedon (father of Alexander of Macedon) conquered it in 342 BC and even renamed it Philippopolis or "the city of Philip". After the decline of the Macedonian state, the city fell into the hands of the Thracian tribes under the name Pulpudeva. Gradually the city grew and became the largest city in Northern Thrace. During the Middle Ages it retained its strategic importance and over the years it fell within the borders of both the Bulgarian and Byzantine Empires. In the 14th century it

IOP Conf. Series: Materials Science and Engineering 1276 (2023) 012008 doi:10.1088/1757-899X/1276/1/012008

was conquered again and fell within the borders of the Ottoman Empire In 1878, Plovdiv was made the capital of the autonomous Ottoman region of Eastern Rumelia, in 1885 it became part of Bulgaria with the unification of that region and the Principality of Bulgaria [1].

At present Plovdiv is one of the most populated industrial and cultural regions of Bulgaria that faces considerable earthquake risk. The earthquakes that mainly influence the hazard in Plovdiv originate near the city. Seismic history of Plovdiv shows that the 1928 earthquakes, Mw6.5 foreshock and the main event with Mw7.1 cause the strongest on-city seismic effects – figure1.

Figure 1. Damages in the buildings after the 1928 earthquakes [2].

2. Creating a database for the building stock

The process of gathering input data in a format that is appropriate for processing and analysis is delicate and responsible. Over the years, several public databases have undergone numerous updates, changes, and even deformations anywhere, as the subject matter of their application varies from one institution to another.

Some databases lack information on the year of construction of buildings, while others lack the type of construction. Databases that are up to date a few years ago are logically missing new buildings, but contain information about the year of construction of buildings for which this information is not available in modern databases, created for another purpose.

After extensive comparative analyses and restructuring of a large part of the data, filtering and combining, the database is combined and harmonized in one general form.

The graphic base is connected in two directions with the textual basis for the building stock through the cadastral identifiers of the individual buildings and properties, thus providing the possibility for the future visualization of different thematic map materials. To help this, an orthogonal coordinate division (zoning) of the graphic parts is constructed - by means of quadrants 600m x 600m – figure 2. Figure 3 shows a quadrants of the grid illustrating the location of the buildings in the city centre.

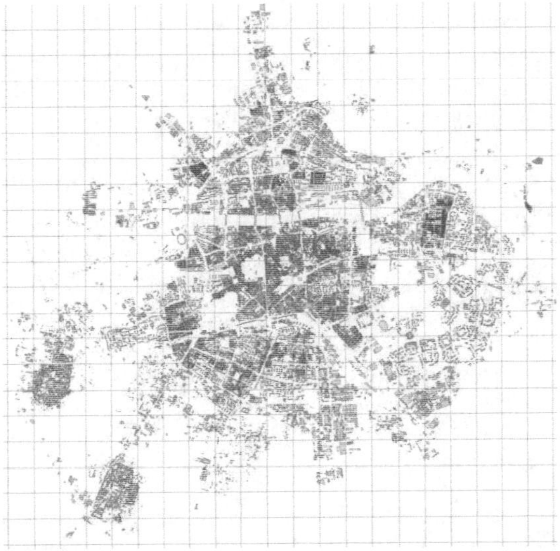

Figure 2. Quadrant network for the city of Plovdiv.

Figure 3. Illustration of the data for quadrants - the central part of the city.

3. Methodology for seismic risk assessment

The present study is based on the "Methodology for Analysis, Evaluation and Mapping of the Seismic Risk of the Republic of Bulgaria", developed by the NIGGG-BAS for Ministry of Regional Development and Public Works, approved by the Minister and having the force of a normative document [3]. The methodology is based on the application of empirical methods for determining the vulnerability of buildings using the approach in EMS-98 [4]. In order to take into account the specific features of the building stock in Bulgaria, a new typology matrix containing 24 types of construction systems has been developed. Buildings are grouped according to their vulnerability class. The next step is determining the level of destruction. For this purpose, the developed DPM (Damage Probability Matrix) is used. Buildings of the same vulnerability class receive the same level of damage (for a given seismic intensity). This is how the number and area of buildings that have been damaged or destroyed are determined. Based on this information, the social losses (number of injured people and deaths) are calculated, depending on what time the earthquake occurred (day or night), as well as the economic losses. The methodology requires that all information is available into a GIS system and is applied separately to each elementary cell of the GIS representation of the studied region.

4. Identification and classification of the building stock

The number of buildings under consideration is 31543 with a total built-up area of 21633194 m^2. This includes all residential and administrative buildings, industrial buildings are not included.

Table 1 illustrates the height distribution of buildings by grouping them into four groups. Most of the buildings (22357 units) can be classified as low-rise up to 3 floors, 6092 buildings are in the range of 4 to 6 floors, 2688 buildings are in the range of 7 to 9 floors and only 406 buildings are above 10 floors. Table 2 illustrates the height distribution of buildings by total floor area (TFA). Low-rise buildings are extremely numerous – 70,9% of all buildings, but their area is only 20,7% of the total area. Medium-height buildings (between 4 and 9 floors) are about 33%, that is approximately as many as those of the second group. High-rise buildings have a relatively small contribution - about 12%.

Table 1. Distribution of buildings (pieces) by number of floors.

Floors	1 to 3	4 to 6	7 to 9	above 10
Pieces	22357	6092	2688	406
%	70,9	19,3	8,5	1,3

Table 2. Distribution of buildings (total floor area – m^2) by number of floors.

Floors	1 to 3	4 to 6	7 to 9	above 10
TFA (m^2)	4482362	7216233	7371973	2562626
%	20.7	33.4	34.1	11.8

One of the important parameters of the building stock is the year of design (construction). It is directly related to the current state of design seismic codes for the year. Bulgarian national seismic codes have been periodically updated and modernized. The timing of these changes is also the boundaries of the respective time periods. In this case 7 time periods are considered. Tables 3 and 4 show the distribution of buildings by construction period. The largest number of buildings (8160 units) were built in the period 1966-1977. This period is followed by 20 years, during which the pace decreases relatively constantly. There is a significant decline in the last decade. In terms of TFA, the maximum is in the period 1978-87.

Table 3. Distribution of number of buildings by construction period.

Period	until 1929	1930-57	1958-65	1966-77	1978-87	1988-2007	after 2007
Pieces	1976	3337	5332	8160	5488	4165	3085
%	6.3	10.6	16.9	25.9	17.4	13.2	9.8

Table 4. Distribution of buildings (total floor area – m^2) by construction period.

Period	until 1929	1930-57	1958-65	1966-77	1978-87	1988-2007	after 2007
TFA (m2)	245018	395505	1490719	4568075	6346159	4676294	3911423
%	1.1	1.8	6.9	21.1	29.3	21.6	18.1

Based on the information about the cadastre design system, each building is classified in a separate type according to the matrix of typologies of the Bulgarian buildings. The total number of typologies is 10. A brief description of the different typologies is given in Table 5. The distribution of buildings by typologies (number of buildings) is given in Table 6.

Table 5. Description of typologies.

Type	Description
M1	Masonry structures, unreinforced masonry with reinforced concrete slabs, beams and belts, not framed or framed by columns
M2	Masonry structures, unreinforced masonry with wooden beams without reinforced concrete belts, not framed by columns
St1	Steel structures, frames with rigid nodes
St2	Steel structures in which seismic forces are absorbed by vertical connections designed as centrically connected in accordance with EN1998-1
RC1	One-storey and multi-storey buildings with single-opening and multi-opening frames
RC2	Skeleton-beamless structures with frames
RC4	Skeleton-beamless structures with reinforced concrete shear walls and / or shear walls framed by columns, non-reinforced masonry
RC5	Skeleton-beam structures with reinforced concrete (or masonry) shear walls
RCp6	Beamless structures with shear walls performed according to the system "lift slabs"
RCp7	Large-panel buildings

Table 6. Distribution of buildings by typologies (number of buildings).

Type	M1	M2	St1	St2	RC1	RC2	RC4	RC5	RCp6	RCp7
Pcs.	10891	8639	22	4	7362	728	482	156	118	3141
%	34.5	27.4	0.1	0.013	23.3	2.3	1.5	0.5	0.4	10.0

The largest number of buildings (10891 pcs.) fall into the M1 typology. In second place is the typology M2 with 8639 pieces. The total number of masonry structures (M1 and M2) is more than half of all buildings - 61.9%. In third place are the reinforced concrete structures of type RC1 with 7362 pieces. Only in fourth place come the large-panel buildings with 3141 pieces.

The distribution of buildings according to the total built-up area is quite different. The area of the reinforced concrete structures RC1 dominates with 32.4%. Immediately after them are large-panel buildings with 27.6%. The total area of masonry structures (M1 and M2) is only 13.6%. Graphically the distribution of buildings by typologies (total floor area m^2) is given in figure 4.

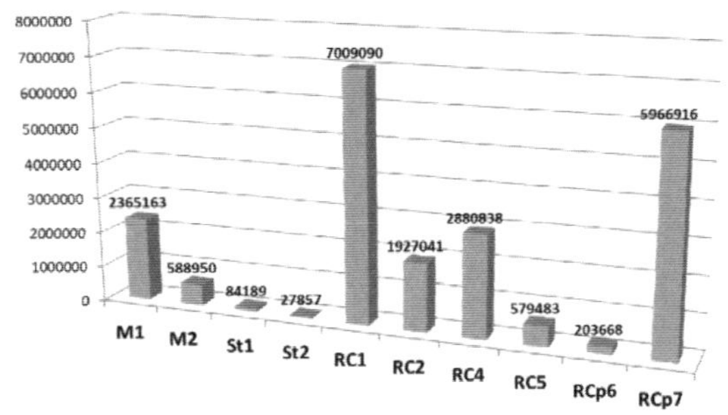

Figure 4. Distribution of buildings by typologies (total floor area m^2).

IOP Conf. Series: Materials Science and Engineering 1276 (2023) 012008 doi:10.1088/1757-899X/1276/1/012008

5. Seismic vulnerability assessment of the building stock

The seismic vulnerability assessment is performed for each building individually, based on information on its load-bearing structure (typology), construction time (design), number of floors and other technical parameters such as the presence of a soft first storey, irregular distribution of stiffness (in plan and height), foundation and type of ground base (soft, hard soils, embankment, rock). Each building is assigned a corresponding vulnerability class. The results obtained are illustrated as follows. Figures 5 and 6 show the distribution of the buildings by number and by TFA - total floor area (m^2).

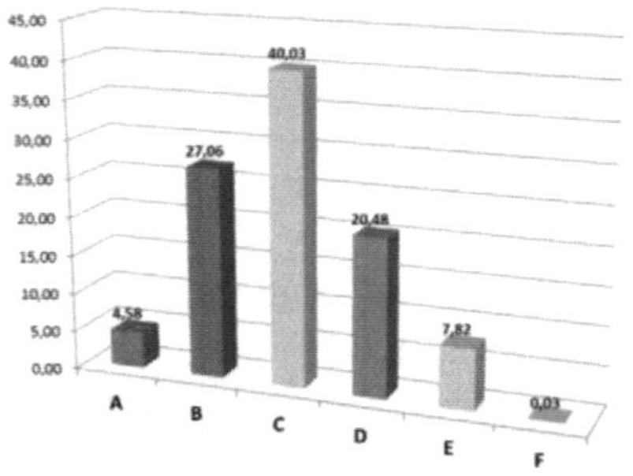

Figure 5. Distribution of number of buildings (%) in vulnerability classes.

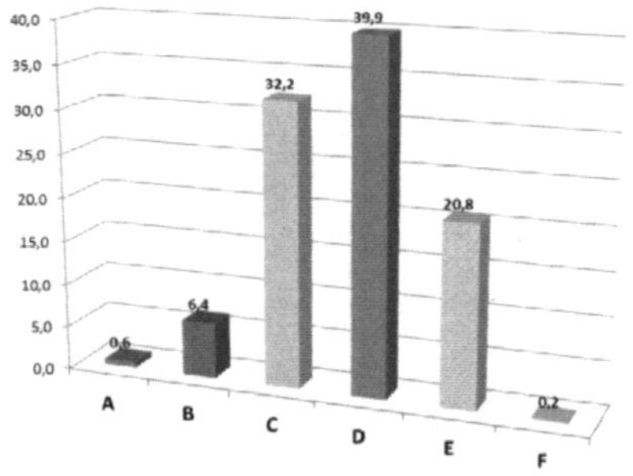

Figure 6. Distribution of total floor area (%) in vulnerability classes.

The largest number of buildings fall into class "C" - 40%. In second and third place are the buildings of class "B" - 27% and class "D" – 20.5%. However, when considering the total built-up area, class "D" dominates with 39.9%. In second place is class "C" with 32%. In third place is class "E" with 20.8% (compared to 7.82% in number of buildings). The most drastic difference is in class "B" - only 6.4% (27% in terms of number of buildings).

6. Deterministic earthquake scenario for the city of Plovdiv in terms of intensity

Deterministic earthquake scenario is performed following the procedure presented in [1]. The city area is enlarged to cover all buildings. The scenario earthquake is 7.1 [1] and scenario is in PGA and PGV. The relations presented in [5] are used to convert PGA (in cm/s^2) and PGV (in cm/s) to intensity:

$$I(PGA) = 3.822 \log(PGA) - 1.361 \quad (1)$$

$$I(PGV) = 2.671 \log(PGV) + 4.018 \quad (2)$$

The final intensity is calculated as:

$$I = (I(PGA) + I(PGV))/2 \quad (3)$$

In figure 7 are presented the Vs30 distribution and evaluated intensities over the city area.

7. Assessment of direct damages and destruction

The building stock of Plovdiv is divided into three main groups, taking into account the specifics of the urban development and the available information on the function and habitation of the buildings - residential buildings, administrative/public buildings and special buildings. The first group includes residential buildings in different parts of the city. The second group includes all administrative and public buildings. The third group is conventionally called "special buildings". It includes buildings with a high concentration of people and those of particular importance in a post-earthquake period. This group includes: hospitals, schools, universities and kindergartens.

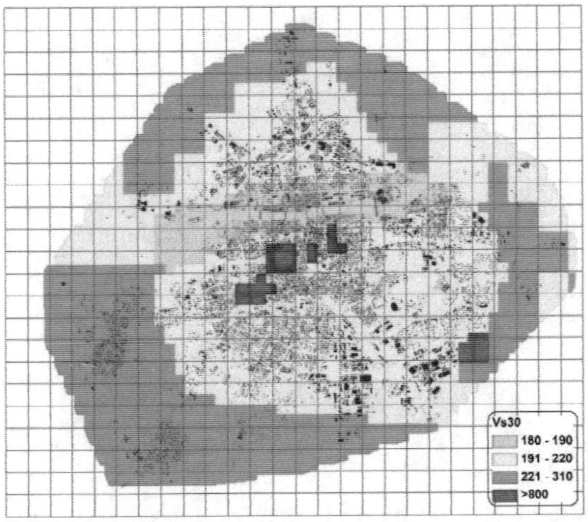

(a)

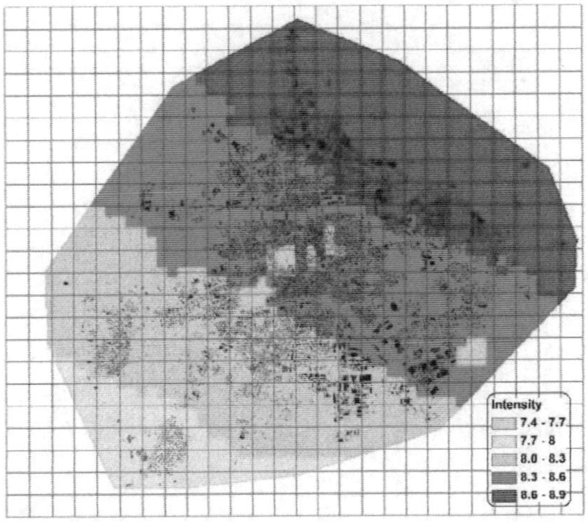

(b)

Figure 7. Vs30 distribution (a) and deterministic earthquake scenario for the city of Plovdiv in intensity (b).

The distribution of buildings, by number and total floor area - TFA (m^2), in the different levels of damage is summarized in Table 7. The same table shows the distribution of buildings as a percentage of the total number and total built-up area.

Table 7. Distribution of buildings (by pieces and TFA) in different levels of damage.

Damage grade/	No damage	Slight	Medium	Heavy	Very heavy	Destruction
Number of buildings	7342	9247	7387	4676	2259	632
Number of build. (%)	23.3	29.3	23.4	14.8	7.2	2.0
Total floor area (m^2)	7345089	6977264	4252592	2088899	797436	171914
TFA (%)	33.9	32.3	19.7	9.6	3.7	0.8

The built-up area of buildings without and with slight damages is 33.9% and 32.3% respectively. Following are buildings with moderate damage – 19.7%. High levels of damage are relatively low – 9.6% for heavy and 3.7% for very heavy damages. The area of the destroyed buildings is minimal – 0.8%. Figures 8 to 11 illustrate some of the results obtained. The most significant damages (in terms of total built-up area) are observed in separate quadrants of the Filipovo and Stolipinovo districts, around the two banks of the Maritsa River, the Old Town and some parts of the Kamenitza district. In these parts of the city there is a combination of two factors. On the one hand, the presence of old buildings with higher vulnerability and on the other very high seismic intensity (I = 8.3 to I = 8.6).

IOP Conf. Series: Materials Science and Engineering 1276 (2023) 012008 doi:10.1088/1757-899X/1276/1/012008

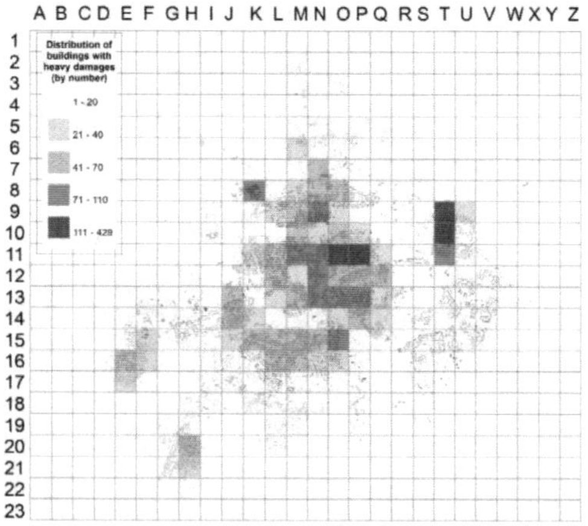

Figure 8. Distribution of buildings with heavy damages – by number.

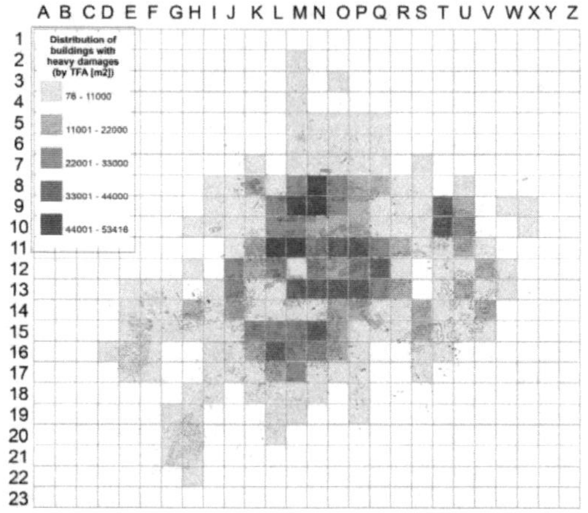

Figure 9. Distribution of buildings with heavy damages - TFA (m^2).

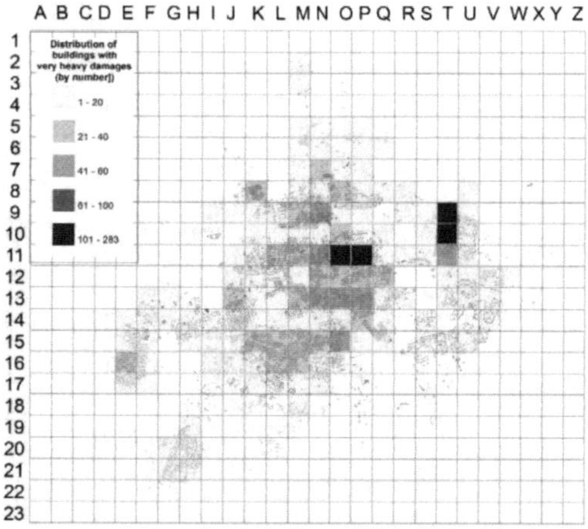

Figure 10. Distribution of buildings with very heavy damages – by number.

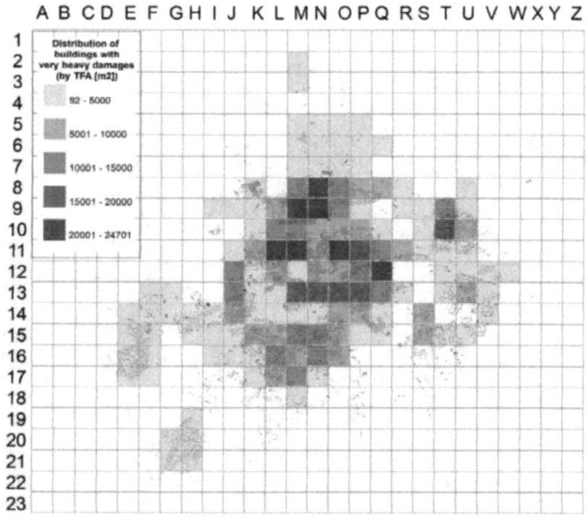

Figure 11. Distribution of buildings with very heavy damages - TFA (m^2).

In assessing the consequences, it is very important to know the distribution of unusable and demolished buildings, so that the volume of restoration work can be planned in advance. Table 8 summarizes the results obtained by detailing the respective groups of buildings - residential, industrial and administrative and special.

The results clearly show that residential buildings are the most vulnerable and with seismic excitation will suffer more damages than the other two groups of buildings.

Table 8. Distribution of unused and demolished buildings (by pieces and TFA).

	Residential	Administrative	Special	Total
Unused buildings (pcs)	4128	536	97	4761
Unused buildings – TFA(m^2)	1545618	171345	87495	1804458
Demolished buildings (pcs)	567	58	8	633
Demolished build. – TFA(m^2)	150060	15147	6707	171914

IOP Conf. Series: Materials Science and Engineering 1276 (2023) 012008 doi:10.1088/1757-899X/1276/1/012008

8. Assessment of social losses - injured people and victims

One of the essential aspects of assessing the consequences of strong earthquakes is the calculation of social losses, since human life is generally priceless. The most victims are observed in the events when there is a total collapse of the structure of a given building. The most important factors affecting social losses are the number of people per square meter and occupation during the earthquake (day or night).

The population of Plovdiv is 342048 people, according to data from the National Statistical Institute (31.12.2020). Regarding the time of realization of the seismic event, there are two options - during the day or at night. The results for affected people - injured and victims are given in Table 9. More detailed information about injured people is shown in Table 10, and they are divided into three categories. After a major earthquake, the homes of a large number of people will be damaged and destroyed to the point of being unusable. The number of people who will need shelter is around 23044 - 25605 people. If the earthquake is in the summer, these people could be temporarily housed in tents. If the earthquake is in winter, they must be housed indoors. At an average rate of 6m^2 per person, the required indoor and heated area is 138266 – 153628m^2, respectively.

On the basis of detailed data on injured people, the need for hospital beds is determined immediately after the seismic event:

- day 2414 - 2683 people;
- 4321 - 4802 people at night.

Table 9. Distribution of the affected people - injured and victims (day and night).

Injured - by day (pcs.)	Victims - by day (pcs.)	Injured - by night (pcs.)	Victims - by night (pcs.)
3721 - 4136	1125 - 1251	6664 - 7406	2024 - 2249

Table 10. Distribution of the injured by category (day and night).

S1-day (pcs.)	S2-day (pcs.)	S3-day (pcs.)	S1-night (pcs.)	S2-night (pcs.)	S3-night (pcs.)
1307- 1453	1166 - 1296	1248 - 1387	2343 - 2604	2080 - 2312	2241 - 2490

Designations: S1 - Minor injuries; S2 - In need of hospital treatment; S3 - Severely injured

9. Estimation of the economic losses due to direct physical damage and destruction

For each of the groups of buildings under consideration, the economic losses are calculated based on the received damages and destructions - Table 11.

Table 11. Economic losses in the main groups of buildings.

	Residential	Administrative	Special	Total
Economic losses (thousands of BGN)	1368037	157340	79381	1604759

Comparison of results for individual groups shows the largest share of buildings in the first group (residential) - BGN 1368 million. The losses in the second group of buildings (administrative) are smaller - approximately 12% than those for the first group. The losses in the third group (special buildings) are much smaller, which is favorable in terms of their importance, immediately after the earthquake.

10. Conclusion

The main question addressed in this article is: if a catastrophic earthquake occurs in the city of Plovdiv, what would be the potential consequences for buildings and people.

The assessment of the vulnerability of buildings shows that the largest number of buildings fall into class "C". In terms of the developed built-up area, the buildings of class "D" dominate. The results clearly show that residential buildings are the most vulnerable and will suffer more damage in an earthquake than the other two groups of buildings.

The number of completely destroyed buildings is 632, which is 0.8% of the total built-up area. The area of the most severely damaged buildings is 9.6% and 3.7% for heavy and very heavy damage grades, respectively.

Based on the detailed data on injured people, the need for hospital beds was determined immediately after the earthquake. In general, social losses are greater at night. The main reason for this is that then people are at home and residential buildings are the most vulnerable and with the greatest damages and destruction. If the earthquake is at night, the number of victims will be about 2 thousand people, while the number of injured people will be about 7 thousand people. The need for hospital beds for severely injured people is about 4500 beds.

The economic losses for the entire city are estimated at about 1604 million BGN, with the largest part falling on residential buildings - 1368 million.

The general conclusion of the study is that the city of Plovdiv is located on an area characterized by a very high level of seismic hazard. In the event of a strong seismic event, the consequences for the city will be very serious. The results show significant damage to a large number of buildings, large social and economic losses. This information should be interpreted in the light of a detailed Preventive Action Program by municipal authorities so that adverse effects can be minimized as much as possible.

Acknowledgements

This work has been carried out in the framework of the National Science Program "Environmental Protection and Reduction of Risks of Adverse Events and Natural Disasters", approved by the Resolution of the Council of Ministers No 577/17.08.2018 and supported by the Ministry of Education and Science (MES) of Bulgaria (Agreement No Д01-363/17.12.2020); and also in the framework of the Contract No D01-404/18.12.2020 (project "National Geoinformation Center (NGIC)" financed by the National Roadmap for Scientific Infrastructure 2020-2027.

References

[1] Solakov D, Simeonova S, Raykova P, Metodiev M 2021 Earthquake scenarios for the city of Plovdiv *Proc. SGEM, 2021, 21(1.1), 701-707*

[2] https://trafficnews.bg/urok-po-istorija/trusat-predi-90-godini-plovdivsko-vzima-23-zhertvi-96900/

[3] *Methodology for analysis, evaluation and mapping of seismic risk of the Republic of Bulgaria* 2018 Ministry of Regional Development and Public Works, *Construction and Architecture, Issue no. 5*, (in Bulgarian)

[4] Grunthal G, Musson R, Schwarz J, Stecci M, 1998 *European Macroseismic Scale 1998-EMS98*; Luxembourg

[5] Caprio M., B. Tarigan, C. Bruce Worden, S. Wiemer, and D. Wald 2015 Ground Motion to Intensity Conversion Equations (GMICEs): A Global Relationship and Evaluation of Regional Dependency. *Bulletin of the Seismological Society of America 105 (3): 1476–1490*

Formulation of parameter limit values for the operation of a large concrete gravity dam

O K Todorov[1], S S Petkova[2] and D S Kisliakov[2*]

[1] Aquapro Engineering EOOD, Block 27, Entr. A, Floor 3, App. 5, Darvenitsa, Sofia, Bulgaria

[2] Department of Hydraulic, Irrigation and Drainage Engineering, University of Architecture, Civil Engineering and Geodesy, 1 Hr. Smirnenski Blvd., 1046 Sofia, Bulgaria

[*] Corresponding author's e-mail: kiss_fhe@uacg.bg

Abstract. This study presents a coupled thermal-structural finite element analysis of a large concrete gravity dam in Bulgaria for exploring the dam response under different actual load combinations. The loads applied to the structure are caused by varying water level in the reservoir and ambient temperature typical for the respective seasons. The developed computational model was calibrated with real plumb measurements. Based on the performed analyses in the ANSYS environment, key limit values for the horizontal displacements measured by the plumbs in the dam as possibly caused by extreme combinations of ambient temperature and water level are formulated for further improvement of the dam operation.

1. Introduction

The technical monitoring and maintenance of a large dam is a string of processes. In addition to the regular visual inspections, a large number of parameters are monitored, and, based on their actual values, corresponding measures are taken. The aim of the whole set of activities is a safe operation of the dam for fulfilling the water management tasks the artificial reservoir is built for. In this sense, the critical limit values of some key parameters are of momentous importance as indicators for the structural behaviour and current state of the facility.

The principles, the scientific base and executive rules of the technical operation, i.e. monitoring and control activities of dams, have been developed and refined over many decades. Currently, they are concisely systematized in an extensive number of different types of resources, e.g. codes, guidelines, recommendations and research studies, for example [1, 2, 3]. In fact, every professional book on dams presents the principles of dam monitoring and surveillance, too [4]. However, the technical operation of a dam, including all monitoring and maintenance activities, is an interdisciplinary field comprising several scientific disciplines. The first one is the systematic knowledge of the necessity of monitoring and maintenance of a civil engineering structure as well as of identifying its actual state [5], i.e. what is to be monitored, how is it to be monitored, which are the key parameters describing both considered structure and its environment, and how the relations among the information flows on structure type and state, environmental impacts and possible damages can be modelled. Further, this knowledge has to be implemented on the particular dam [1, 2, 3]. A sub-case constitutes the concrete dams with their specific features, and especially the concrete gravity dams [6], as the one considered in the present work.

Another technical field is the development and industrial implementation of technologies and technical devices for data acquisition related to particular structural parameters of concrete dams. This extensive area remains outside of the scope of this work.

A next related problem is the necessity of increasing both the efficiency and reliability of the monitoring and maintenance process for concrete dams. To this end, the implementation of different automation and data transfer concepts and systems has been already widely used for decades [7], as has been the development and application of modern, highly sophisticated methods for data processing. As for the latter, some more recent developments should be outlined here, besides the classical statistical approaches. In [8], for structural health monitoring of concrete dams, an inverse analysis method based on a novel hybrid fireworks algorithm (FWA) and the radial basis function (RBF) model was proposed. A damage of concrete dams was diagnosed by identifying the elastic modulus of materials using the displacement changes at different reservoir water levels. FWA is a global optimization intelligent algorithm, and the proposed hybrid algorithm combines the FWA with the pattern search algorithm, which has a high capability for local optimization.

The procedure proposed in [9] was based on identification of zone-wise moduli of elasticity as representative for the diagnosis of possible damage in aged large concrete dams. Changes of reservoir water level with annual periodicity were considered as diagnostic experiments. A number of displacements related to the water level changes were measured on the downstream surface of the dam by radar instruments. Seasonal thermal effects were also taken into account by temperature measurements inside the dam. Finally, damage diagnosis was carried out by minimization of a batch discrepancy function between measured and computed displacements with respect to the sought Young's moduli, as an inverse analysis in a linear thermoelasticity problem.

The work [10] was especially dedicated to the uncertainties in the dam deformation monitoring and analysis. A novel cloud-Verhulst hybrid prediction model was established by combing a cloud model with the Verhulst model.

Artificial neural networks are used for decades in the monitoring data analysis of dams. As a recent application in this field, the paper [11] can be mentioned here presenting a singular value diagnostic method for concrete dam deformation monitoring. It was shown that due to particular disadvantages, a shallow neural network was not appropriate. Instead, a convolutional neural network model was implemented with the advantages of deep learning, automatic feature extraction, good model fitting and strong generalizability.

A problem of particular importance is the stability of a concrete dam with respect to the rock foundation. For identification of the limit state equilibrium, the foundation conditions have to be modelled as realistically as possible. In the paper [12], the rigid-body equilibrium of a wedge representing the dam and a downstream rock wedge was analyzed according to the large displacement regime. Failure mechanisms were identified, analytically described and numerically validated. Application to two Portuguese large concrete gravity dams confirmed the applicability of the developed approach.

In this work, a study is presented on the technical monitoring and maintenance of a large concrete gravity dam in Bulgaria. It was put into operation more than 60 years ago, and during this period, considerable experience has been gained already regarding the dam behaviour under different operational conditions. The subject of the present study was formulation of limit values for the horizontal displacements of the dam at representative points as measured by the plumbs for critical combinations of the water levels in the reservoir and ambient temperatures. These values should further serve the safe operation of the dam.

2. Problem formulation and used methodology

2.1. Problem formulation and structural parameters of the dam

The considered concrete gravity dam was put into operation in 1958, figure 1. As mentioned above, there is already a long-term experience on the structural performance of the dam, gained during its operation. The dam is ageing, and due to possible climate change, an increase of the peak values of some environmental parameters, e.g. ambient temperature, might be expected. In the process of dam monitoring, the formulation of limit values for the most critical parameters, e.g. the horizontal displacements measured by the two plumbs in the dam in blocks No.13 and No.15, would be of crucial importance for its further safe operation. Thus, if such defined values are exceeded in corresponding load cases, this will be a clear sign for an emergency situation with known consequences already identified in the study for defining these limit values.

For solving this task, a computational model of the dam was developed and carefully calibrated. Then, the decisive load cases possibly generating the extreme deformations of the dam and the maximum possible measurement point displacements, respectively, were formulated in the particular case. The calculated displacement values for these extreme load cases were analysed with respect to the related stress and strain conditions of the dam for further use in the process of dam monitoring and maintenance for the upgrade of the safe operation procedures.

Figure 1. The modelled dam (photo by D. Kisliakov).

The dam was built of conventional mass concrete. It consists of 25 blocks, and each of them has a width of 13.00 m. There are two additional blocks at both ends with a width of 6.50 m each. The spillway is located on the central blocks No.9 to No.17, and the total spillway length is 94.50 m, figure 1 and figure 2. The upstream slope of the dam is 1:0.09, and the downstream one – 1:0.75. The non-overflown blocks have a triangle profile. There are three control galleries in the dam body. The grouting gallery has dimensions (width/height) of 2.00 m / 3.50 m and passes over the whole contour of the dam body. There are 9 spillway fields at the dam crest equipped with flap gates, figure 1. Based on the carried out structural testing of the materials and on the study of available sources, the identified computational parameters of the dam concrete also are shown in table 1.

Table 1. Main parameters of the dam.

Parameter	Value
Maximum height of the dam	67.5 m
Length over the crest	338.0 m
Crest width	6.00 m
Maximum water level in the reservoir	227.40 m.a.s.l.
Minimum water level in the reservoir	177.00 m.a.s.l.
Modulus of elasticity of concrete	2.10^{10} N/m^2
Unit weight of concrete	2300 N/m^3

Poisson's ratio	0.20
Thermal conductivity of concrete [13, 14]	2.00 W/mK
Specific heat of concrete [13, 14]	0.2
Concrete thermal expansion coefficient [13, 14]	4.10^{-6}

The physical parameters of the rock foundation domain were considered according to the results of the geological investigations performed during the dam construction period.

2.2. Development and calibration of the computational model

The computational model for solution of the formulated problem was developed in the ANSYS environment [15]. First, a geometry model of the dam and its foundation was created. Then, a finite element discretization of the different domain volumes was carried out with assignment of corresponding material properties to the applied finite elements.

The geometry of the dam was spatially modelled as realistically as possible, figure 2.

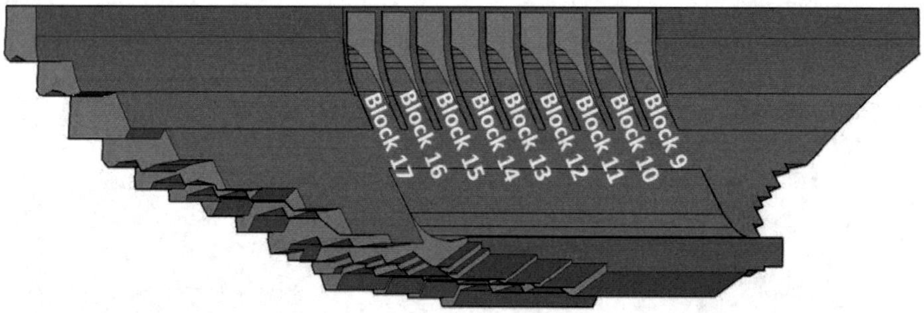

Figure 2. Geometry model of the dam and numbers of central blocks.

Thereafter, the developed geometry model of the dam was smoothed for avoiding subsequent problems of the numerical solution resulting from the properties of the performed discretization of the structural domain. Thus, the smoothed and slightly idealized geometry model of the dam was obtained, figure 3.

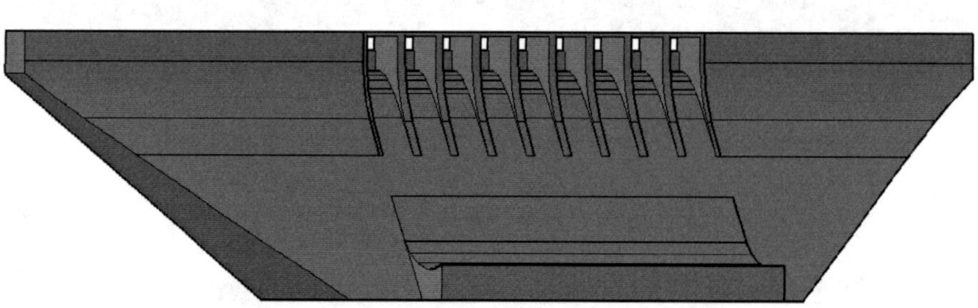

Figure 3. Smoothed geometry of the dam.

In the next step, discretization of the geometry volumes of the different physical domains was performed, figure 4. As it can be seen, the bridge structure at the dam crest was not explicitly modelled further on. This approach was applied as a first approximation. However, the obtained results, as it will be discussed further below, showed that the applied level of sophistication with respect to this structure was entirely sufficient. Moreover, this approximation is on the safety side related to the formulated problem. If the computed horizontal displacements by this model for the decisive extreme load cases are not a cause for concern, then additional structural constraints at the crest imposed by the bridge will only reduce these displacements.

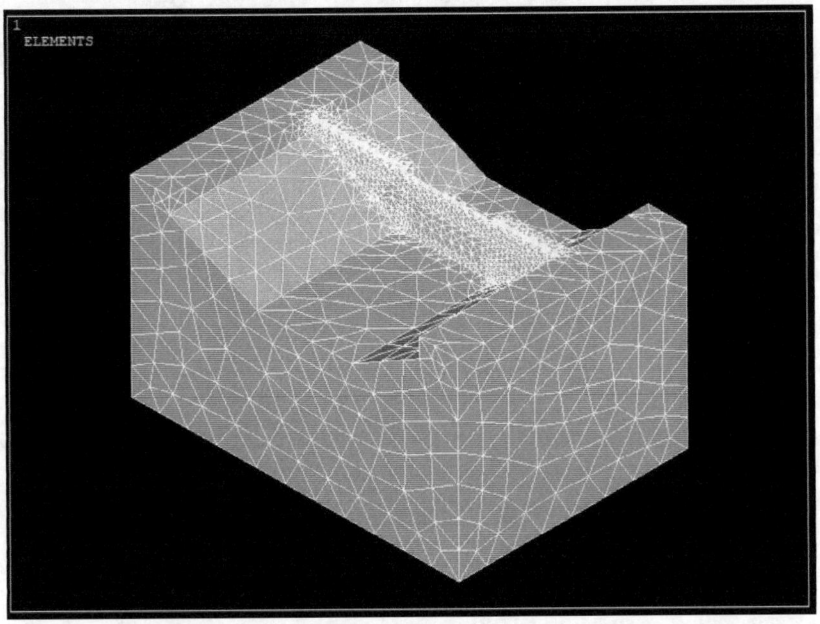

Figure 4. Finite element mesh of the dam and foundation domains.

Further, the boundary conditions of the model and the load cases for model calibration had to be formulated. Here, it should be noted that with respect to the formulated particular task, a coupled thermal-structural analysis of the dam was performed, given the appropriate boundary conditions in the rock foundation domain and the loading by the ambient conditions.

In general, the coupled thermal-structural analysis required is a complicated task. It can have a varying formulation, different related methodologies can be applied depending on the particular problem formulation, and different computational resources can be implemented, respectively. The last decades have marked a substantial progress in this field, especially in modelling of the coupled thermal-mechanical behaviour of mass concrete structures. Currently, highly sophisticated methods already exist for transient thermal-structural analyses of concrete dams allowing for modelling of these coupled processes during both construction and subsequent operation periods of a dam [16]. Case studies also were successfully covered by such analyses, e.g. [17]. Some of these analysis methodologies are summarized in technical regulations in this field [18]. In the present study, the guidelines [19, 20] were predominantly followed. Of course, the fundamental principles of Scientific computing also were met [21].

In general, the coupled thermal-structural problem consists of two parts. The first one is the thermal analysis, and, as a result, the temperature field in the dam body is obtained under the thermal impact of specific/given ambient conditions. Then, the structural problem is solved, whereas the already known thermal field, together with particular structural load combination, delivers the resulting stress-strain fields for this coupled load case.

The investigated dam has already been in operation for a long time. Hence, it was not necessary to apply a transient model for the thermal load, i.e. for modelling of the impact of the ambient air and water temperatures. First, calibration of the developed computational model was performed. For this purpose, days with measured air temperatures in a relatively broad range were selected. The corresponding water level elevations also were known. Since water temperature was not measured, vertical temperature distribution in the reservoir was assumed according to the data [22] for the neighbouring reservoir just upstream of the one considered here. Based on available sources [14], it was assumed that in the kernel of the dam, a zone with the average multi-annual ambient temperature

(12.9°C) has been built for the long operation period so far. The temperature of the foundation rock also was assumed equal to this multi-annual average. Numerous preliminary computational studies confirmed this assumption. Thus, the corresponding thermal load cases were formulated, and the corresponding analyses were run.

At the second stage of each particular analysis, the dam was loaded with the obtained temperature fields, the dead weight and the corresponding hydrostatic load, figure 5. The horizontal displacements transverse to the dam axis in the control points were calculated and compared with the measured values on those days by the plumbs. The results of excellent agreement are presented in table 2.

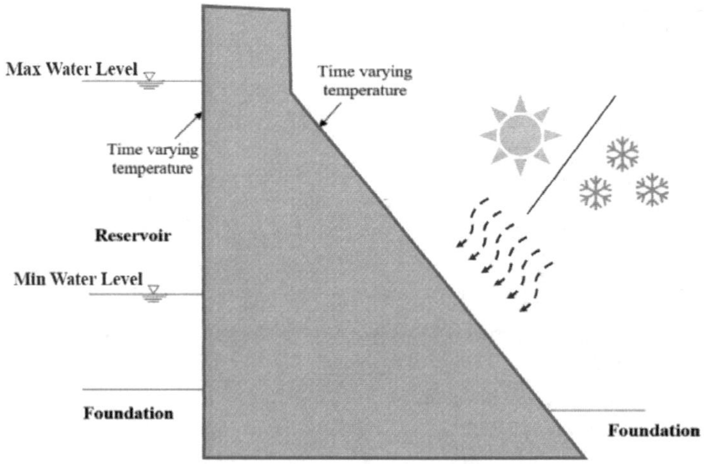

Figure 5. Formulation of the coupled thermal-structural load cases.

Finally, after the model development and calibration process was successfully completed, the critical extreme load cases representing the most unfavourable combinations of extreme water level elevations and ambient temperatures were formulated, figure 5. For these coupled thermal-structural load cases, the already calibrated computational model was applied. The calculated transversal (with respect to the dam axis) horizontal displacements were the aimed results of this study. They should further serve as limit values whose exceeding would cause a concern in the operation of the dam.

Table 2. Results from the calibration of the computational FE.

Date	Air temperature (°C)	Water level elevation (m.a.s.l.)	Measured displacement block No.13 (mm)	Calculated displacement block No.13 (mm)
02.04.2018	16	224.82	4.73	4.58
18.03.2019	10	221.47	3.86	3.80
08.09.2020	22	218.80	-2.38	-2.30

3. Results and discussion

The formulation of limit values of the monitored dam parameters for the safe operation of the dam is an extremely responsible task. Even when a working and calibrated computational model of the facility is available, the set of monitored parameters of dam behaviour is extensive, and the real performance of such a complicated system, like a dam-foundation-reservoir system, in the real environment is very far from being describable and/or controlled by a single parameter. The monitored parameters, such as reservoir water level, ambient temperatures, displacements of the dam crest (in a geodetic way), displacements in block No.13 and block No.15 by means of plumbs, uplift pressure at different points, etc, vary. The measurements of these parameters are of different periodicity, and the accuracy is affected by many factors. The horizontal displacements of the dam, as measured by the plumbs in blocks No.13 and No.15, were assumed as integral parameters of the structural performance of the dam, because at these points, the influence of the atmospheric conditions is minimal, and the accuracy of the measurement devices is the highest possible. This was the reason why these parameters were used for calibration of the developed computational model. This also was the reason why limit values exactly of these parameters had to be formulated as critical for the dam operation. The two extreme load cases used for calculating these limit displacement values were realistic combinations of water level elevations and ambient temperatures, which were supposed to deliver the maximum displacements.

We mentioned here the horizontal displacements in blocks No.13 and No.15 as integral parameters whose limit values were determined in the present study for the formulated limit load cases. In fact, these displacements are directly related to the corresponding stress field in the dam and in the considered blocks, respectively. Thus, the parameter which possibly can cause concern is the tensile stress in some zone of the concrete dam. Since the stress field in the dam body cannot be measured and monitored, the horizontal displacements measured by the plumbs indirectly represent the limit values for the resulting stresses in a particular load case. It is in this sense that the formulation of limit values of the displacements for the dam operation was the subject of the present computational study.

3.1. Limit load case 1

The assumed parameters of the limit load case 1 are presented in table 3. The limit load case is characterized by the lowest ambient temperatures, here – long term average values for February and highest water level elevation in the reservoir. Thus, the maximum possible displacements towards the downstream slope of the dam would be caused.

As an illustration, the calculated thermal field and FE node displacement field in block No.13 are shown in figure 6 and figure 7. The maximum calculated stresses for this limit load case are presented in table 4. These are vertical normal stresses normal to the contact foundation joint of the dam. In this case, there are only compressive normal stresses at the foundation joint.

Table 3. Characteristics of the limit load case 1.

Air temperature (°C)	Water level elevation (m.a.s.l.)	Water depth upstream at the dam (m)	Water temperature (°C)
4	227.40	67.40	bottom: 7, middle layer: 5, surface: 4

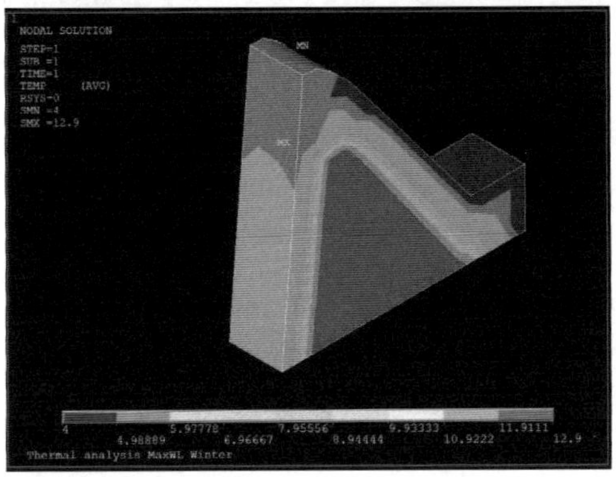

Figure 6. Results as thermal field in block No.13 for limit load case 1.

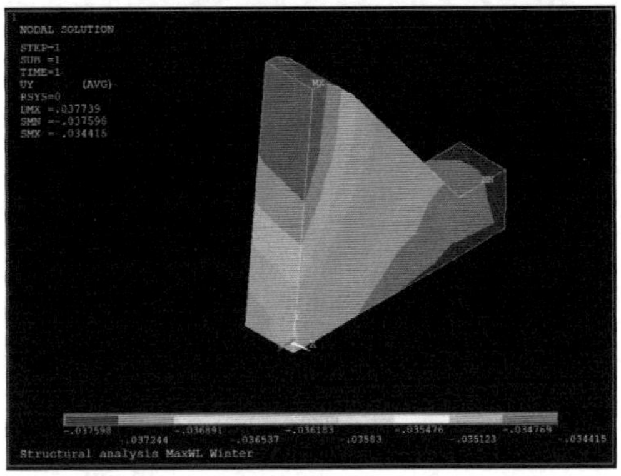

Figure 7. Results as FE node displacements in block No.13 for limit load case 1.

Table 4. Resulting maximum vertical normal stresses at the foundation contact joint of block No.13 for limit load case 1.

Limit load case 1	Upstream toe of the dam (MN/m^2)	Downstream toe of the dam (MN/m^2)
Maximum stress	-1.7 (compressive)	-2.0 (compressive)

3.2. Limit load case 2

The assumed parameters of the limit load case 2 are presented in table 5. This load case is characterized by the highest ambient temperatures, here – long term average values for August and lowest water level elevation in the reservoir. Thus, the maximum possible displacements towards the upstream slope of the dam would be caused.

Table 5. Characteristics of the limit load case 1.

Air temperature (°C)	Water level elevation (m.a.s.l.)	Water depth upstream at the dam (m)	Water temperature (°C)
25	204.00	44.00	bottom: 7, middle layer: 12, surface: 17

As an illustration, the calculated thermal field and FE node displacement field in block No.13 are shown in figure 8 and figure 9. The maximum calculated stresses for this limit load case are presented in table 6. Again, these are vertical normal stresses normal to the contact foundation joint of the dam.

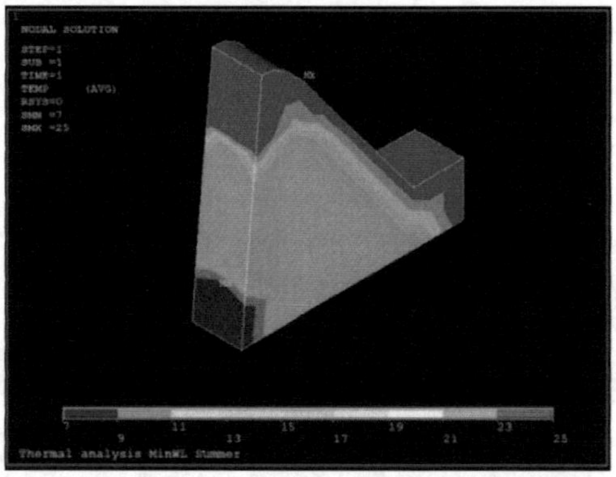

Figure 8. Results as thermal field in block No.13 for limit load case 2.

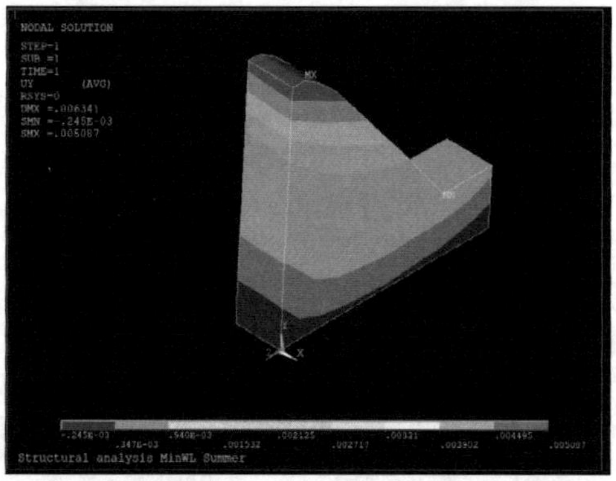

Figure 9. Results as FE node displacements in block No.13 for limit load case 2.

Table 6. Resulting maximum vertical normal stresses at the foundation contact joint of block No.13 for limit load case 1.

Limit load case 2	Upstream toe of the dam (MN/m^2)	Downstream toe of the dam (MN/m^2)
Maximum stress	-0.2 (compressive)	-1.7 (compressive)

In this load case, there are no tensile stresses, either.

All results for block No.15 in terms of displacements and stresses were very closed to the cited values for block No.13.

Finally, it should be noted that the obtained results for the formulated limit load cases were in good agreement with both simplified independent control calculations performed according to [23] and the developed statistical relations built by the dam owner, based on the monitoring during the operational period of the dam so far.

4. Conclusions

The main conclusion drawn from the carried out computational study is that in the beginning of the 1950s, the dam has been appropriately designed with sufficient reserve with respect to its bearing capacity. This is of particular importance because of further possible impacts, such as an earthquake.

The results of this study clearly shows that even an arriving at the obtained limit values for the horizontal displacements measured by the plumbs would not be a cause for concern with respect to the developed stresses in the dam body and indirectly indicated by these displacements. The particular values of the compressive stresses would remain far below the compressive strength of the dam concrete.

References

[1] Georgiev L 1970 *Operation of Hydraulic Engineering Systems (in Bulgarian)* (Tehnika, Sofia) p 400

[2] ICOLD Bulletin 68 1989 *Monitoring of dams and their foundations. State of the art* (International Commission on Large Dams, Paris, France)

[3] ICOLD Bulletin 168 2017 *Recommendations for operation, maintenance and rehabilitation* (International Commission on Large Dams, Paris, France)

[4] Tanchev L 2014 *Dams and Appurtenant Hydraulic Structures* 2nd Edition (CRC Press, Taylor & Francis Group, London, UK) p 1096

[5] Siu-Seong Law and Xin-Qun Zhu 2009 *Damage Models and Algorithms for Assessment of Structures under Operating Conditions* (CRC Press, Taylor & Francis Group, London, UK) p 349

[6] ICOLD Bulletin 117 2000 *The gravity dam. A dam for the future. Review and recommendations* (International Commission on Large Dams, Paris, France)

[7] ICOLD Bulletin 118 2000 *Automated dam monitoring systems. Guidelines and case histories* (International Commission on Large Dams, Paris, France)

[8] Si-qi Dou, Jun-jie Li and Fei Kang 2019 Health diagnosis of concrete dams using hybrid FWA with RBF-based surrogate model *Water Science and Engineering* **12**(3) 188-195

[9] Ardito R, Maier G and Massalongo G 2008 Diagnostic analysis of concrete dams based on seasonal hydrostatic loading, *Engineering Structures* **30** 3176-85

[10] Jin-ping He, Zhen-xiang Jiang, Cheng Zhao, Zheng-quan Peng and Yu-qun Shi 2018 Cloud-Verhulst hybrid prediction model for dam deformation under uncertain conditions, *Water Science and Engineering*, **11**(1) 61-7

[11] Jie Yang, Xu-dong Qu and Meng Chang 2019 An intelligent singular value diagnostic method for concrete dam deformation monitoring *Water Science and Engineering* **12**(3) 205-12

[12] Pereira R, Batista A L, Neves L C and Lemos J V 2022 Deduction of ultimate equilibrium limit states for concrete gravity dams, keyed into rock mass foundations based on large displacement analysis *Structures* **38** 1180-90

[13] Amin M N, Jeong-Su Kim, Yun Lee and Jin-Keun Kim 2009 Simulation of the thermal stress in mass concrete using a thermal stress measuring device *Cement and Concrete Research* **39** 154-64

[14] Žvanut P, Turk G and Kryžanowski A 2021 Thermal analysis of a concrete dam taking into account insolation, shading, water level and spillover *Appl. Sci.*, **11** 705

[15] ANSYS® Academic Research Mechanical, Release 2018 R1 (ANSYS, Inc.)

[16] Mingchao Li, Wen Si, Shengli Du, Mengxi Zhang, Qiubing Ren and Yang Shen 2020 Thermal deformation coordination analysis of CC-RCC combined dam structure during construction and operation periods, *Eng. Structures* **213** 110587

[17] Noorzaei J, Bayagoob K H, Thanoon W A and Jaafar M S 2006 Thermal and stress analysis of Kinta RCC dam *Eng. Structures* **28** 1795–1802

[18] TL 1110-2-542 1997 *Thermal Studies of Mass Concrete Structures* (Dept. of the Army, U.S. Army Corps of Engineers, Washington, DC 20314-1000)
[19] Malm R *Guideline for FE analyses of concrete dam* Report 2016:270 (Energiforsk) p 160
[20] Nguyen L and Salamon J 2018 *Guide for Analysis of Concrete Dam Structures using Finite Element Methods*, DSO-2018-09, Dam Safety Technology Development Program (US Dept. of the Interior, Bureau of Reclamation, Technical Service Center) Denver, Colorado p 72
[21] Wilson G, Bryan J, Cranston K, Kitzes J, Nederbragt L and Teal T K, *Good Enough Practices in Scientific Computing*, PLOS Submission, 31. Aug. 2016, https://arxiv.org/abs/1609.00037v1 (last visited June 02, 2022)
[22] Ivanova A 2017 Vertical Temperature Profile and Dissolved Oxygen in Kardzhali Dam, Bulgaria, *Bulg. J. of Animal Husbandry*, **54**(3) 79-86 (in Bulgarian)
[23] Herzog M A M 1999 *Practical dam analysis* (Thomas Telford Publishing, London, UK) p 241

Modified Anderson-Bjork's method for solving non-linear equations in structural mechanics

N Ganchovski [1], **A Traykov** [2]

[1] Proektsoft EOOD, 34-36 Peyo Yavorov blvd, Sofia, Bulgaria
[2] Department of Structural Mechanics, University of Architecture, Civil Engineering and Geodesy, Sofia, Bulgaria

proektsoft.bg@gmail.com

Abstract. One of the greatest problems in numerical methods is to provide high speed and stability at the same time. In this paper, we propose a new hybrid root finding algorithm, based on the appropriate combination of bisection, false position and the Anderson-Bjork's method. In the general case, the achieved convergence is compatible to the fastest available methods, while stable performance in the worst case is provided. A comparative analysis versus popular root finding algorithms is performed as well. For that purpose, the respective procedures are implemented in the C# programming language. The proposed algorithm is included into the Calcpad engineering calculations platform.

1. Introduction

There are a lot of root finding algorithms available today. The reason for this is that no one of them is universally superior. Some are faster, but more unstable. Some behave well for certain functions, but fail for others. Our goal is to find or develop a general purpose algorithm that is fast and stable enough for any equations, used in structural mechanics. It will be implemented into the open source calculation platform Calcpad. For that purpose, we will review the properties of existing methods, first.

2. Overview of existing methods

In general, root finding algorithms are divided into three main groups: bracketing, iterative and hybrid.

2.1. Bracketing methods

Bracketing methods encloses the root into successively shrinking intervals. They are more stable, but slower than others. Some of the most popular bracketing methods are:

2.1.1. Bisection. This is one of the most popular algorithms, mostly for its simplicity. The initial interval is split into two equal parts by: $x_3 = (x_1 + x_2) / 2$. Then, we select those sub-interval in which the function has different signs at the ends: If $\mathbf{sign}(f(x_1)) \neq \mathbf{sign}(f(x_3))$ then $x_2 = x_3$ else $x_1 = x_3$ (Figure 1). The iterations continue until the root gets enclosed in a sufficiently small interval.

Bisection is the most stable of all algorithms. According to the Bolzano's theorem, if $f(x)$ is continuous in the interval $[a, b]$ and $\mathbf{sign}(f(a)) \neq \mathbf{sign}(f(b))$ then at least one root exists. Only the bisection method will always find the root in a constant time. We can even pre-calculate the required number of iterations to reach a given precision ϵ:

$$n = \log_2 \frac{|x_2 - x_1|}{2\epsilon} \tag{1}$$

Then, why not always use bisection? The problem is that it is also one of the slowest algorithms, with linear convergence. Later, we will show that it is possible to use faster methods and still retain the stability of bisection.

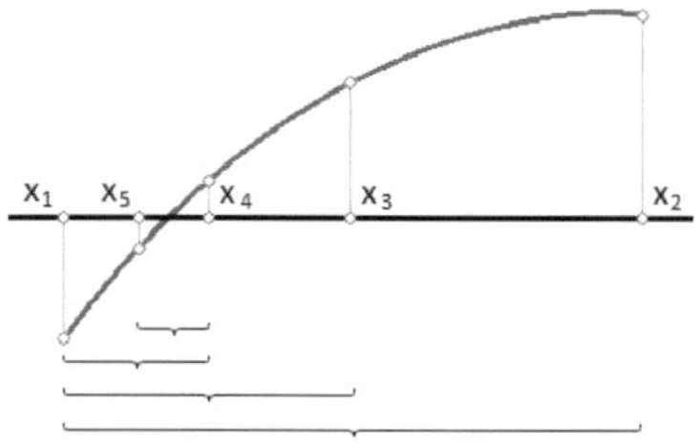

Figure 1. Bisection method.

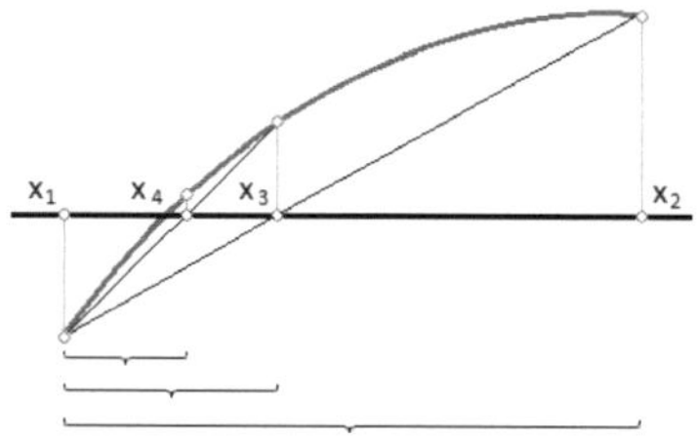

Figure 2. False-position.

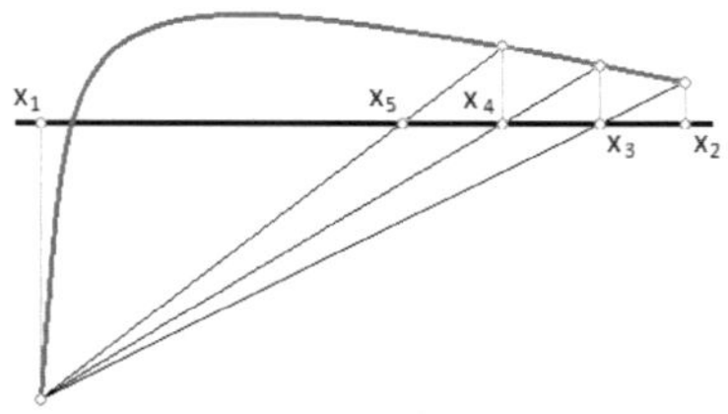

Figure 3. Poor behaviour of the false-position method.

2.1.2. False-position. This method is similar to bisection but the intermediate point is calculated by linear interpolation (Figure 2):

$$x_3 = \frac{x_1 \cdot f(x_2) - x_2 \cdot f(x_1)}{f(x_2) - f(x_1)} \tag{2}$$

It also has linear convergence, and is supposed to work faster for well behaved functions. Unfortunately, it is not so reliable. For functions like the one of Figure 3, its performance can degrade far bellow bisection.

It also has another problem. For most functions, the interval shrinks only at one side, while the other remains fixed. However, this problem has already been solved and the derived algorithms appear to be among the fastest available.

2.1.3. Illinois method. It is an improvement of the false-position method [1]. It uses similar formula for the intermediate point, but the value of the function is multiplied by 0.5 for the end that remained fixed at the previous iteration. In this way, the Illinois method achieves superlinear convergence with factor of 1.442. Another similar method is the Pegasus method [2].

2.1.4. Anderson-Bjork's method. Anderson and Bjork [3] propose, instead of a fixed value of 0.5 as in the Illinois method, to use a reduction factor m as follows:

- $m = 1 - \dfrac{f(x_3)}{f(x_2)}$ if $m > 0$,
- otherwise: $m = 0.5$

In this way, they achieve significant improvement in performance, providing superlinear convergence with factor of 1.7.

2.1.5. ITP method. Its name is abbreviation of "Interpolate, truncate, project". It is the first algorithm that achieves superlinear convergence by preserving the minmax optimality of the bisection method [4]. The convergence factor is the same as the secant method, equal to 1.618. It seems that this is exactly what we need. However, we believe that we will manage to outperform the ITP method, as discussed further.

2.1.6. Ridders' method. It is similar to false position, but instead of linear interpolation, it uses an exponential function [7]. For that purpose, the midpoint of the interval is determined first. Then, a new intermediate point is evaluated by intersecting the interpolation function with the abscissa:

$$x_4 = x_3 + (x_3 - x_1)\frac{sign[f(x_2)]\sqrt{f(x_3)^2 - f(x_1)f(x_2)}}{f(x_1)} \tag{3}$$

The Ridders's method has superlinear convergence with factor of $\sqrt{2} \approx 1.414$, which is less than the Anderson-Bjork's method.

2.2. Iterative methods

Iterative methods use the previous point to successively estimate the next (hopefully better) approximation of the root. They are usually faster than bracketing methods, but much more unstable.

2.2.1. Newton-Raphson. This is one of the fastest and most popular root finding methods with quadratic convergence. It starts from a point that is near to the root. The next point is evaluated by intersecting the tangent with the abscissa. For that purpose, the following equation is used:

$$x_{n+1} = x_n - \frac{f(x_n)}{f'(x_n)} \tag{4}$$

Graphical representation of the algorithm is shown on Figure 4.

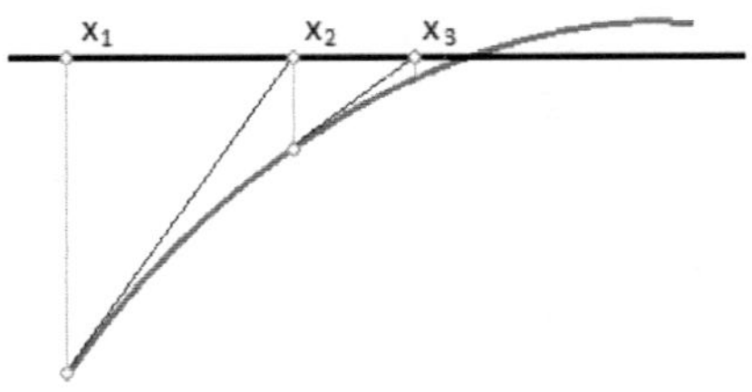

Figure 4. Newton-Raphson's method.

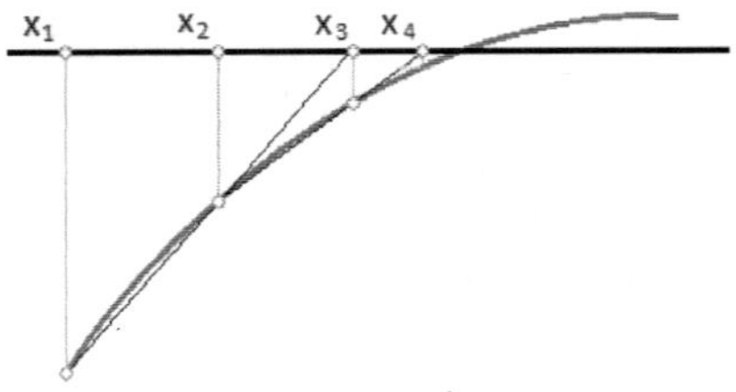

Figure 5. Secant method.

However, this method is not suitable for a general purpose algorithm. It has to satisfy plenty of requirements. Firstly, the function must have a continuous non-zero derivative. An analytical representation of the derivative should be available as well. Otherwise, it would be computationally too expensive to evaluate it numerically. If the function does not comply with the requirements, the solution may easily fail. Some common examples are shown on Figure 6:

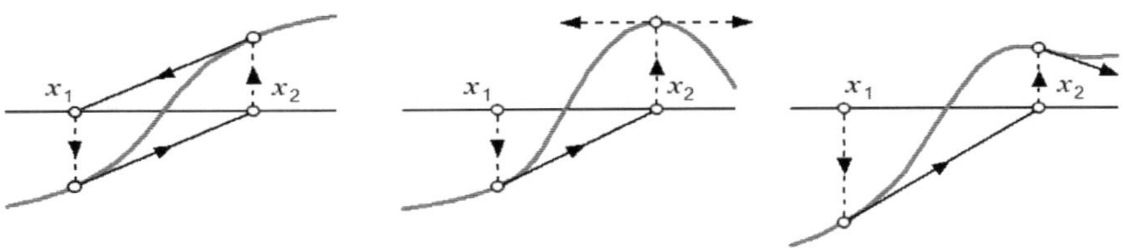

Figure 6. Examples for failure of Newton-Raphson's method.

2.2.2. Secant method. It is similar to the Newton-Raphson's method, but does not require the derivative to exist. Instead, two successive points are used to construct the secant (Figure 5):

$$x_3 = x_2 - f(x_2) \cdot \frac{x_2 - x_1}{f(x_2) - f(x_1)} \tag{5}$$

Unfortunately, it inherits all other drawbacks of the Newton-Raphson's method. Additionally, we pay a little trade-off with the performance. It has superlinear convergence with factor of 1.618. But it is the same as the ITP method, which is stable, and less than Anderson-Bjork's method.

2.2.3. Haley's method. It is similar to Newton-Raphson but of higher order. It uses the following equation to obtain the next approximation:

$$x_2 = x_1 - \frac{2 f(x_1) f'(x_1)}{2 f'(x_1)^2 - f(x_1) f''(x_1)} \tag{6}$$

It has a cubic convergence, but being even more demanding, it is not suitable for our purposes.

2.2.4. Muller's method. It is the higher order analogue of the secant method, but it constructs a parabola trough three successive points [6]:

$$x_3 = x_2 - \frac{2c}{b + sign(b)\sqrt{b^2 - 2c}}, \tag{7}$$

where a, b and c are calculated as follows:

$a = \frac{\delta_2 - \delta_1}{h_2 - h_1}$, $b = a h_1 + \delta_1$, $c = f(x_2)$, where

$h_1 = x_2 - x_1$, $h_2 = x_3 - x_2$, $\delta_1 = \frac{f(x_2) - f(x_1)}{h_1}$, $\delta_2 = \frac{f(x_3) - f(x_2)}{h_2}$

It has superlinear convergence with factor of 1.84. However, like the other iterative methods, it is not reliable enough for a general purpose use.

2.3. Hybrid methods

Hybrid methods combine two or more simple methods, trying to get the most out of them and overcome the drawbacks. In this way, we obtain both reliable and fast algorithm, such as the Brent's method.

2.3.1. Brent's method. It incorporates bisection, secant method and inverse quadratic interpolation [8]. It is an improvement of the earlier Decker's method [9]. The implementation is a little bit complicated and we are not going into details here. It is important that it manages to achieve a superlinear convergence with a factor of 1.839, while retaining the reliability of bisection. That makes it a good candidate for our purposes. However, it has been shown by Suhadolnik [10] that the performance of Brent's method can fall far under bisection even for very simple functions like x^3 and x^5.

2.4. Comparison table

Analytical study of the convergence rate for some of the above methods was made by Kumar and Vipan [12]. They presented the results in a comparison table. An extended version of this table, including all of the above method is presented bellow:

Table 1. Comparison of root-finding methods.

Method	Type	Base equation order	q-convergence	Order	Reliability
Bisection	Bracketing	1-st order	linear	1	stable
False-position		1-st order	linear	1	volatile
Illinois		1-st order	superlinear	1.442	nearly stable
ITP		1-st order	superlinear	1.618	stable
Anderson-Bjork		1-st order	superlinear	1.7	nearly stable
Ridders		2-nd order	superlinear	$\sqrt{2}$	nearly stable
Newton-Raphson	Iterative	1-st order	quadratic	2	unstable
Secant		1-st order	superlinear	1.618	unstable
Muller		2-nd order	superlinear	1.84	unstable
Halley		2-nd order	cubic	3	unstable
Brent	Hybrid	1-st and 2-nd	superlinear	1.839	nearly stable

We also added a reliability estimation as follows:

- **stable** – has a guarantied performance in worst case equal or not much bellow bisection;
- **nearly stable** – performance can drop bellow bisection but still a root can be found within a reasonable number of iterations;
- **volatile** – a root will be found even in the worst case but after unreasonable number of iterations (as much as to make the software to "freeze");
- **unstable** – the algorithm is possible to diverge or finish after unreasonable number of iterations;

Based on the above analysis, we can conclude that available root finding methods are always a compromise between speed and stability. That is why, the quest for better algorithms is still contemporary and the possibilities for further improvements are not exhausted. The most straightforward and easiest way to do that is to properly combine existing methods into new hybrid algorithms. We will demonstrate bellow that this can improve stability, speed and general applicability for the incorporated methods.

3. Enhancing the false-position method

As we discussed above, false-position is usually faster than bisection, but for some functions, its performance can drop significantly bellow. We can improve both methods by combining them into a single hybrid algorithm. Instead of blindly using a single method alone, we can estimate which method will be faster for the next step and switch between methods.

Similar approach was proposed by Sabharwal [11] in his „Blended" algorithm. On each iteration he computes the function for both methods and compares the absolute values. The method with smaller absolute value (closer to zero) is relevant for the next iteration.

However, this approach has two drawbacks:

- It requires two function computations at each iteration. So, even if the number of iterations is the same as bisection, the algorithm will be nearly as twice as slower.
- The smaller absolute value does not properly indicate the better method. For example, for the function of Figure 3, the algorithm will always select the worse method (false-position) until it gets close enough to the root.

That makes the above algorithm efficient mostly for monotonic functions.

IOP Conf. Series: Materials Science and Engineering 1276 (2023) 012010 doi:10.1088/1757-899X/1276/1/012010

We will implement a different criterion for switching methods. False-position is based on linear interpolation, so the closer is the function to a straight line, the faster will be the solution. Most functions "look" straighter as the interval gets smaller. It is because the relative curvature becomes smaller as well. So, by narrowing the interval preliminary with bisection, we can boost the performance of false-position for ill-behaved function. In this way, we will force it to start with better initial conditions. This will result in steeper line in the convergence chart, as shown on Figure 7.

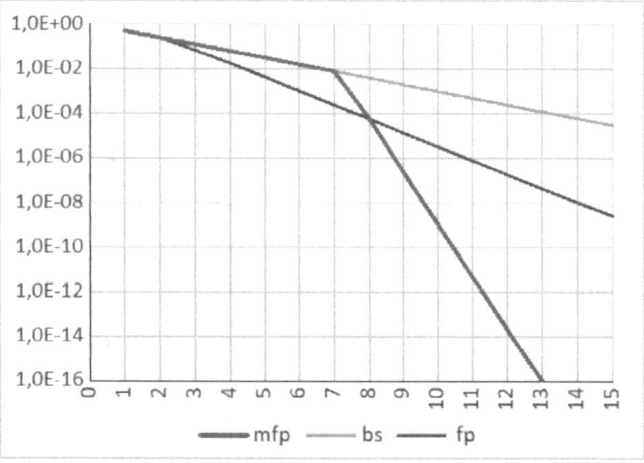

Figure 7. Bisection, false-position and combined algorithm convergence:
bs – bisection,
fp – false-position,
mfp – modified false-position.

Both methods are linear, but the combined algorithm has superlinear (or bi-linear) convergence. Even if bisection is slower at the beginning, the overall performance will be faster due to the steeper false-position line.

The question is: how to select the appropriate moment for switching. If we switch earlier than necessary, we will have less iterations along the bisection line, but flatter false-position line. We propose to measure the relative deviation of the function from a straight line at the middle point. For that purpose, the following expression will be applied:

$$|y_3 - y_m| < k\left(|y_3| + |y_m|\right), \tag{8}$$

where y_3 is the function value and y_m is the ordinate of the chord in the middle of the interval, as shown on Figure 8.

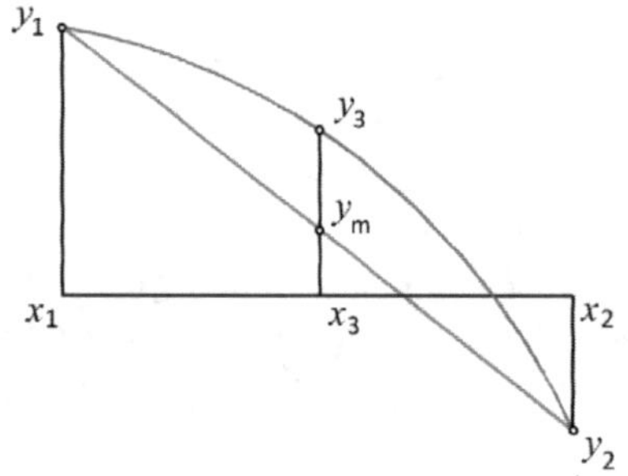

Figure 8. Relative deviation of a function.

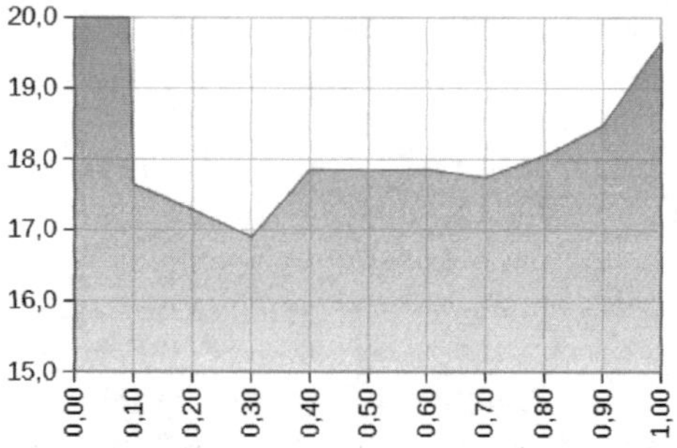

Figure 9. Average number of iterations, in dependence of *k*.

The optimal value of *k* is not the same for different functions and initial intervals. However, it would be acceptable to find some average value that works for most cases. For that purpose, we performed numerical experiments with different types of functions (see §7. further). Then, we calculated the average number of iterations in dependence of k, as displayed on Figure 9.

The results show that there is no significant difference (about one iteration) for $k \in (0.1, 0.9)$. The optimal value is located between 0.1 – 0.4. We will select the middle value of 0.25. If we plot the convergence charts, we will see how k affects the behavior of the algorithm (Figure 10).

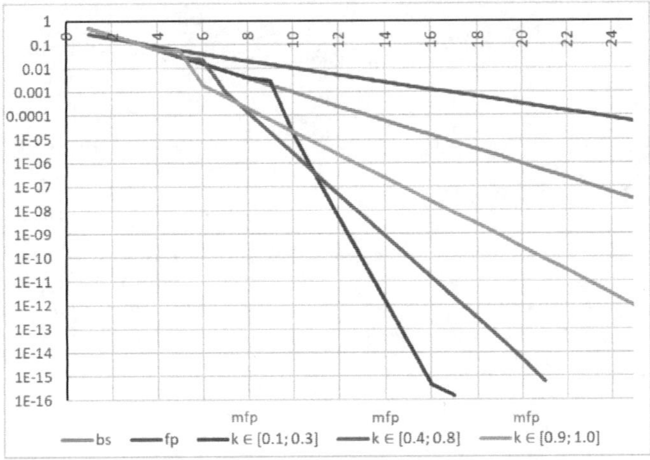

Figure 10. Convergence charts for different values of k.

The above graphics are obtained for the function $f(x) = 4 - \tan(x)$ in the interval [0. 1.5].

4. Do we need to switch back?

In some rare cases, the above algorithm can be tricked as shown on Figure 11:

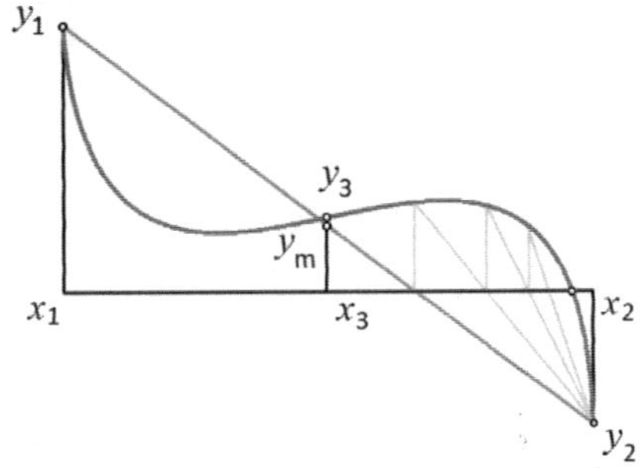

Figure 11. Example for improper switching to false-position method.

In other cases, the function can have a local bulge near the root. That is why we will need some way to switch back to bisection. The simplest way is to reset the algorithm to bisection, if it fails to converge after the expected number of iterations, defined by equation (1).

5. Overcoming the fixed end problem

As we mentioned above, the false-position method has another major drawback. For some functions, only one of the endpoints is moved, while the other remains fixed. There are several algorithms, such as Pegasus, Illinois and Anderson-Bjork [3], that were created to solve this problem and thus, managed to significantly accelerate the solution.

A detailed comparison of their performance was made by Galdino [13], using numerical testing. For the selected set of functions, the Anderson-Bjork's algorithm showed the best average score. So, we can use it to further improve the performance of our algorithm. For that purpose, we will replace the false-position method with the Anderson-Bjork's one. The convergence diagrams for $f(x) = 4 - \tan(x)$, for both methods, compared to bisection and false-position are presented bellow:

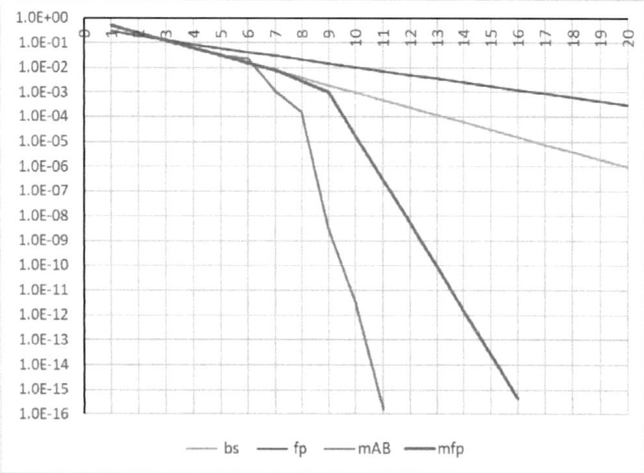

Figure 12. Convergence of different root-finding methods:
bs – bisection,
fp – false-position,
mfp – modified false-position,
mAB – modified Anderson-Bjork.

6. The final algorithm

The final algorithm, implemented in the C# language, is presented bellow:

```
internal static double? ModAB(Func F, double x1, double x2, double eps)
{
  double y1 = F(x1), y2 = F(x2);               //Evaluate the function at both ends
  int side = 0;                    //For tracking the side that has moved at the previous iteration
  int N = -(int)(Math.Log2(Precision) / 2) + 1; //Expected number of iterations
  double x0 = x1;                              //For storing the abscissa from the previous iteration
  bool Bisection = true;
  for (int i = 1; i <= n; ++i)
  {
    double x3, y3;
    if (Bisection)                             //Bisection step
    {
      x3 = (x1 + x2) / 2; y3 = F(x3);          //Midpoint abscissa and function value
      double ym = (y1 + y2) / 2;               //Ordinate of chord at midpoint
      if (Math.Abs(ym - y3) < 0.25 * (Math.Abs(ym) + Math.Abs(y3))
        Bisection = false;                     //Switch to false-position
    }
    else                                       //False-position step
      x3 = (x1 * y2 – y1 * x2)/(y2 - y1); y3 = F(x3)

    if (y3 == 0 || Math.Abs(x3 - x0) <= eps)   //Convergence check
      return x3;                               //Return the result

    x0 = x3;                                   //Store the abscissa for the next iteration
    if (side == 1)                             //Apply Anderson-Bjork modification for side 1
    {
      double m = 1 - y3 / y1;
```

```
        if (m <= 0) y2 *= 0.5; else y2 *= m;
    }
    else if (side == 2)                       //Apply Anderson-Bjork modification for side 2
    {
        double m = 1 - y3 / y2;
        if (m <= 0) y1 /= 2; else y1 *= m;
    }
    if (Math.Sign(y1) == Math.Sign(y3))       //If the left interval does not change sign
    {
        if (!Bisection) side = 1;             //Store the side that move
        x1 = x3; y1 = y3;                     //Move the left end
    }
    else                                      //If the right interval does not change sign
    {
        if (!Bisection) side = 2;             //Store the side that move
        x2 = x3; y2 = y3;                     //Move the right end
    }
    if (i % N == 0)                           //If the expected number of iterations is exceeded
    {
        Bisection = true;                     //Reset to bisection
        side = 0;
    }
  }
  return null;                                //If the algorithm failed to converge for n iterations
}
```

7. Numerical tests

In order to select the most appropriate algorithm, we performed numerical tests with the potential candidates: Illinois, Anderson-Bjork, ITP, Ridders, Brent and the newly proposed method. Bisection and false-position were also included for comparison. For our tests, we selected some of the functions that Oliveira and Takahashi used for their ITP method [4], with initial interval [-1, 1] and $\epsilon = 10^{-15}$. The results are summarized in the table bellow:

Table 2. Results from numerical tests.

Function	Equation	Number of iterations for method[1)]							
		bs	fp	ill	AB	ITP	mAB	Rid	Bre
f_{01}	$xe^x - 1$	50	30	10	8	10	9	8	9
f_{02}	$\tan\left(x - \frac{1}{10}\right)$	50	40	7	6	10	7	10	8
f_{03}	$\sin(x) + \frac{1}{2}$	50	10	7	7	9	7	14	7
f_{04}	$4x^5 + x^2 + 1$	50	33	12	10	14	11	14	10
f_{05}	$x + x^{10} - 1$	50	51	11	9	15	12	10	11
f_{06}	$\pi^x - e$	50	15	8	6	9	7	8	7
f_{07}	$\ln\left\|x - \frac{10}{9}\right\|$	50	65	11	8	3	8	10	9

***f*₀₈**	$\frac{1}{3}+sign(x)\lvert x\rvert^{1/3}+x^3$	50	13	12	10	15	10	14	10
***f*₀₉**	$\left(x+\frac{2}{3}\right)\div\left(x+\frac{101}{100}\right)$	50	200	15	5	52	8	14	12
***f*₁₀**	$(x\,10^6-1)^3$	50	2	2	2	53	50	4	138
***f*₁₁**	$e^x(x\,10^6-1)^3$	50	200	96	116	53	50	4	140
***f*₁₂**	$\left(x-\frac{1}{3}\right)^2\tan^{-1}\left(x-\frac{1}{3}\right)$	50	200	95	115	52	50	84	142
***f*₁₃**	$sign(3x+1)\left(1-\sqrt{1-\frac{(3x+1)^2}{9^2}}\right)$	25	200	26	35	27	25	36	65
***f*₁₄**	$\left(x>\frac{1-10^6}{10^6}\right)\frac{1+10^6}{10^6}-1$	50	50	39	39	51	50	54	49
***f*₁₅**	$\frac{1}{21x-1}\left(x\neq\frac{1}{21}\right)$	50	13	94	94	53	50	72	52
***f*₁₆**	$\left(\frac{x}{2}\right)^2+\left\lceil\frac{x}{2}\right\rceil-1$	50	51	56	56	51	50	40	50
***f*₁₇**	$\lceil 10x-1\rceil+\frac{1}{2}$	50	48	52	52	49	13	36	46
***f*₁₈**	$x+\frac{\sin(x\,10^6)}{10}+\frac{1}{10^3}$	50	23	21	21	24	23	24	20
***f*₁₉**	$(x>-1)\left(1+\sin\left(\frac{1}{x+1}\right)\right)-1$	50	54	15	13	52	18	40	51
***f*₂₀**	$202x-2\left\lfloor\frac{2x+10^{-2}}{2\cdot10^{-2}}\right\rfloor-\frac{1}{10}$	50	4	4	4	12	3	4	3
***f*₂₁**	$202x-2\left\lfloor\frac{2x+10^{-2}}{2\cdot10^{-2}}\right\rfloor-\frac{1}{10}$	50	200	75	73	52	50	70	125
Sum	Number of iterations	1025	1502	668	689	666	511	570	964
Ave	Average number	48,8	71,5	31,8	32,8	31,7	24,3	27,1	45,9
Max	Maximal number	50	200	96	116	53	50	84	142
Func.	Method	bs	fp	ill	AB	ITP	mAB	Rid	Bre

[1] Legend:

bs	– Bisection	ITP	– Interpolate, truncate, project
fp	– False-position	mAB	– Modified Anderson-Bjork (new)
ill	– Illinois	Rid	– Ridders
AB	– Anderson-Bjork	Bre	– Brent

IOP Conf. Series: Materials Science and Engineering 1276 (2023) 012010 doi:10.1088/1757-899X/1276/1/012010

8. Conclusions

Based on the above results, we can conclude that we managed to improve the average and worst case performance of both false-position and Anderson-Bjork's algorithms. The proposed method preserves the stability of bisection for poorly behaved functions, similar to ITP method. At the same time, it provides the best performance in a lot of cases.

We can also say that Illinois, Anderson-Bjork's, Ridders', ITP and Brent's methods showed similar performance and there was a different "winner" each time. In all cases, the proposed algorithm was very close to the best one, with only a few iterations behind. This makes it suitable for implementation as a general purpose method for solving non-linear equations in structural mechanics. So far, we successfully applied this method for different mechanical problems:

- finding the neutral line for an arbitrary cross section with material nonlinearity;
- ultimate load calculation for elements with variable cross section with arbitrary shape, initial imperfections, material and geometric nonlinearity;
- capacity of reinforced concrete beams with prestressing tendons, etc.

References

[1] Dowell M and Jarratt P 1971 A modified regula falsi method for computing the root of an equation *BIT* **11** pp 168–174

[2] Dowell M and Jarratt P 1972 The "Pegasus" method for computing the root of an equation *BIT* **12** pp 503–508

[3] Anderson N and Björck Å 1973 A new high order method of regula falsi type for computing a root of an equation *BIT* **13** pp 253–264

[4] Oliveira I and Takahashi R 2021 An Enhancement of the Bisection Method Average Performance Preserving Minmax Optimality *ACM Trans. on Mathematical Software* **47**, 1 Article 5

[5] Halley E 1694 A new and general method of finding the root equations *Philos. Trans. Roy. Soc.* London, 18

[6] Muller D 1956 A Method for Solving Algebraic Equations Using an Automatic Computer *Mathematical Tables and Other Aids to Computation* vol 10 no 56 pp 208-215

[7] Ridders C 1979 A new algorithm for computing a single root of a real continuous function *IEEE Trans. on Circuits and Systems* vol 26 no 11 pp 979-980

[8] Brent R 1971 An algorithm with guaranteed convergence for finding a zero of a function *The Computer Journal* vol 14 Issue 4 pp 422–425

[9] Dekker T 1969 Finding a zero by means of successive linear interpolation *Constructive Aspects of the Fundamental Theorem of Algebra*, ed B Dejon and P Henrici (London: Wiley-Interscience) ISBN 978-0-471-20300-1

[10] Suhadolnik A 2012 Combined bracketing methods for solving nonlinear equations *Applied Mathematics Letters* vol 25 Issue 11 pp 1755-1760, ISSN 0893-9659

[11] Sabharwal C 2019 Blended Root Finding Algorithm Outperforms Bisection and Regula Falsi Algorithms *Mathematics* vol 7 Issue 11 article 1118

[12] Kumar R and Vipan 2015 Comparative Analysis of Convergence of Various Numerical Methods *Journal of Computer and Mathematical Sciences* vol 6 issue 6 pp 290-297

[13] Galdino S 2011 A family of regula falsi root-finding methods P*roceedings of 2011 World Congress on Engineering and Technology* (Shanghai, China: IEEE Press) pp 514-517, ISBN 978-1-61284-365-0

[14] Press W, Teukolsky S, Vetterling W and Flannery B 2007 *Numerical Recipes 3rd Edition: The Art of Scientific Computing* Cambridge University Press USA

Capacity of axially loaded RC columns with arbitrary cross sections to Eurocode 2

M. Traykova [1], **N. Ganchovski** [2], **A. Traykov** [3]

[1] Department of RC Structures, University of Architecture, Civil Engineering and Geodesy, Sofia, Bulgaria
[2] Proektsoft EOOD, 34-36 Peyo Yavorov blvd, Sofia, Bulgaria
[3] Department of Structural Mechanics, University of Architecture, Civil Engineering and Geodesy, Sofia, Bulgaria

alex_fce@uacg.bg

Abstract. Unlike older design codes, Eurocode 2 does not provide a procedure for calculation of the capacity of axially loaded columns. For that purpose, a general algorithm is developed, using the principles of structural mechanics and mathematics: It starts with constructing the P-Δ relationship (buckling curve) for the element, accounting for initial imperfections, minimum eccentricity, material nonlinearity and second order effects. Then, the respective bending moments are determined from the displacements and the P-M relationship is obtained. Finally, the axial load capacity in planar domain is determined by intersecting the P-M diagram with the interaction curve for the RC section. In spacial domain, the P-Δ relationship is defined by the respective "buckling surface". Its intersection with the interaction surface represents a spacial curve in the (P, Mx, My) coordinate system. The axial capacity is determined as the minimum of P values along the curve. The solution is performed by numerical methods.

1. Introduction

Usually, design checks of RC columns for bending with axial force are performed by comparing the external bending moment with the ultimate bending capacity ($M_{Ed} \leq M_{Rd}$), providing that the internal and external axial loads are in equilibrium ($N_{Ed} = N_{Rd}$). However, in most cases, columns are loaded only with axial force and there are no external bending moments. Then, minimal eccentricities, initial imperfections and second order effects are applied besides the axial load. In such cases, it is more convenient to determine the ultimate axial force capacity $N_{Rd,II}$ and compare it to the imposed external load N_{Ed}. This approach provides more valuable information for the structural designer.

2. Capacity of RC sections for bending with axial load

Unfortunately, Eurocode 2 does not provide detailed methods or formulas for calculation of the capacity for bending with axial load. Only general principles are described in section 6.1. of EN 1992-1-1. They can be used by different authors to develop formulas, numerical models or computer programs. On their bases. we have created a numerical procedure that is applicable for cross sections with arbitrary shapes and reinforcement.

Parabolic-linear and linear plastic stress-strain relationships are used for the concrete and reinforcement, respectively (Figures 1 and 2). Stain is distributed linearly along the section height.

Strain diagram is varied from pure tension to pure compression through stages 1, 2 and 3 as shown on Figure 3.

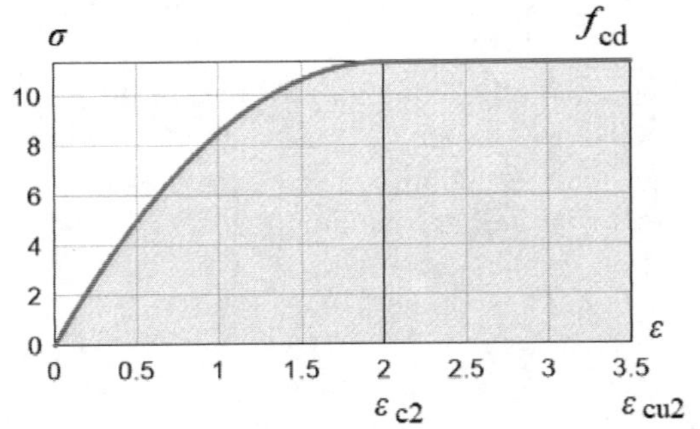

Figure 1. Stress-strain relationship for concrete.

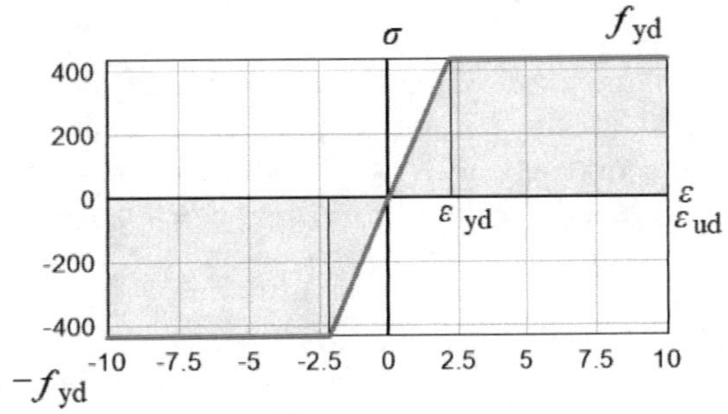

Figure 2. Stress-strain relationship for steel.

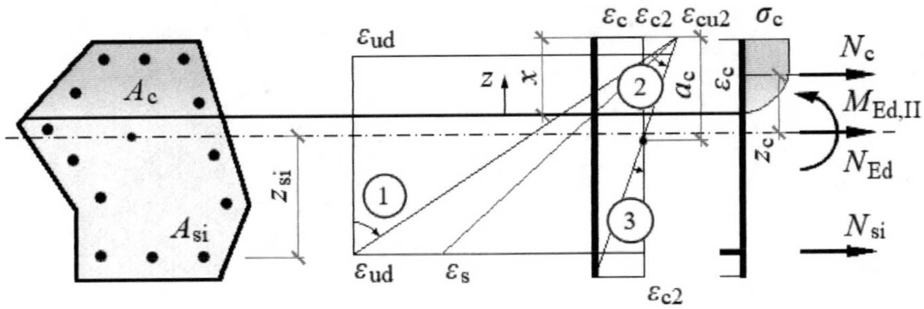

Figure 3. Theoretical basis for bending-with-axial-force design.

For a given strain diagram, the respective stresses in concrete and reinforcement bars are evaluated. The resulting compression force in concrete is calculated via numerical integration using adaptive Gauss-Lobatto quadrature [1]. The equilibrium of the internal and external axial forces is found by solving the equation:

$$N_{Rd}(\varepsilon)+N_{Ed}=0 \tag{1}$$

,where ε is a general strain parameter that describes all possible positions of the strain diagram. The equation is solved using the modified Anderson-Bjork root finding method [2]. The axial cross section resistance is determined as follows:

$$N_{Rd}=N_c+\sum N_{si}, \tag{2}$$

where:

$N_c=\int_0^x \sigma_c(z)\cdot b(z)\,dz$ - the resultant force in the concrete; (3)

$$\sigma_c(z)=\begin{cases} f_{cd}\left[1-\left(1-\dfrac{\varepsilon_c(z)}{\varepsilon_{c2}}\right)^n\right] \text{for } 0\le\varepsilon_c(z)\le\varepsilon_{c2} \\ f_{cd} \text{ for } \varepsilon_{c2}\le\varepsilon_c(z)\le\varepsilon_{cu2}\end{cases} \text{ - concrete stress;} \tag{4}$$

$N_{si}=A_{si}\cdot\sigma_{si}$ is the force in reinforcement bar "*i*"; (5)

$$\sigma_s(\varepsilon)=\max\left(-f_{yd};\min\left(\varepsilon\cdot E_s;f_{yd}\right)\right)$$ - stress-strain relationship for steel. (6)

The bending resistance of the section is determined accordingly:

$$M_{Rd}=N_c\cdot z_c+\sum N_{si}\cdot z_{si} \quad (7)$$

For the case of biaxial bending, the above equation is applied along the direction of the total external moment. Then, besides the strains in the extreme edges, we need to determine the angle of the neutral line so that the total internal moment acts in the same direction as the external one. We may notice that the neutral line is not perpendicular to the vector of the total bending moment, so again we should use numerical methods to find the angle.

3. N-M interaction diagram/surface

If we sweep the strain diagram from pure tension to pure compression and calculate N_{Rd} and M_{Rd} for each position, we will obtain a closed curve in the M–N plane(Figure 4). For biaxial bending, we also need to rotate the neutral line and calculate the respective N_{Rd}, $M_{Rd,x}$ and $M_{Rd,y}$. The resulting object is a closed surface in the M_x–M_y–N coordinate system which we will call "interaction surface" (Figure 5).

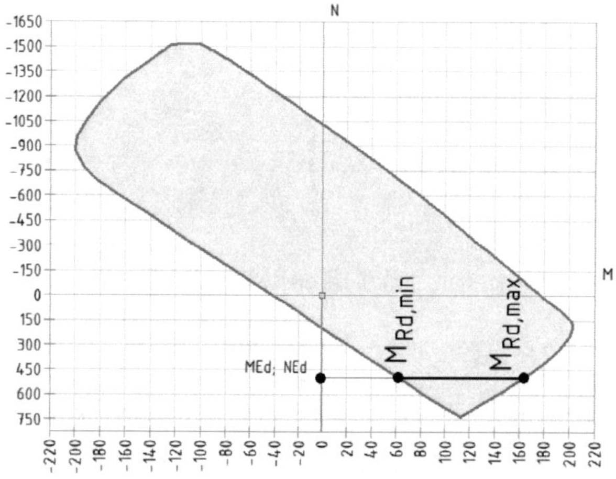

Figure 4. Interaction diagram.

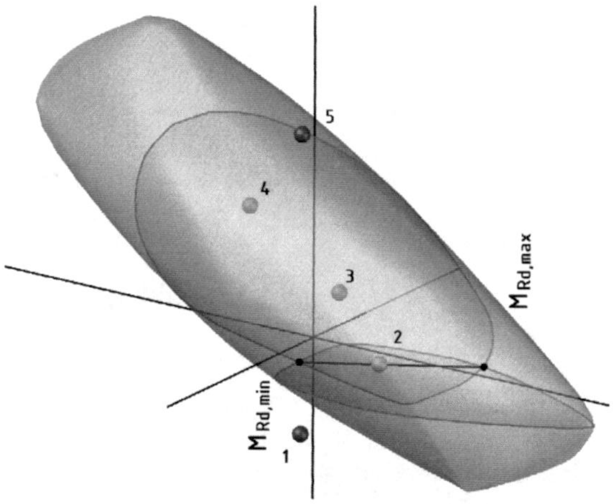

Figure 5. Interaction surface.

4. Design criteria

As we mentioned above, most authors propose the equation $M_{Ed} \leq M_{Rd}$ for design check of RC sections, where M_{Rd} is obtained for the equilibrium of internal and external axial forces. However. this approach may not be fully correct for unsymmetrical sections and/or reinforcement. For example, the asymmetrically reinforced section, shown on Figure 6 will fail for the applied load even though $M_{Ed} \leq M_{Rd}$. The correct approach for all cases is to check if the load is inside the interaction curve (Figure 4) which can be expressed by the equation:

$$M_{Rd,min} \leq M_{Ed} \leq M_{Rd,max} \quad (8)$$

The same approach is applicable for biaxial bending, as shown on Figure 5. Here, $M_{Rd,min}$, M_{Ed} and $M_{rd,max}$ are located on a horizontal line in the plane of the external moment. In this way, we provide that the load point is inside the interaction surface. Once we have the surface constructed, we can use it to check many load cases as displayed on Figure 5.

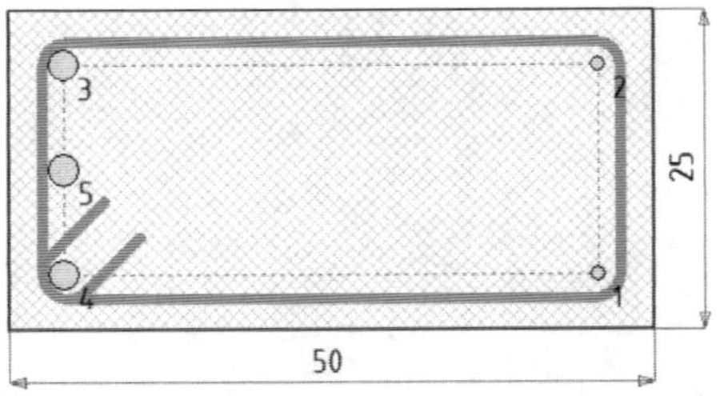

Figure 6. Sample cross section.

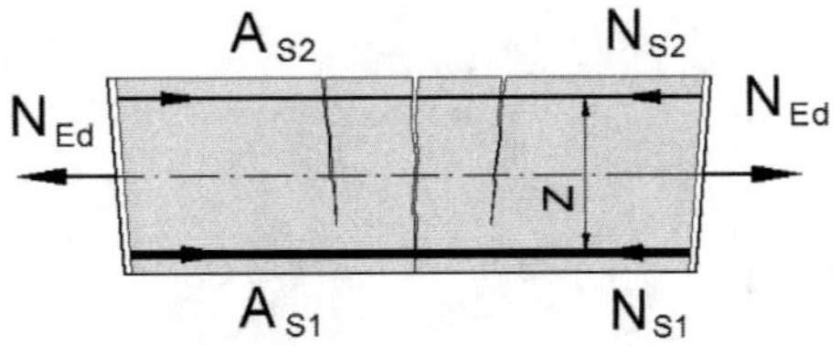

Figure 7. Sample load.

5. Capacity of columns

Besides the cross section and material properties, the capacity of the element depends on the minimum eccentricity, initial imperfections and second order effects, that should be also taken into account.

5.1. Minimum eccentricity

According to EN 1992-1-1, section 6.1 (4) [3], the minimum eccentricity is calculated by the equation:

$$e_0 = \max(h/30;\ 20\text{mm}), \tag{9}$$

where h is the section depth. Eurocode does not provide information how to proceed for arbitrary shaped sections, subjected to biaxial bending. The logical approach is to take the distance between the extreme points, projected in the plane of interest. Since the eccentricity may act in any direction, it can be represented by a closed curve. For simplicity, we can calculate the eccentricities in the two main directions and construct an ellipse.

The eccentricity results in a bending moment that is obtained by the equation: $M_{Ed,0} = N \cdot e_0$. For plane bending, it can be represented by two lines in the $M–N$ plane (Figure 8). If the external moment M_{Ed} is less than $M_{Ed,0}$ the latest is relevant. And since the eccentricity can act in both directions, we should make two checks: for $-M_{Ed,0}$ and $+M_{Ed,0}$, regardless of the sign of M_{Ed}. For unsymmetrical sections, it can happen that the opposite side is the relevant one as shown on Figure 8. The point where the line intersects the interaction diagram defines the axial capacity of a section with minimum eccentricity: $N_{Rd,0}$.The minimum value from both sides should be taken as relevant.

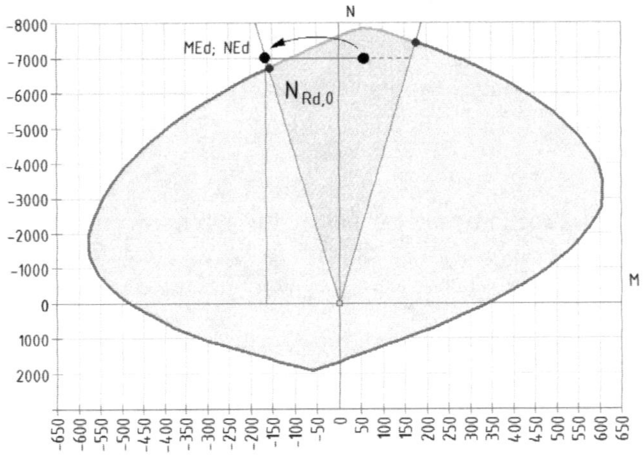

Figure 8. Minimum eccentricity check with $M–N$ interaction diagram.

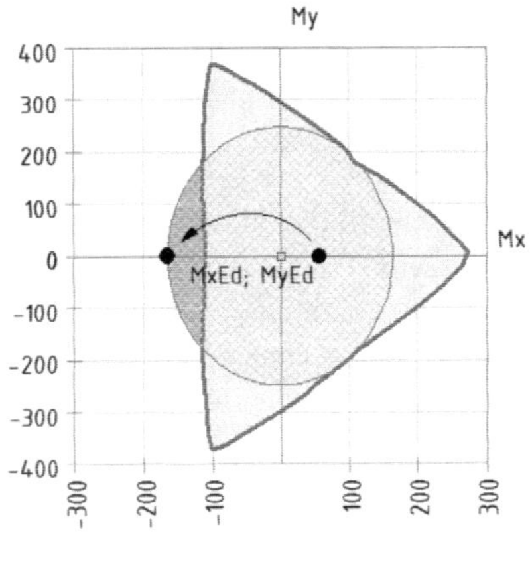

Figure 9. Min eccentricity check with M_x - M_y interaction diagram.

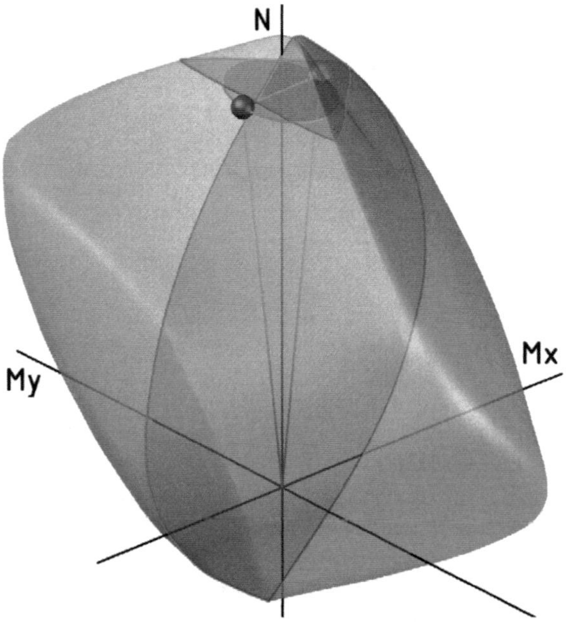

Figure 10. Minimum eccentricity check with interaction surface.

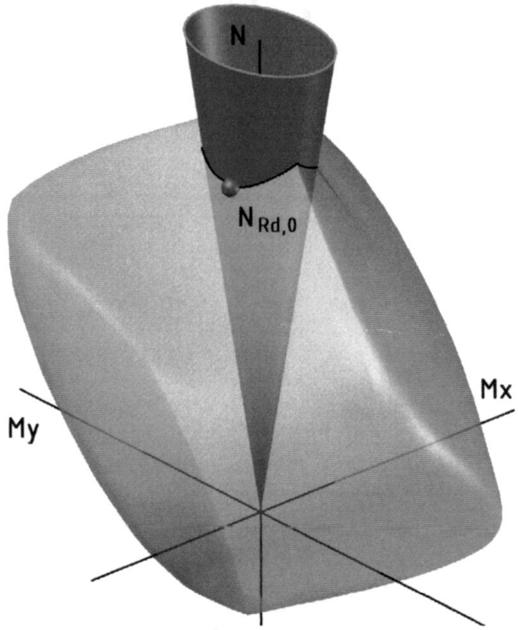

Figure 11. Axial capacity for minimum eccentricity.

For the spatial formulation, the eccentricity can be represented by inverted elliptic cone as displayed on Figure 11. Since, it can act in any direction, the entire cone will be the relevant load whenever the external moment falls inside the cone/pyramid, regardless of its sign and direction. Then the check is reduced to a mere geometrical problem: We intersect the pyramid and the interaction surface with a horizontal plane at $N = N_{Ed}$ and compare both shapes. If the eccentricity polygon falls entirely inside the M_x-M_y interaction diagram in the $N = N_{Ed}$ plane, the check is satisfied (Figure 10).

The definition of the axial load capacity $N_{Rd,0}$ is a bit more complicated. If we intersect the inverted pyramid with the interaction surface, we will obtain a spatial closed curve. The capacity is the minimum of the curve in respect to the N coordinate as displayed on Figure 11

5.2. *Initial imperfections*

Initial imperfections are defined in EN 1992-1-1, section 5.2. For isolated members, they may be taken as an eccentricity:

$e_i = l_0/400$, where

l_0 is the effective length of the column.

Since they do not depend on the shape of the cross section, they can be equally applied in any direction. If we draw the eccentricity in plan, it will represent a circle with radius e_i. The respective bending moments can be represented as an inverted cone in the M_x-M_y-N coordinate system like the one on Figure 11.

5.3. *Second order effects*

They are result from bucking and induce additional eccentricities. Unlike initial imperfections and minimum eccentricity, the respective bending moments increase hyperbolically with the compressive force. When the latest is close to the buckling load, the bending moment theoretically grows to infinity. However, the column will brake long before that.

There are three methods in Eurocode 2 to account for second order effects: general method, nominal stiffness and nominal curvature. For the purpose of our research, we will use the nominal stiffness method, according to EN 1992-1-1, section 5.8.7.2. However, any of the other methods may be applied with equal success, without affecting the final conclusions.

The following procedure is used:

Bending moment with initial imperfections - $M_{0,Ed} = M_{Ed} + N_{Ed} \cdot e_i$

Factor depending on the concrete grade - $k_1 = \sqrt{\frac{f_{ck}}{20\,MPa}}$ [EN 1992-1-1 (5.23)]

Relative axial force $n = \frac{N_{Ed}}{A_c \cdot f_{cd}}$

Factor depending on axial force and slenderness - $k_2 = \frac{n \cdot \lambda}{170} < 0.20$ [EN 1992-1-1 (5.24)]

Maximum value - $k_2 < 0.20$

Effective creep ratio - $\varphi_{ef} = \varphi(\infty, t_0) \cdot M_{0,Eqp} / M_{0,Ed}$

Factor for effects of cracking, creep, etc. $K_c = \frac{k_1 \cdot k_2}{1 + \varphi_{ef}}$ [EN 1992-1-1 (5.22)]

Factor for contribution of reinforcement – $K_s = 1$

Design value of the modulus of elasticity of concrete – $E_{cd} = \frac{E_{cm}}{\gamma_{cE}}$ [EN 1992-1-1 (5.20)]

Nominal stiffness – $EI = K_c \cdot E_{cd} \cdot I_c + K_s \cdot E_s \cdot I_s$

Buckling load – $N_B = \frac{\pi^2 \cdot EI}{L_o^2}$

Bending moment with second order effects – $M_{Ed,II} = \frac{M_{0,Ed}}{1 - \frac{N_{Ed}}{N_B}}$ [EN 1992-1-1 (5.30)]

Minimum eccentricity – $e_o = max\left(\frac{h}{30}; 20\,mm\right)$ [EN 1992-1-1, § 6.1(4)]

Design bending moment – $M_{Ed} = max\left(M_{Ed,II}; N_{Ed} \cdot e_o\right)$

5.4. *Buckling capacity*

For the planar formulation, the second order effects can be represented by a hyperbolic curve that asymptotically approaches the buckling load. The intersection of the hyperbola with the interaction diagram defines the buckling resistance of the column as shown on Figure 12 a).

For the spatial formulation, the second order effects can be displayed as a surface that looks like gramophone funnel. Its intersection with the interaction diagram represents a closed spatial curve. The minimal "*N*" coordinate on the curve is the column buckling capacity as on Figure 12 b).

It is not necessary that buckling capacity point will occur in either of the main directions. Especially for columns with complex, non-symmetric shapes and different boundary conditions in both directions.

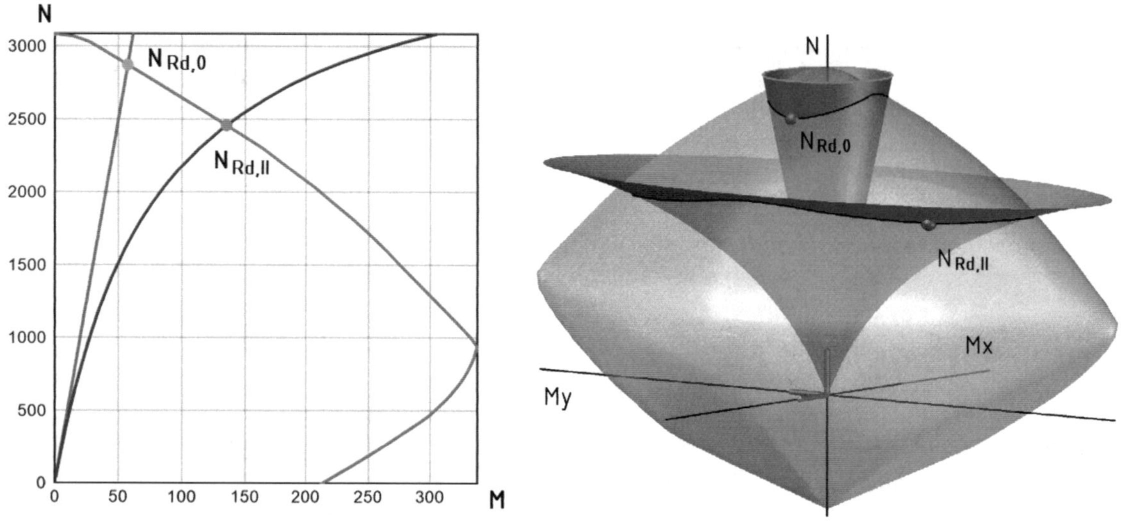

a) 2D solution b) 3D solution

Figure 12. Buckling capacity of columns.

6. Proposed algorithm

Based on the above discussion, we propose the following algorithm for calculation of the axial load capacity of arbitrary shaped RC columns, with account to minimum eccentricity, initial imperfections and second order effects:

6.1. *Planar solution*

1. Construct the interaction diagram by variation of the strain diagram.
2. Intersect the minimum eccentricity line with the interaction diagram = $N_{Rd,0}$.
3. Find the buckling load and construct the buckling curve.
4. Intersect the buckling curve with the interaction diagram = $N_{Rd,II}$.
5. Calculate the final column resistance $N_{Rd} = \min(N_{Rd,0}, N_{Rd,II})$.

6.2. *Spatial solution*

1. Construct the interaction surface by variation of the neutral line direction and the strain diagram.
2. Construct the minimum eccentricity inverted cone.
3. Intersect the cone with the interaction surface and obtain the intersection curve.
4. Find the minimum coordinate on the curve in respect to its N-coordinates = $N_{Rd,0}$.
5. Find the buckling loads and construct the buckling funnel surface.
6. Intersect the buckling surface with the interaction surface and obtain the intersection curve.
7. Find the minimum coordinate on the curve in respect to its N- coordinates = $N_{Rd,II}$.
8. Calculate the final column resistance as $N_{Rd} = \min(N_{Rd,0}, N_{Rd,II})$.

7. Additional considerations

7.1. *Selection of a coordinate system*

For arbitrary shaped cross sections, it is not always obvious how to locate the section in the *Oxy* plane. However the laws of mechanics require us to perform the calculations about the principle axes through the geometric center. That is why we will locate the geometric center at point $O(0, 0)$ and rotate the section so that the principle axes coincide with x and y axes.

7.2. *Spatial distribution of buckling effects*

Buckling analyses are performed about the two principle axes separately. For that purpose, the respective boundary conditions or effective lengths should be specified in both directions. Two buckling loads are determined for the two directions and the respective second order effects are calculated. For all other directions, the influence of the second order effects is approximated as an ellipse.

7.3. *Performance*

At a first glance, the proposed 3D algorithm may look computationally intensive. However it can be efficiently implemented on the contemporary hardware. All surfaces can be stored as 3D meshes (polygons). Once they are constructed, the problem is reduced to pure geometric algorithms. The proposed method is suitable for parallelization, so we can take advantage of the today's multi-core processors. The other possibility is to benefit from the computational power of 3D graphic cards, if available. In all cases, we will be able to calculate hundreds or even thousands of columns in a second.

8. Software and implementation

The 2D version of the algorithm is implemented as a Calcpad worksheet. It it is available for free, under the MIT license. The 3D version is included into the PMM Expert professional software, developed by Proektsoft. It was used for preparation of all 3D examples in the current paper.

9. Worked example

To illustrate how the proposed method works, we developed a simple example for a rectangular section with different boundary conditions in both directions. We calculated the buckling resistance using both methods – planar and spatial and compared the results.

Cross Section – R50x30 + 4N28

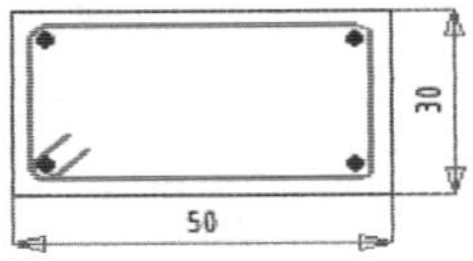

Concrete area - A_b =1500 cm^2
Bar count - n_b = 4
Bar diameter - d_b = 28mm
Reinforcement area - A_s = 24 cm^2
Reinforcement ratio - ρ = 1.6%

Buckling Data

Column length – L = 420cm
Buckling length – L_{ox} = 2*L
Buckling length – L_{oy} = 1*L
Creep factor - $\varphi(\infty,t_0)$ = 2.5
Permanent load ratio - K_G = 75%

Concrete grade C25/30

Characteristic compressive cylinder strength - $f_{ck}=25$ MPa
Partial safety factor for concrete - $\gamma_c=1.5$, $\alpha_{cc}=0.85$
Design compressive cylinder strength - $f_{cd}=\dfrac{\alpha_{cc}\cdot f_{ck}}{\gamma_c}=\dfrac{0.85\cdot 25}{1.5}=14.17$ MPa
Mean value of cylinder compressive strength - $f_{cm}=f_{ck}+8=25+8=33$ MPa
Secant modulus of elasticity - $E_{cm}=22\cdot\left(\dfrac{f_{cm}}{10}\right)^{0.3}=22\cdot\left(\dfrac{33}{10}\right)^{0.3}=31.48$ GPa
Ultimate compressive strain - $\varepsilon_{cu2}=0.0035$
Strain at the end of parabolic part of the diagram - $\varepsilon_{c2}=0.002$

Reinforcement grade B500

Characteristic yield strength - $f_{yk}=500$ MPa
Partial safety factor for steel - $\gamma_s=1.15$
Design yield strength - $f_{yd}=\dfrac{f_{yk}}{\gamma_s}=\dfrac{500}{1.15}=434.78$ MPa
Modulus of elasticity - $E_s=200$ GPa

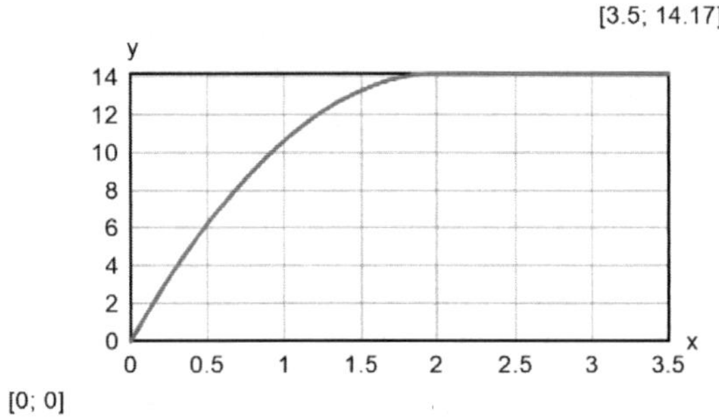

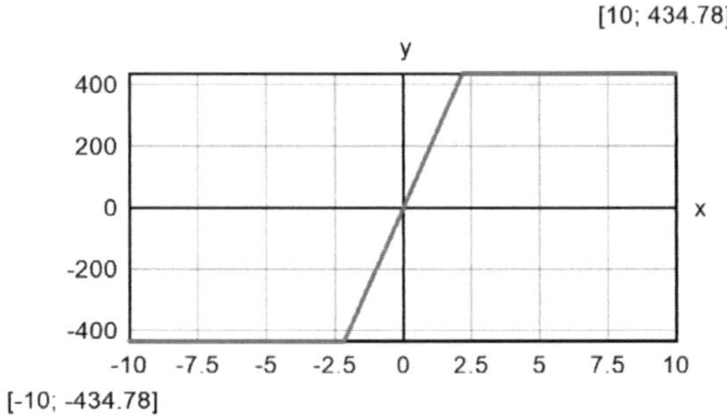

Planar solution

Geometric imperfections and accidental eccentricity

Number of vertical members contributing to the total effect - $m=1$

Reduction factor for height - $\alpha_h=\dfrac{2}{\sqrt{L\cdot 10^{-3}}}=\dfrac{2}{\sqrt{4200\cdot 10^{-3}}}=0.976$

This value is limited within the interval: $2/3 \le \alpha_n \le 1$

$$\alpha_h=max\left(\frac{2}{3};min\left(\alpha_h;1\right)\right)=max\left(\frac{2}{3};min\left(0.976;1\right)\right)=0.976$$

Reduction factor for number of members - $\alpha_m=\sqrt{0.5\cdot\left(1+\dfrac{1}{m}\right)}=\sqrt{0.5\cdot\left(1+\dfrac{1}{1}\right)}=1$

Basic inclination value - $\theta_o=\dfrac{1}{200}=0.005$

Inclination - $\theta_i=\theta_o\cdot\alpha_h\cdot\alpha_m=0.005\cdot 0.976\cdot 1=0.00488$ [EN 1992-1-1 (5.1)]

Eccentricity - $e_i=\dfrac{\theta_i\cdot L_o}{2}=\dfrac{0.00488\cdot 8400}{2}=20.49$ mm [EN 1992-1-1 (5.2)]

Minimum eccentricity [EN 1992-1-1, § 6.1(4)]

$e_{ox}=max\left(\dfrac{h}{30};20\right)=max\left(\dfrac{500}{30};20\right)=20$ mm $\quad e_{oy}=max\left(\dfrac{b}{30};20\right)=max\left(\dfrac{300}{30};20\right)=20$ mm

Second order effects based on the nominal stiffness method

Relative axial force - $n=\dfrac{N_{Ed}}{A_c\cdot f_{cd}\cdot 10^{-3}}=\dfrac{2271}{150000\cdot 14.17\cdot 10^{-3}}=1.07$

Factor depending on the concrete strength class - $k_1=\sqrt{\dfrac{f_{ck}}{20}}=\sqrt{\dfrac{25}{20}}=1.12$ [EN 1992-1-1 (5.23)]

Factor depending on axial force and slenderness [EN 1992-1-1 (5.24)]

$k_{2x}=\dfrac{n\cdot\lambda_x}{170}=\dfrac{1.07\cdot 58.2}{170}=0.366$ $\quad k_{2y}=\dfrac{n\cdot\lambda_y}{170}=\dfrac{1.07\cdot 48.5}{170}=0.305$

Maximum value - $k_2 < 0.20$ $\quad$ Maximum value - $k_2 < 0.20$

$k_{2x}=min(0.366;0.2)=0.2$ $\quad k_2=min\left(k_{2y};0.2\right)=min(0.305;0.2)=0.2$

Effective creep ratio

$\varphi_{ef}=\varphi(\infty,t_0)\cdot K_G=2.5\cdot 0.75=1.88$

Factor for effects of cracking, creep etc. [EN 1992-1-1 (5.22)]

$$K_c=\frac{k_1\cdot k_2}{1+\varphi_{ef}}=\frac{1.12\cdot 0.2}{1+1.88}=0.0778$$

$K_s=1$ – factor for contribution of reinforcement

$\gamma_{cE}=1.2$ [BS EN 1992-1-1 (5.20), NA.1]

Design value of the modulus of elasticity of concrete [EN 1992-1-1 (5.20)]

$$E_{cd}=\frac{E_{cm}}{\gamma_{cE}}=\frac{31.48}{1.2}=26.23 \text{ GPa}$$

Moment of inertia of the concrete cross section

$$I_{cx}=\frac{b\cdot h^3}{12}=\frac{300\cdot 500^3}{12}=3125000000 \text{ mm}^4 \qquad I_{cy}=\frac{h\cdot b^3}{12}=\frac{500\cdot 300^3}{12}=1125000000 \text{ mm}^4$$

Radius of gyration

$$i_x=\sqrt{\frac{I_{cx}}{A_c}}=\sqrt{\frac{3125000000}{150000}}=144.34 \text{ mm} \qquad i_y=\sqrt{\frac{I_{cy}}{A_c}}=\sqrt{\frac{1125000000}{150000}}=86.6 \text{ mm}$$

Slenderness ratio

$$\lambda_x=\frac{L_{ox}}{i_x}=\frac{8400}{144.34}=58.2 \qquad \lambda_y=\frac{L_{oy}}{i_y}=\frac{4200}{86.6}=48.5$$

Second moment of area of the reinforcement

$$I_{sx}=A_{s1}\cdot\left(\frac{h}{2}-d_1\right)^2+A_{s2}\cdot\left(\frac{h}{2}-d_2\right)^2 \qquad I_{sy}=A_{s1}\cdot\left(\frac{b}{2}-d_1\right)^2+A_{s2}\cdot\left(\frac{b}{2}-d_2\right)^2$$

$$I_{sx}=2\cdot 1200\cdot\left(\frac{500}{2}-50\right)^2=96000000 \text{ mm}^4 \qquad I_s=2\cdot 1200\cdot\left(\frac{300}{2}-50\right)^2=24000000\text{mm}^4$$

Nominal stiffness [EN 1992-1-1 (5.21)]

$$EI_x=K_c\cdot E_{cd}\cdot I_c+K_s\cdot E_s\cdot I_s \qquad EI_y=K_c\cdot E_{cd}\cdot I_c+K_s\cdot E_s\cdot I_s$$

$$EI_x=0.0778\cdot 26.23\cdot 3125\cdot 10^6+1\cdot 200\cdot 96\cdot 10^6 \qquad EI_y=0.0778\cdot 26.23\cdot 1125\cdot 10^6+1\cdot 200\cdot 24\cdot 10^6$$

$$EI_x=25575184994\text{kN}\cdot\text{mm}^2 \qquad EI_y=7095066598\text{kN}\cdot\text{mm}^2$$

Buckling load [EN 1992-1-1 (5.17)]

$$N_{Bx}=\frac{\pi^2\cdot EI_x}{L_{ox}^2}=\frac{3.14^2\cdot 25575184994}{8400^2}=3577.34 \text{ kN} \qquad N_{By}=\frac{\pi^2\cdot EI_y}{L_{oy}^2}=\frac{3.14^2\cdot 7095066598}{4200^2}=3969.7 \text{ kN}$$

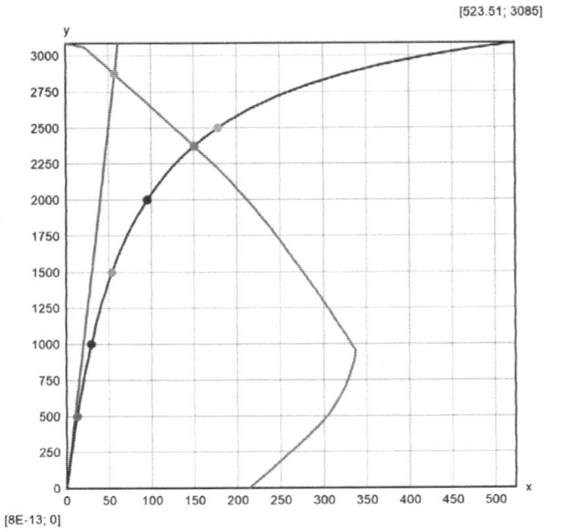

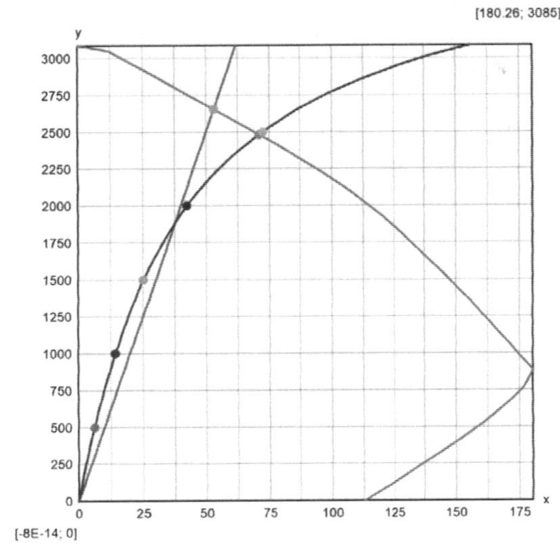

Ultimate capacity for axial force

Cross section only - $N_{Rd,max}=3168.48$ kN

Capacity for minimum eccentricity

$N_{Rd,0x}=2872.74$ kN $\qquad N_{Rd,0y}=2653.59$ kN

Capacity for geometric imperfections

$N_{Rd,eix}=2865.99$ kN $\qquad N_{Rd,eiy}=2879.7$ kN

Capacity for imperfections and II order effects

$N_{Rd,IIx}=2375.23$ kN s $\qquad N_{Rd,IIy}=2482.87$ kN

Buckling factor

$\varphi_x=\frac{N_{Rd,IIx}}{N_{Rd,max}}=\frac{2375.23}{3168.48}=0.75 \qquad \varphi_y=\frac{N_{Rd,IIy}}{N_{Rd,max}}=\frac{2482.87}{3168.48}=0.784$

Spatial solution

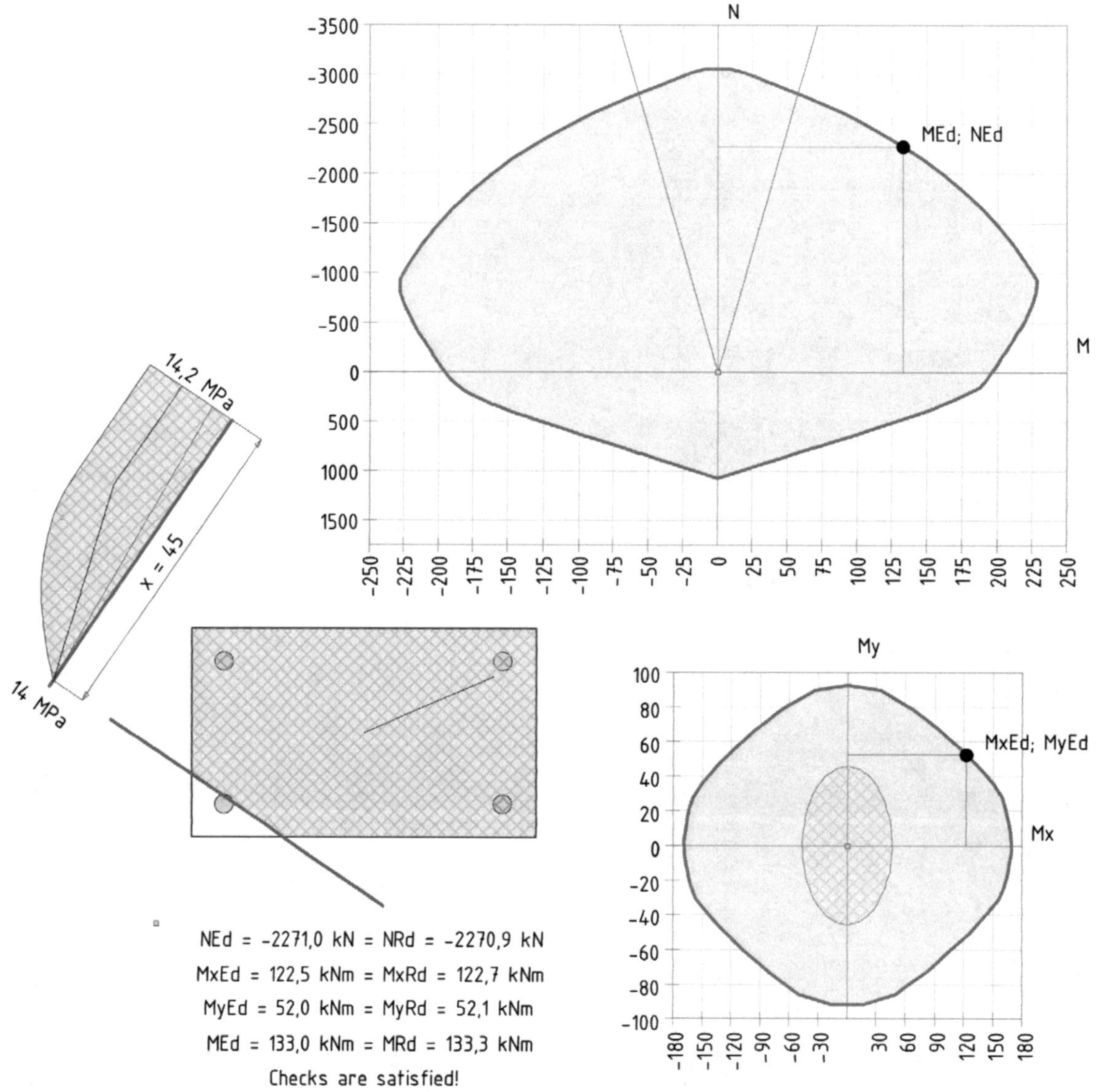

Axial force - N = -2271 kN, Bending moments - M_x = 0 kNm, M_y = 0 kNm

Capacity with imperfections and II order effects - $N_{Rd,II}=2271$ kN

Buckling factor - $\varphi=\frac{N_{Rd,II}}{N_{Rd,max}}=\frac{2271}{3168.48}=0.717$

Spatial buckling resistance is a bit lower than the orthogonal one and it is located at slightly diverted direction, as expected. This is due to the fact that the column is stronger along the orthogonal directions which can be seen in the shape of the interaction surface. But if we restrict the column to

IOP Conf. Series: Materials Science and Engineering 1276 (2023) 012011 doi:10.1088/1757-899X/1276/1/012011

buckle only along one of the two orthogonal directions, the spatial method gives the same results as the planar one.

10. Conclusions

Based on the above discussion, we may draw the following conclusions:

1. Design codes do not provide sufficient recommendations about the buckling capacity of arbitrary shaped columns. Numerical methods and computer software/simulations are required.

2. Using one planar solution only may not be enough for certain shapes. On the other hand, it may be too conservative to account for the minimum eccentricities and initial imperfections simultaneously in both directions, as recommended in sections 5.8.9 of EN 1992-1-1 [3]. For example, for circular sections, it will be simply equivalent to increase the eccentricities by approx. 1.4 times.

3. Even for simple cross sections, it may be difficult to predict the most unfavorable direction to apply the minimal eccentricity and initial imperfections, if the boundary conditions of the element are different in both directions.

4. The proposed algorithm gives a robust and solid mathematical approach to find the buckling resistance of RC columns with arbitrary shapes. Additionally, it provides a clear and illustrative graphical representation of the solution.

11. Limitations and future work

1. The above method can be applied for relatively compact sections. RC cores may have much more complicated behavior like warping, local buckling, flexural-torsional buckling etc. The assumption for the planarity of sections may not be not valid for them and they have to be designed by different methods.

2. Although we choose the most unfavorable direction to apply the minimum eccentricity and initial imperfections, the statistical probability for that may not be very high. A lot of laboratory tests are needed for verification. Additional correction factor may be included to account for that circumstance.

References

[1] Gander W and Gautschi W 2000 Adaptive Quadrature - Revisited *BIT Numerical Mathematics* **40** pp 84–101

[2] Anderson N and Björck Å 1973 A new high order method of regula falsi type for computing a root of an equation *BIT* **13** pp 253–264

[3] EN 1992-1-1 (2004): *Eurocode 2: Design of concrete structures. Part 1-1: General rules and rules for buildings*

Analysis of a column with variable, arbitrary shaped cross section, including initial imperfections, geometric and material nonlinearity

A. Traykov [1], **N. Ganchovski** [2]

[1] Department of Structural Mechanics, University of Architecture, Civil Engineering and Geodesy, Sofia, Bulgaria
[2] Proektsoft EOOD, 34-36 Peyo Yavorov blvd, Sofia, Bulgaria

alex_fce@uacg.bg

Abstract. In this paper, we propose a general method for solving structural mechanics problems by developing a "functional model". It is implemented by defining the required geometrical and physical relationships in their general form via analytical and numerical functions. For that purpose, a special programming language Calcpad is used. The obtained data structure is processed by a core engine, which generates the computer code, required for the solution. This model can be used for calculation of different partial cases, after specifying the specific input functions and parameters. In this way, the need for developing a different procedure for each separate case is eliminated. The proposed method is applied for the analysis of a column with variable, arbitrary shaped cross section, accounting for initial imperfections, geometric and material nonlinearity.

1. Introduction

There are many different programming paradigms that exist in computer science. The most common of them are the imperative (procedural, object orientated) and the declarative (functional, logical, etc) ones. They correspond to the different ways we think, organize and perform calculations. Currently, most calculations are made in the imperative way: as a sequence of formulas in a predefined order. In fact, that is how most design codes are implemented. On the other hand, it is convenient to think about structures as hierarchy of objects with properties. For example, a structural element has type, material, cross section, etc. A cross section has shape and dimensions, and so on. This logically corresponds to object oriented programming. For examples, different sections can be represented by classes, derived from an abstract base class "Shape". Area properties will be calculated by different methods, depending on the shape, that override the respective abstract base methods (A, Av, Sy (Cz), Iy, Wel, Wpl, etc.).

2. Functional modeling

However, there is another approach, which has not been well investigated so far, but may bring some unexpected benefits. And it is to thing about a structure as a set of functions. The idea is to represent all structural properties as functions, e.g. geometry, cross sections, materials, loads, etc. Then, we can define the corresponding laws of physics in a general form – as a hierarchy of functional relationships. In this way, we will describe our solution algorithm. It will depend only the on general functional representations of the input data. We will call this "**Functional model**". Later, we can use this model, to solve a variety of specific problems, by plugging in different input functions.

For example, lets take the cross section properties, that we discussed previously. Instead of classes with properties and methods, we can represent a cross section by a function of the type $b(z,h), z \in [0,h]$. Then, we can define the area properties by their common definitions from structural mechanics:

$$A(h) = \int_0^h b(z,h)\,dz,\; S(h) = \int_0^h b(z,h)\cdot z\,dz,\; I(h) = \int_0^h b(z,h)\cdot z^2\,dz, \text{ etc.}$$

Further, we can use this code to calculate the properties of any sections and this will save us the efforts to write separate methods for each shape. Also, we will need to preliminary develop a general procedure for finding definite integrals by numerical methods. One may say that this will work much slower and that is correct. But let us also look at the advantages:

- simpler and faster development;
- smaller code base that is easier to support and maintain;
- cleaner code and better separation of responsibilities;

Actually, contemporary computers are really fast, so speed is normally not a big issue today. In future, new technologies, such as quantum computing and new superconductors can even make them hundreds to million times faster [1]. In fact, the above method is analogous to functional programming paradigm in computer science, but in respect to structural mechanics. It actually allows us to effectively implement functional programming languages for structural analysis.

3. Functional programming

Unlike the imperative one, in functional programming the order of calculations is not defined. It depends on the internal implementation of the compiler/interpreter platform behind the specific language. Instead, we describe our problem in terms of chained functions that call each other and pass arguments in a declarative manner. Then, the input is parsed by the language processing core. It is stored internally as a data structure, called PFG (Program Flow Graph) which is a special type of directed graph. Figure 1 shows a sample code for calculation of normal stresses along a tapered T-shaped beam and the corresponding PFG structure.

```
b1 = bf - b; Af = b1* hf
h(x) = h1 + i*x
zf(h) = h - hf/2; Aw(h) = b*h
A(h) = Aw(h) + Af
S(h) = Aw(h)*h/2 + Af*zf(h)
zc(h) = S(h)/A(h)
I(h) = b*h^3/12 + b1*hf^3/12 +
       Aw(h)*(h/2 - zc(h))^2 +
       Af*(zf(h) - zc(h))^2
σ1(x) = N/A(h(x)) +
        M(x)/I(h(x))*zc(h(x))
σ2(x) = N/A(h(x)) -
        M(x)/I(h(x))*(h(x) -
        zc(h(x)))
```

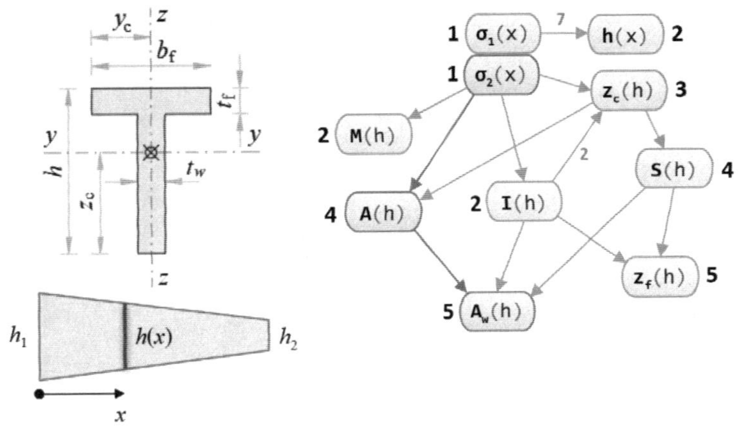

a) Sample code b) Problem scheme c) PFG with topological sorting

Figure 1. Sample code with the corresponding program flow graph.

Functions are located in the vertices of the graph and edges represent the connections defined by the function calls. The math core engine performs the calculations by processing the above graph. It may perform BFS with topological sorting, or DFS walking. Also different optimization methods are applied to reduce the number of evaluations, such as dynamic programming (memoization), folding (rewriting), pruning, etc. This approach has been implemented in the Calcpad engineering calculation platform. It provides a simple declarative language that can be used to describe a structural analysis problem. Then, it parses the source code, performs the calculations and generates a calculation report. It also includes some general-purpose and effective numerical algorithms:

- $Area – numerical integration by an adaptive Gauss-Lobatto [2] quadrature;
- $Root – nonlinear equation solver, using a modified Aderson-Bjork method [3];
- $Slope – numerical differentiation by Richardson's extrapolation [4] on a symmetric two node stencil, etc.

4. Worked example

To demonstrate the proposed method, we have developed a sample functional model, by using Calcpad. It is intended for the analysis of columns with arbitrary shaped cross sections, that are variable along column height. We also takes into account initial imperfections, geometric and material nonlinearity. Cross section and material data are represented in the form of general functions:

$$b(z, h), @ z \in [0, h]$$

$$\sigma(\varepsilon, \varepsilon_{min}, \varepsilon_{max}), @ \varepsilon \in [\varepsilon_{min}, \varepsilon_{max}].$$

All the required mathematical and mechanical relationships are defined parametrically in respect to section shape and height, as follows:

4.1. *Section properties*

They are specified by the general integral relationships:

Area - $A(h) = \text{\$Area}\{b(h; z); z \in [0\,mm; h]\}$

First moment of area - $S(h) = \text{\$Area}\{b(h; z) \cdot z; z \in [0\,mm; h]\}$

Center of gravity- $z_c(h) = \dfrac{S(h)}{A(h)}$

Second moment of area - $I(h) = \text{\$Area}\{b(h; z) \cdot (z - z_c(h))^2; z \in [0\,mm; h]\}$

Section moduli - $W_{el,1}(h) = \dfrac{I(h)}{z_c(h)}, W_{el,2}(h) = \dfrac{I(h)}{h - z_c(h)}$

4.2. *Moment-curvature relationship*

Mean section strain - $\varepsilon_N(h) = \dfrac{-N_{Ed}(h)}{A(h) \cdot E}$ (section cannot be plastic for axial load only)

Stain at section top edge - $\varepsilon_t(h; \varepsilon_b) = \dfrac{\varepsilon_b \cdot (z_c(h) - h) + \varepsilon_N(h) \cdot h}{z_c(h)}$

Bending - $M_{pl}(h; \varepsilon_b) = \text{\$Area}\left\{\sigma\left(\varepsilon_z\left(\dfrac{z}{h}; \varepsilon_b; \varepsilon_t(h; \varepsilon_b)\right)\right) \cdot b(h; z) \cdot (z_c(h) - z); z \in [0\,mm; h]\right\}$

Curvature - $\theta(h; \varepsilon_b) = \dfrac{\varepsilon_b - \varepsilon_N(h)}{z_c(h)}$

Parametric plot - $\text{\$Plot}\{\theta(h; \varepsilon_b) | M_{pl}(h; \varepsilon_b) @ \varepsilon_b = \varepsilon_N(h) : \varepsilon_{ud}\}$

4.3. *Nonlinear stiffness* - $K(h; \varepsilon_b) = \dfrac{M_{pl}(h; \varepsilon_b)}{\theta(h; \varepsilon_b)}$

4.4. Initial imperfections

They are code specific and depend on the type of the structure. We will use structural Eurocodes for steel design, but similar approach can be used for any other case. According to EN 1993-1-1 [5], initial imperfections are defined as follows:

- global inclination to eq. (5.5):

$\varphi = \varphi_0 \cdot \alpha_h \cdot \alpha_m$, where:

$\varphi_0 = \dfrac{1}{200}$ - basic value of inclination;

$\alpha_h = max\left(\dfrac{2}{3}; min\left(\sqrt{\dfrac{4}{L}}; 1\right)\right)$ - reduction factor for column height;

$\alpha_m = \sqrt{0.5\left(1 + \dfrac{1}{m}\right)}$ - reduction factor for the number of columns in a row m;

- local bow to eq. (5.6) and table 5.1:

$e_0 = \dfrac{L}{150}$ - for plastic analysis and buckling curve "C"

Both imperfections can be combined in a single function that will represent the resulting eccentricity of the column axis along height:

$$e_0(x) = \varphi \cdot x - \frac{4 \cdot x \cdot (L - x)}{L^2} \cdot e_0$$

4.5. Geometric nonlinearity

The solution will be performed by the fixed point iterative method. We will use the displacement at the top of the column for the target parameter - $\Delta_{max} = 0.001$ (0.1%). Displacements will be calculated by the Maxwell-Mohr integral.

Bending moment due to a unit force at height x above column base - $\overline{M}(\xi; x) = x - \xi$

Deformation at the bottom edge of the section in dependence of the bending moment:

$\varepsilon(h; M) = \text{\$Root}\left\{M_{pl}(h; \varepsilon_b) - M; \varepsilon_b \in [0; 2 \cdot \varepsilon_{ud}]\right\}$

Initial iteration:

Bending moment diagram from external loads and initial imperfections:

$M_{e0}(x) = H \cdot (L - x) + P \cdot (e_0(L) - e_0(x))$

Maxwell-Mohr integrand:

$$f_0(\xi; x) = \frac{M_{e0}(\xi) \cdot \overline{M}(\xi; x)}{K\left(h(\xi); \varepsilon\left(h(\xi); M_{e0}(\xi)\right)\right)}$$

Initial deflected shape:

$d_0(x) = \text{\$Area}\left\{f_0(\xi; x); \xi \in [0\,m; x]\right\}$

Every other iteration „i“:

Bending moment diagram - $M_i(x) = H \cdot (L - x) + P \cdot \left(d_{i-1}(L) - d_{i-1}(x) + e_0(L) - e_0(x)\right)$

We can see from the above expression that it depends on the deflected shape from the previous iteration. To avoid building a recursive call chain with increasing depth, we will “freeze” the bending diagram using Chebyshev approximation of the first kind:

$$M_{i,Ch}(x) \approx \sum_{k=0}^{n-1} c(k) T_k\left(2\frac{x}{L} - 1\right) - \frac{1}{2}c(0) \quad (1)$$

where $T_k(x)$ is the k-th Chebyshev polynomial;

$$c(k)=\frac{2}{n}\sum_{j=0}^{n-1} M_i(x(j))\cdot\cos\left(\frac{(j+0.5)\cdot k}{n}\cdot\pi\right)$$ - the *k*-th interpolation factor;

$$x(k)=\frac{L}{2}\cdot\left(1+\cos\left(\frac{k+0.5}{n}\cdot\pi\right)\right)$$ - the abscissa of the k-th interpolation node.

We will use the first five polynomials, which are defined as:

$$T_0(x)=1;\quad T_1(x)=x;\quad T_2(x)=2\cdot x^2-1;\quad T_3(x)=4\cdot x^3-3\cdot x;\quad T_4(x)=8\cdot x^4-8\cdot x^2+1.$$

We can optimize equation (1) in respect to its computational effectiveness. For that purpose, we will present it in the following form:

$$P_{Ch}(x)=a_0+x\cdot(a_1+x\cdot(a_2+x\cdot(a_3+x\cdot a_4)))\tag{2}$$

where the new factors a_k are calculated as follows:

$$a_0=\frac{c_0}{2}-c_2+c_4;\quad a_1=c_1-3\cdot c_3;\quad a_2=2\cdot c_2-8\cdot c_4;\quad a_3=4\cdot c_3;\quad a_4=8\cdot c_4.$$

So, further calculations are continued as follows:

Approximated bending moment diagram - $M_{i,Ch}(x)=P_{Ch}\left(\frac{2\cdot x}{L}-1\right)$

Maxwell-Mohr integrand - $f_i(\xi;x)=\dfrac{M_{i,Ch}(\xi)\cdot\overline{M}(\xi;x)}{K(h(\xi);\varepsilon(h(\xi);M_{i,Ch}(\xi)))}$

Displacement diagram - $d_i(x)=\$\text{Area}\{f_i(\xi;x);\xi\in[0\,m;x]\}$

Displacement at top - $d_{i,L}=d_i(L)$

Relative error - $\Delta_i=\dfrac{d_{i,L}-d_{i-1,L}}{d_{i,L}}$

If $\Delta_i<\Delta_{max}$, then the convergence is achieved and the iterations terminate. The obtained internal forces are used further to check the plastic capacity of the section. If $\Delta_i>\Delta_{i-1}$, the solution diverges and the program exits with an error message. It normally occurs when we are very close or over the buckling load. This can be also an approximate method for finding the buckling force by variation of the axial load.

4.6. Section capacity

Strain along the section height - $\varepsilon_z(\xi;\varepsilon_1;\varepsilon_2)=\varepsilon_1+(\varepsilon_2-\varepsilon_1)\cdot\xi$, where

$\xi=z/h$ is the relative height of the neutral line;

ε_1 and ε_2 are the strains at the top and bottom edge of the section, respectively.

In this way we introduced the Euler–Bernoulli hypothesis that sections will remain plain after deformations.

Bending capacity:

$$M_{pl,Rd}(h)=\$\text{Area}\left\{\sigma\left(\varepsilon_z\left(\frac{z}{h};\varepsilon_{ud};-\varepsilon_{ud}\right)\right)\cdot b(h;z)\cdot(z_c(h)-z);z\in[0\,mm;h]\right\}$$

Axial force capacity:

Tension - $$N_{pl,Rd,t}(h)=\$\text{Area}\left\{\sigma\left(\varepsilon_z\left(\frac{z}{h};\varepsilon_{ud};\varepsilon_{ud}\right)\right)\cdot b(h;z);z\in[0\,mm;h]\right\}$$

Compression - $$N_{pl,Rd,c}(h)=\$\text{Area}\left\{\sigma\left(\varepsilon_z\left(\frac{z}{h};-\varepsilon_{ud};-\varepsilon_{ud}\right)\right)\cdot b(h;z);z\in[0\,mm;h]\right\}$$

4.7. P-M interaction diagram

For that purpose, we will define the strains at top and bottom edge as functions on a single strain parameter $\varepsilon = [\varepsilon_{max} ; -\varepsilon_{max}]$, as follows:

Bottom edge strain - $\varepsilon_1(\varepsilon) = min(\varepsilon + \varepsilon_{ud} ; \varepsilon_{ud})$, where ε_{ud} is the ultimate deformation

Top edge strain - $\varepsilon_2(\varepsilon) = max(-\varepsilon_{ud} ; \varepsilon - \varepsilon_{ud})$

Along the section height - $\varepsilon_{zz}(\xi ; \varepsilon) = \varepsilon_z(\xi ; \varepsilon_1(\varepsilon) ; \varepsilon_2(\varepsilon))$

Parametric equations for bending and axial force:

Bending - $M_{pl,Rd}(h ; \varepsilon) = \$\text{Area}\left\{\sigma\left(\varepsilon_{zz}\left(\frac{z}{h} ; \varepsilon\right)\right) \cdot b(h ; z) \cdot (z_c(h) - z) ; z \in [0\,mm ; h]\right\}$

Axial force - $N_{pl,Rd}(h ; \varepsilon) = \$\text{Area}\left\{\sigma\left(\varepsilon_{zz}\left(\frac{z}{h} ; \varepsilon\right)\right) \cdot b(h ; z) ; z \in [0\,mm ; h]\right\}$

Diagram plot - $\$\text{Plot}\{M_{pl,Rd}(h ; \varepsilon) | N_{pl,Rd}(h ; \varepsilon) @ \varepsilon = -\varepsilon_{max} : \varepsilon_{max}\}$

4.8. *Capacity check for bending with axial load*

Strain parameter for the applied axial load:

$\varepsilon_N(h) = \$\text{Root}\{N_{pl,Rd}(h ; \varepsilon) = -N_{Ed}(h) ; \varepsilon \in [0 ; \varepsilon_{max}]\}$

Ultimate bending moment - $M_{pl,N,Rd}(h) = M_{pl,Rd}(h ; \varepsilon_N)$

Capacity check - $M_{Ed}(h) \le M_{pl,N,Rd}(h)$

5. Input data

Now, our functional model is completed and we can use it to solve columns with different shapes and materials. For this example, we will use elliptical pipe with variable height along the column axis. At the top of the column, it will finish in a circular shape. The material will be steel grade S355 as per EN 1993-1-1 with linear-plastic model.

Column length - $L = 3\,m$

5.1. Loads

Vertical force - $P = 3700\,kN$, Horizontal force - $H = 10\,kN$

5.2. Material

Modulus of elasticity - $E = 210\,GPa$

Characteristic yield strength - $f_{yk} = 355\,MPa$

Partial safety factor - $\gamma_s = 1.05$

Design yield strength - $f_{yd} = \frac{f_{yk}}{\gamma_s} = \frac{355\,MPa}{1.05} = 338.1\,MPa$

Yield strain - $\varepsilon_{yd} = \frac{f_{yd}}{E} = \frac{338.1\,MPa}{210\,GPa} = 0.00161$

Ultimate strain - $\varepsilon_{ud} = \frac{10}{1000} = 0.01$

Stress-strain relationship - $\sigma(\varepsilon) = max(-f_{yd} ; min(\varepsilon \cdot E ; f_{yd}))$

5.3. Cross section – elliptical-to-circular hollow pipe

Width - $b = 200\,mm$

Height:

- at the bottom of the column - $h_b = 400\,mm$
- at the top of the column - $h_t = 200\,mm$

- difference - $\Delta h = h_t - h_b = 200\ mm - 400\ mm = -200\ mm$
- as a function along column height - $h(x) = h_b + \frac{\Delta h \cdot x}{L}$

We assumed linear distribution, but any other can be defined in a similar way.
Wall thickness - $t = 20\ mm$
Inner width - $b_2 = b - 2 \cdot t = 200\ mm - 2 \cdot 20\ mm = 160\ mm$
Inner height - $h_2(h) = h - 2 \cdot t$

Function for the outer perimeter - $b_1(h; z) = \frac{2 \cdot b}{h} \cdot \sqrt{\left(\frac{h}{2}\right)^2 - \left(z - \frac{h}{2}\right)^2}$

Function for the inner perimeter:

$$b_2(h; z) = if\left((z > t) \cdot (z < h - t); \frac{2 \cdot b_2}{h_2(h)} \cdot \sqrt{\left(\frac{h_2(h)}{2}\right)^2 - \left(z - \frac{h}{2}\right)^2}; 0\ m\right)$$

Final function - $b(h; z) = b_1(h; z) - b_2(h; z)$

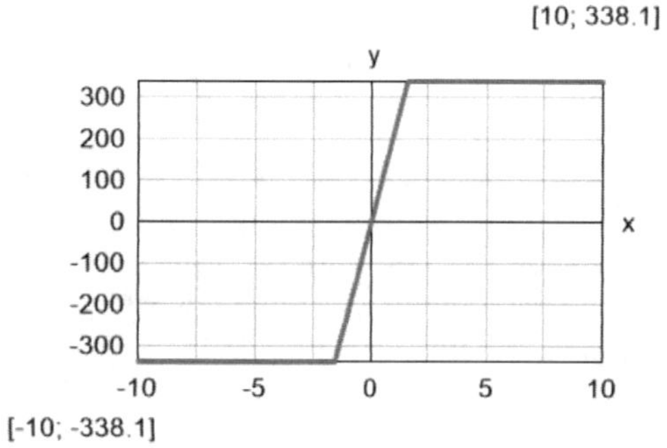

Figure 2. Stress-strain relationship.

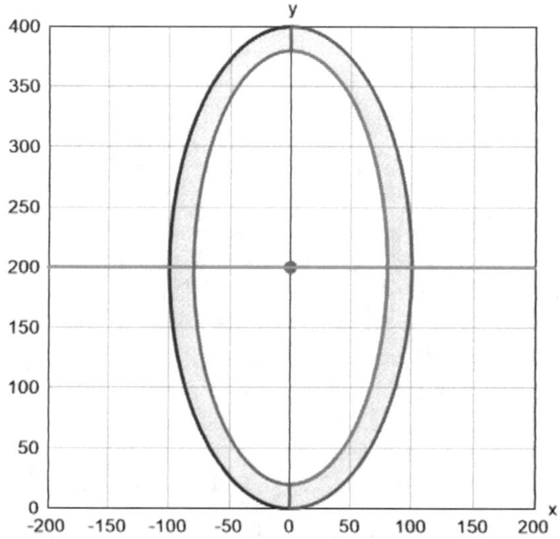

Figure 3. Cross section shape at column base.

6. Results

6.1. Nonlinear stiffness

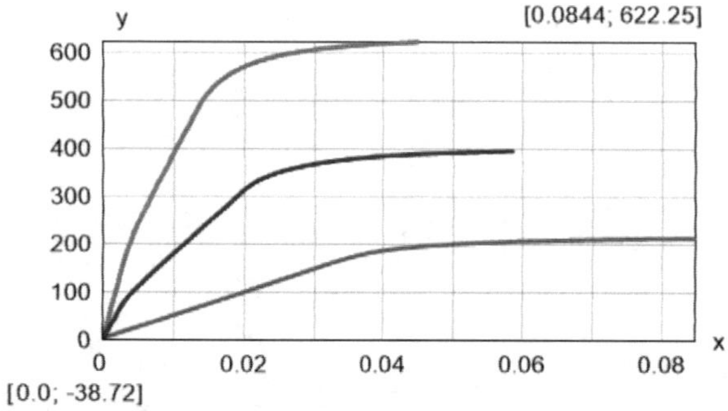

Figure 4. Moment-curvature diagrams.

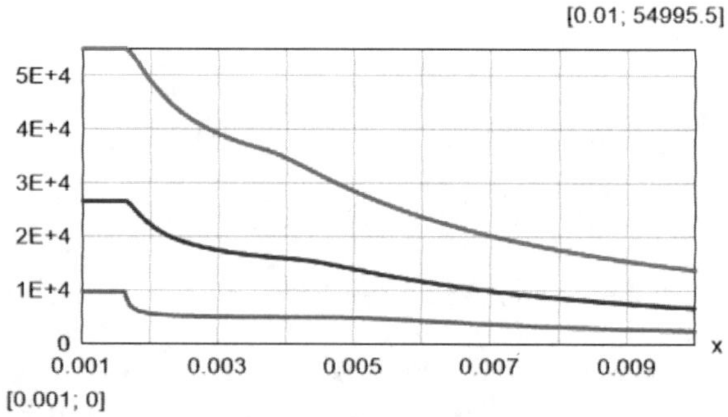

Figure 5. Nonlinear stiffness diagrams.

6.2. Initial imperfections

Global inclination to eq. (5.5) from EN 1993-1-1:

Basic value for inclination - $\varphi_0 = \frac{1}{200} = 0.005$

Reduction factor for height - $\alpha_h = 1$

Reduction factor for number of columns in a row - $\alpha_m = 1$

Initial inclination - $\varphi = \varphi_0 \cdot \alpha_h \cdot \alpha_m = 0.005 \cdot 1 \cdot 1 = 0.005$

Eccentricity at the top of the column - $e_\varphi = \varphi \cdot L = 0.005 \cdot 3\,m = 15\,mm$

Initial local bow to eq, (5.6) and table 5.1:

Bow arrow - $e_0 = \frac{L}{150} = \frac{3\,m}{150} = 20\,mm$ - for buckling curve "C"

The deflected axis shape is displayed on Figure 6.

6.3. Geometric nonlinearity

Iteration results are presented in the following table:

Table 1. Iterative solution results.

Iteration	Control displacement, mm	Relative error
0	16.71	
1	27.99	0.4030
2	36.49	0.2330
3	43.14	0.1540
4	48.39	0.1080
5	52.55	0.0792
6	55.85	0.0593
7	58.49	0.0451
8	60.6	0.0348
9	62.28	0.0270
10	63.63	0.0212
11	64.71	0.0167
12	65.58	0.0133
13	66.28	0.0106
14...	66.86	0.0086
...47	69.75	< 0.001

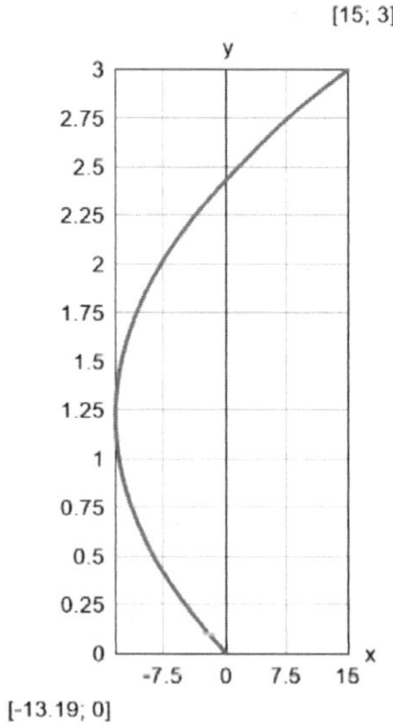

Figure 6. Initial imperfections.

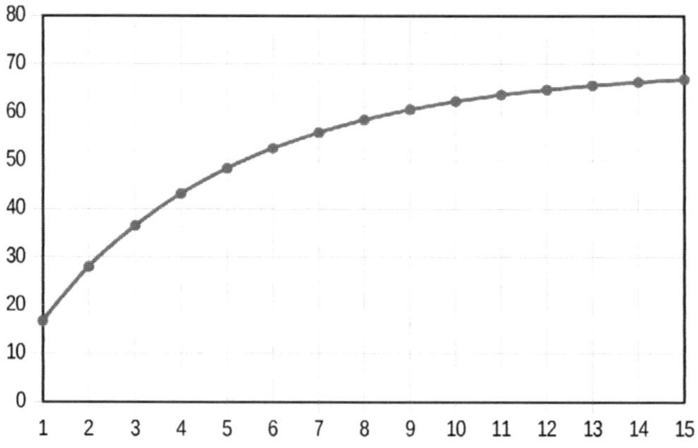

Figure 7. Control displacement vs solution steps.

6.4. Displacements

Displacement at the top from linear-elastic solution without imperfections:

$$d_{el}(x)=\$\text{Area}\left\{\frac{M(\xi)\cdot\overline{M}(\xi;x)}{EI(h(\xi))};\xi\in[0\,m;x]\right\};\quad d_{el,L}=d_{el}(L)=d_{el}(3\,m)=2.45\,mm$$

Displacement on top from linear-elastic solution with initial imperfections:

$$d_0(x)=\$\text{Area}\{f_0(\xi;x);\xi\in[0\,m;x]\};\qquad d_{0,L}=d_0(L)=d_0(3\,m)=16.71\,mm$$

Displacement on top from non-linear solution with initial imperfections:

$$d_{pl,L}=d_1(L)=d_1(3\,m)=69.75\,mm$$

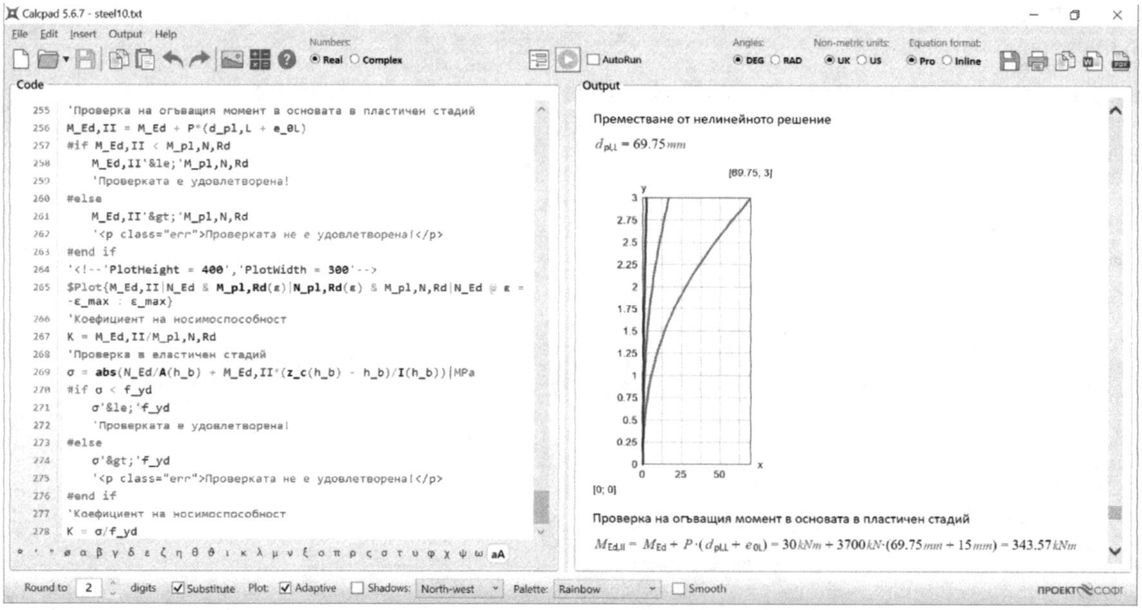

Figure 8. Displacements in Calcpad.

6.5. Design internal forces

Axial force - $N_{Ed} = -P = -3700\ kN$

Bending moments from applied loads - $M_{Ed} = H \cdot L = 10\ kN \cdot 3\ m = 30\ kNm$

Bending moment from initial imperfections:

$M_{Ed,e0} = M_{Ed} + P \cdot e_{0L} = 30\ kNm + 3700\ kN \cdot 15\ mm = 85.5\ kNm$

Bending moment from non-linear solution with initial imperfections:

$M_{Ed,II} = M_{Ed} + P \cdot (d_{pl,L} + e_{0L}) = 30\ kNm + 3700\ kN \cdot (69.75\ mm + 15\ mm) = 343.57\ kNm$

6.6 Section capacity check at column base

Ultimate bending capacity:

$$M_{pl,Rd} = \$\text{Area}\left\{\sigma\left(\varepsilon_z\left(\frac{z}{h}; \varepsilon_{ud}; -\varepsilon_{ud}\right)\right) \cdot b(h;z) \cdot (z_c(h) - z); z \in [0\ mm; h]\right\} = 630.04\ kNm$$

Ultimate compression force capacity:

$$N_{pl,Rd,c} = \$\text{Area}\left\{\sigma\left(\varepsilon_z\left(\frac{z}{h}; -\varepsilon_{ud}; -\varepsilon_{ud}\right)\right) \cdot b(h;z); z \in [0\ mm; h]\right\} = -5948.08\ kN$$

Strain parameter for the applied axial load:

$\varepsilon_N = \$\text{Root}\{N_{pl,Rd}(\varepsilon) = -N_{Ed}; \varepsilon \in [0; \varepsilon_{max}]\} = 0.00803$

Corresponding inelastic bending capacity:

$M_{pl,N,Rd} = M_{pl,Rd}(\varepsilon_N) = M_{pl,Rd}(0.00803) = 366.88\ kNm$

Capacity check:

$M_{Ed,II} = 343.57\ kNm \le M_{pl,N,Rd} = 366.88\ kNm$

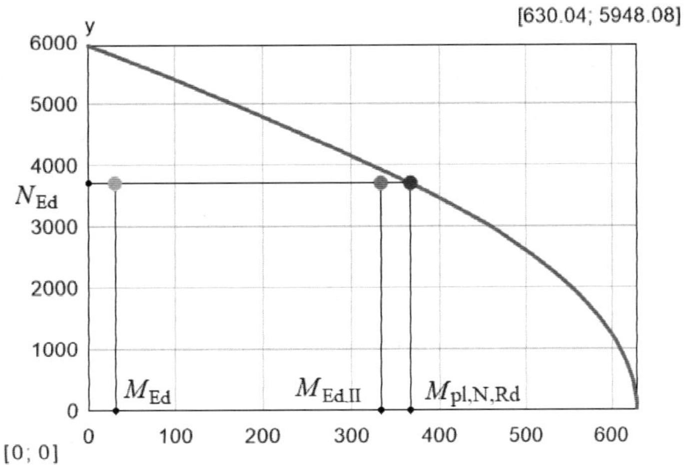

Figure 9. Interaction diagram and capacity check.

7. Numerical verification

For the verification of the above results, we performed nonlinear FEA using Ansys numerical simulation software. For that purpose, we created a solid model with initial imperfections that were incorporated into the geometry, as shown on Figure 10. The same linear-plastic material model was applied, as in the above calculations.

Figure 10. Solid model.

Figure 11. Mechanical model with boundary conditions.

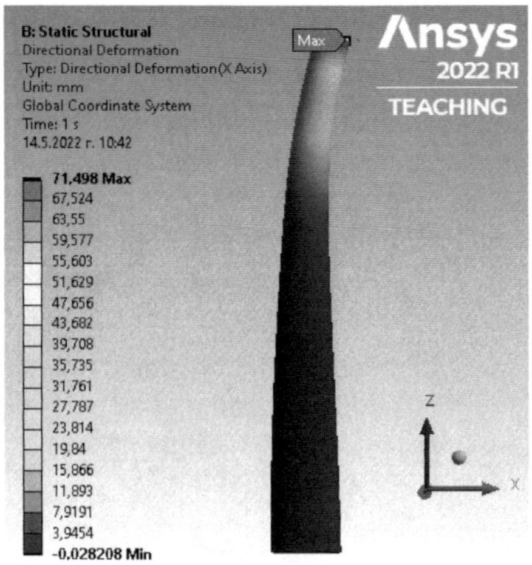

Figure 12. Displacements, mm.

The obtained maximal displacement at top is d_{FEA} = 71.5 mm, which is not too far from our solution. The relative difference is:

$$\Delta = \frac{d_{FEA} - d_{pl,L}}{d_{FEA}} = \frac{71.5\,mm - 69.75\,mm}{71.5\,mm} = 0.0245 = 2.45\,\%.$$

8. Conclusions

Based on the above results, we can conclude that mathematical models, based on chained functional relationships (called "functional models"), can be successfully applied for the solution of nonlinear mechanical problems. They can be organized in a library with highly reusable code blocks that can be combined or plugged one into another for rapid development of structural analysis programs.

They can be used in structural design instead of the imperative formulas, defined for the separate shapes in the design codes. For other cases, that are not covered by the design codes, they can be compatible to numerical simulations.

It could be convenient to use interaction diagrams for plastic design of steel sections, instead of the interactions formulas, provided in Eurocode 3. They can be obtained numerically, even for sections that are not included in the code.

Calcpad math core engine is capable of processing complex functional models that include numerical methods.

9. Future work

With some modifications, the above method can be used to obtain buckling curves. It will be good to be compared against the nominal buckling curves in Eurocode 3 for standard sections. Then, the procedure can be used to develop buckling curves for non-standard sections.

By variation of the vertical load, we can obtain the P-δ (P-M) curve for the element. Then, we can intersect it with the interaction diagram and find the ultimate compression capacity which is the maximum axial load that the column can take without horizontal loads (only with initial imperfections).

P-M interaction curves will be developed for quick plastic design of steel shapes for bending with axial load.

We will create a library of ready made functional models that can be used by Calcpad developers for solving different kind of mechanical problems.

Ready made programs will be shared via the Calcpad cloud platform to be used by practicing engineers in their everyday work.

References

[1] Wu H, Wang Y, Xu Y *et al.* 2022 The field-free Josephson diode in a van der Waals heterostructure. *Nature* **604** pp 653–656

[2] Gander W and Gautschi W 2000 Adaptive Quadrature – Revisited. *BIT Numerical Mathematics* **40** pp 84–101

[3] Anderson N and Björck Å 1973 A new high order method of regula falsi type for computing a root of an equation *BIT* **13** pp 253–264

[4] Richardson L 1911 The approximate arithmetical solution by finite differences of physical problems including differential equations, with an application to the stresses in a masonry dam *Philosophical Transactions of the Royal Society* vol 210 Issue 459–470 pp 307–357

[5] EN 1993-1-1:2005/A1:2014. *Eurocode 3: Design of steel structures. Part 1-1: General rules and rules for buildings*

Determination of the natural frequencies and periods of a body with a fixed point

T Todorov [1]

[1] Department of Technical Mechanics, University of Architecture, Civil Engineering and Geodesy, 1 "Hristo Smirnenski" Blvd., Sofia 1164, Bulgaria

tstodorov_fhe@uacg.bg

Abstract. In the current paper the determination of the natural frequencies and periods of a body with a fixed point is given. Two main approaches are considered. The first one is an analytical solution based on the principles of Lagrangian mechanics and the second is a finite element analysis. Although some equations are provided, the implementation of the analytical solution is achieved via MATLAB codes for simplicity and easier calculations. A finite element program (SAP2000) is used for the numerical solution. Different loading patterns are considered and the results of the two methods are compared. As the geometry of the problem is three-dimensional, some peculiarities of the model are discussed as well.

1. Introduction

The most important characteristics of a dynamic system are the natural frequencies and periods. These are characteristics that are independent of the external loading and are inherent of the structure [1]. The determination of the natural frequencies and periods is crucial because of the resonance phenomena, which can be quite dangerous for the integrity of the system. Whenever the frequency of the external load approaches any of the natural frequencies of the system (the number of the natural frequencies of any system is equal to the number of its degrees of freedom) the amplitude of vibration increases considerably [2]. Therefore, the frequencies of the external loads are to differ from the natural frequencies of the structure.

It is vital for the engineer to be able to assess these physical quantities. That is why the current paper focuses on their determination. Moreover, the various loading patterns used for the FEM (finite element method) analysis show that the definition of the point masses has a non-negligible effect when a three dimensional (3D) problem is considered. Nowadays, such problems are mostly modelled.

2. Problem statement

A body with a fixed point is considered. It consists of three rods – one alongside each axis of a Cartesian coordinate system. Point masses can be applied to random positions of the body (see figure 1). The oscillations of the system are maintained by linear springs. Such a body is chosen because the three dimensional (3D) nature of the problem is to be emphasized. Furthermore, the usage of rods allows for frame elements to be applied in the FEM analysis and, as the shape functions of frame elements are exact, this leads to a higher level of accuracy when comparing results. Such a model can be solved analytically as well which can be used as a verification.

As the analytical solution is achieved via MATLAB codes, various parameters are used to describe different sizes and masses. The problem can be defined and solved with any geometry of the body but for a more "observable" and verifiable from a theoretical point of view result a body with planes of material symmetry is used. That way the assembled matrices – the matrix of inertial characteristics and the stiffness matrix are symmetric [3]. The matrix of inertial characteristics is even diagonal. For the stiffness matrix not be diagonal springs with different stiffness are applied. If both matrices were diagonal, the differential equations (DE) describing the oscillations would be independent. As the classical problem of a body with a fixed point is three dimensional (3D) with three degrees of freedom (DOF), it is preferred to work with a system of DE.

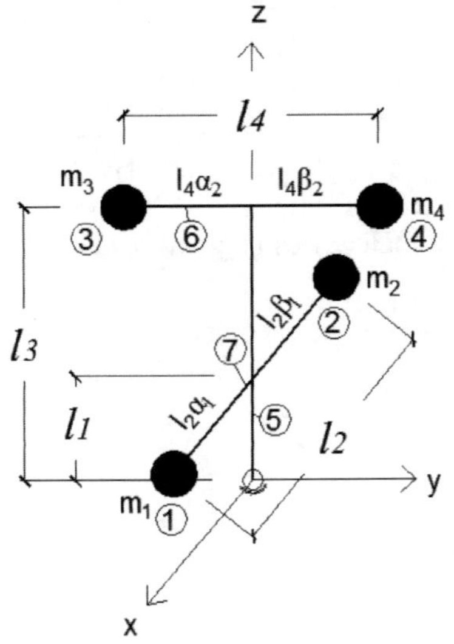

Figure 1. Scheme of the geometry of the considered body. The origin of the Cartesian coordinate system is at the fixed point. Point masses m_1 to m_4 are placed along axes x and y. Linear springs can be attached to random points of the body, however, they are usually parallel to given axes (most commonly – to x and y). Parameters allow for easier calculations of the principal mass moments of inertia.

In order to find the natural frequencies and periods of the defined body the aforementioned system of differential equations (DE) is to be generated [4].

3. Theoretical background

Any mechanical system can be analysed in a number of ways. For more complex systems the usage of more formal approaches is recommended. Such an approach is given by Lagrangian mechanics which is applied to derive the DE of motion of the body with a fixed point. It is known [5] that the problem at hand has three degrees of freedom (DOF) and three generalised coordinates are to be used to define the motion. Therefore, the matrices written below are three by three by default. The sizes of the matrices will only be specifically denoted if a possibility of misinterpretation exists.

The general form of the Lagrange's equations (of the second kind) is [6]:

$$\frac{d}{dt}\left(\frac{\partial E_k}{\partial \dot{q}_i}\right)-\frac{\partial E_k}{\partial q_i}+\frac{\partial E_p}{\partial q_i}+\frac{\partial \Phi}{\partial \dot{q}_i}=Q_j. \tag{1}$$

In equation (1) E_k is the kinetic energy, E_p is the potential energy of the system, Φ is the dissipative function and Q_j is the vector of generalized forces. The system of differential equations (DE), given by equation (1) consists of three second order linear nonhomogeneous DE. Such a system can be solved by an appropriate substitution [7] that reduces the order of the derivatives with respect to time. That system can be written in matrix form:

$$[a]\{\ddot{q}\}+[b]\{\dot{q}\}+[c]\{q\}=\{Q\} \tag{2}$$

In equation (2) $[a]$, $[b]$ and $[c]$ are the matrix of inertial characteristics, the damping matrix and the stiffness matrix respectively. As the simple harmonic oscillations are of interest, the damping matrix is assumed to be the zero matrix. The generalized coordinates are denoted by the vectors $\{\ddot{q}\}$, $\{\dot{q}\}$ and $\{q\}$. The rotations about the x, y and z axes are chosen as generalized coordinates and derivatives with respect to time are classically represented with dots .Finally, the vector of the generalized forces is given by $\{Q\}$. In the case of simple harmonic oscillations $\{Q\}$ is a zero vector. The equation of the simple harmonic oscillations is [8]:

$$[a]\{\ddot{q}\}+[c]\{q\}=\{0\} \tag{3}$$

All matrices in equation (3) are three by three and all vectors are three by one in size. The solution of (3) can be presented as a linear combination of exponential functions. However, for the determination of the natural frequencies and periods of a n-degree of freedom (DOF) system a specific procedure is to be implemented [1]. If the matrix of inertial characteristics is given by:

$$[a]_{nxn}=\begin{bmatrix} a_{11} & a_{12} & \dots & a_{1n} \\ a_{21} & a_{22} & \dots & a_{2n} \\ \dots & \dots & \dots & \dots \\ a_{n1} & a_{n2} & \dots & a_{nn} \end{bmatrix}_{nxn} \tag{4}$$

and the stiffness matrix is given by:

$$[c]_{nxn}=\begin{bmatrix} c_{11} & c_{12} & \dots & c_{1n} \\ c_{21} & c_{22} & \dots & c_{2n} \\ \dots & \dots & \dots & \dots \\ c_{n1} & c_{n2} & \dots & c_{nn} \end{bmatrix}_{nxn} \tag{5}$$

, then the equation that determines the natural frequencies is:

$$\Delta(\omega^2)=\begin{vmatrix} c_{11}-\omega^2 a_{11} & c_{12}-\omega^2 a_{12} & \dots & c_{1n}-\omega^2 a_{1n} \\ c_{21}-\omega^2 a_{21} & c_{22}-\omega^2 a_{22} & \dots & c_{2n}-\omega^2 a_{2n} \\ \dots & \dots & \dots & \dots \\ c_{n1}-\omega^2 a_{n1} & c_{n2}-\omega^2 a_{n2} & \dots & c_{nn}-\omega^2 a_{nn} \end{vmatrix}_{nxn}=0. \tag{6}$$

Equation (6) is an equation of nth power with respect to ω^2. The number of natural frequencies ω_i is equal to the number of degrees of freedom (DOF) of the system. If the system is at state of stable equilibrium, the roots $\omega_1^2,\omega_2^2,...\omega_n^2$ of equation (6) are positive. For a 3 DOF system (see figure 1), the equation that gives the natural frequencies is:

$$\Delta(\omega^2)=\begin{vmatrix} c_{11}-\omega^2 a_{11} & c_{12}-\omega^2 a_{12} & c_{13}-\omega^2 a_{13} \\ c_{21}-\omega^2 a_{21} & c_{22}-\omega^2 a_{22} & c_{23}-\omega^2 a_{23} \\ c_{31}-\omega^2 a_{31} & c_{32}-\omega^2 a_{32} & c_{33}-\omega^2 a_{33} \end{vmatrix}=0. \tag{7}$$

The solutions of equation (7) are obtained by developing MATLAB codes for simplicity and easier manipulation.

4. Solving equation (7) with MATLAB codes

Prior to the solution of equation (7), the matrices $[a]$ and $[c]$ are to be derived. Based on figure 1, a code for determining the principal mass moments of inertia can be written. For simplicity the springs that maintain the oscillations of the system are considered at the same positions where the point masses are. The springs are also parallel to the axes of the Cartesian coordinate system. That way a symmetric stiffness matrix is obtained.

A code for calculating the elements of matrix $[a]$ is given below (figure 2). By changing the input values of the parameters, different geometries of the body can be studied [9].

```
%% Calculation of the matrix [a]
clear, clc
l1=50;l2=50;l3=120;l4=60;% Lengths in cm !
R5=1;r5=0.9;R6=1;r6=0.9;R7=1;r7=0.9; %Radii of pipes cm !
alpha1=0.5;alpha2=0.5;beta1=0.5;beta2=0.5;%alpha and beta from 0 to 0,5
%to determine the position of the point mass
p=7850; %density of steel kg/m^3
A5=pi()*(R5^2-r5^2)/10000;A6=pi()*(R6^2-r6^2)/10000;A7=pi()*(R7^2-
r7^2)/10000;
m1=0;m2=0;m3=0;m4=0;% masses in kg
% No information is required below - only calculations
m5=A5*p*l3/100;m6=A6*p*l4/100;m7=A7*p*l2/100;
%Determination of the distances d
dx1=l2*alpha1;dy1=0;dx2=-l2*beta1;dy2=0;dx3=0;dy3=-l4*alpha2;
dx4=0;dy4=l4*beta2;dx5=0;dy5=0;dx6=0;dy6=0;dx7=0;dy7=0;
dz1=l1;dz2=l1;dz3=l3;dz4=l3;dz5=l3/2;dz6=l3;dz7=l1;
Ixy=0;
Iyz=(m3*dy3*dz3)+(m4*dy4*dz4); %Products of inertia
Ixz=(m1*dx1*dz1)+(m2*dx2*dz2);
% Determination of axial mass moments of inertia
%Mass moments of inertia with respect to x
Ix7=(m7/2)*(R7^2+r7^2);Ix5=(m5/12)*(3*R5^2+3*r5^2+l3^2);
Ix6=(m6/12)*(3*R6^2+3*r6^2+l4^2);
Ix=Ix7+Ix6+Ix5+m1*(dz1^2+dy1^2)+m2*(dz2^2+dy2^2)+m3*(dz3^2+dy3^2)+m4*(dz4
^2+dy4^2)+m5*(dz5^2+dy5^2)+m6*(dz6^2+dy6^2)+m7*(dz7^2+dy7^2);
%Mass moments of inertia with respect to y
Iy6=(m6/2)*(R6^2+r6^2);Iy5=(m5/12)*(3*R5^2+3*r5^2+l3^2);
Iy7=(m7/12)*(3*R7^2+3*r7^2+l2^2);
Iy=Iy5+Iy6+Iy7+m1*(dz1^2+dx1^2)+m2*(dz2^2+dx2^2)+m3*(dz3^2+dx3^2)+m4*(dz4
^2+dx4^2)+m5*(dz5^2+dx5^2)+m6*(dz6^2+dx6^2)+m7*(dz7^2+dx7^2);
%Mass moments of inertia with respect to z
Iz5=(m5/2)*(R5^2+r5^2);Iz6=(m6/12)*(3*R6^2+3*r6^2+l4^2);
Iz7=(m7/12)*(3*R7^2+3*r7^2+l2^2);
Iz=Iz5+Iz6+Iz7+m1*(dx1^2+dy1^2)+m2*(dx2^2+dy2^2)+m3*(dx3^2+dy3^2)+m4*(dx4
^2+dy4^2)+m5*(dx5^2+dy5^2)+m6*(dx6^2+dy6^2)+m7*(dx7^2+dy7^2);
A=[Ix -Ixy -Ixz
    -Ixy Iy -Iyz
    -Ixz -Iyz Iz];%Matrix of inertia
a=A*(1/10000)% dimension kgm^2
Mtot=m1+m2+m3+m4+m5+m6+m7%total mass
zc=(m1*l1+m2*l1+m3*l3+m4*l3+m5*l3*0.5+m7*l1+m6*l3)/(100*Mtot)%coordinate
of center of mass [m]
```

Figure 2. MATLAB code for determination of the elements of matrix $[a]$.

The stiffness matrix $[c]$ is also manipulated by developing a MATLAB code (figure 3).

```
%% stiffness matrix [c]
clear, clc
ki1a=100;ki1b=150;
ki2a=110;ki2b=120;%spring stiffness at a point [N/m]
kj3a=115;kj3b=125;%input
kj4a=130;kj4b=145;
m=1.077;% total mass [kg]
zc=0.7348;%coordinate of center of mass [m]
g=9.81;%grav. acceleration
xki1=0.25;zki1=0.5;%coordinates of springs [m]
xki2=-0.25;zki2=0.5;%input
ykj3=-0.3;zkj3=1.20;
ykj4=0.3;zkj4=1.20;
% determination of coefficients cij - no input required
ki1=ki1a+ki1b;ki2=ki2a+ki2b;kj3=kj3a+kj3b;kj4=kj4a+kj4b;
Cai=((zki1)^2)*ki1+((zki2)^2)*ki2;
c11=Cai+m*g*zc;
c12=0;c21=0;
Cagi=ki2*abs(zki2)*abs(xki2)-ki1*abs(xki1)*abs(zki1);
c13=Cagi;
Cbj=((zkj4)^2)*kj4+((zkj3)^2)*kj3;
c22=Cbj+m*g*zc;
Cbgj=abs(ykj3)*abs(zkj3)*kj3-abs(ykj4)*abs(zkj4)*kj4;
c23=Cbgj;
Cgi=ki1*((xki1)^2)+ki2*((xki2)^2);
Cgj=((ykj4)^2)*kj4+kj3*((ykj3)^2);
c33=Cgi+Cgj;c31=c13;c32=c23;
c=[c11 c12 c13; c21 c22 c23; c31 c32 c33] %stiffness matrix
```

Figure 3. MATLAB code for determination of the elements of matrix $[c]$.

As the considered problem has planes of material symmetry, a simpler code can be used for the solution of (7):

```
v=inv(A).*c;
v(1);v(5);v(9);
T1=2*pi()/sqrt(v(1));f1=1/T1;
T2=2*pi()/sqrt(v(5));f2=1/T2;
T3=2*pi()/sqrt(v(9));f3=1/T3;
T=[T1 T2 T3];
f=[f1 f2 f3];
Period=T';Frequency=f';
table(Period,Frequency)
```

Figure 4. MATLAB code for calculation of the natural frequencies and periods of the considered body.

The codes given above are implemented for the calculation of the natural frequencies and periods of the body with a fixed point .Two main loading patterns are considered. The first pattern assumes that no point masses are applied (the mass of the system is generated by the frame elements only), and the second pattern defines point masses (in such a way that the material symmetry is preserved). Point masses of 1.25 kg are used. The results of the numerical experiment are presented in tables. They can be found in section 7.

5. Finite element analysis

A finite element analysis of the considered body is performed. The program SAP2000 is used (figure 5). The body itself is modelled with frame elements that represent the rods in each of the three directions. The cross sections of the frame elements are defined as pipes with the same diameters as the ones used in the MATLAB codes. Springs are applied at the proper points with respective stiffness coefficients. The fixed point is modelled via restraints (displacements along x, y and z axes). For the purposes of the current research only modal analysis is enabled. As the considered problem has three degrees of freedom (DOF) the analysis is limited to the first three modes and corresponding natural frequencies.

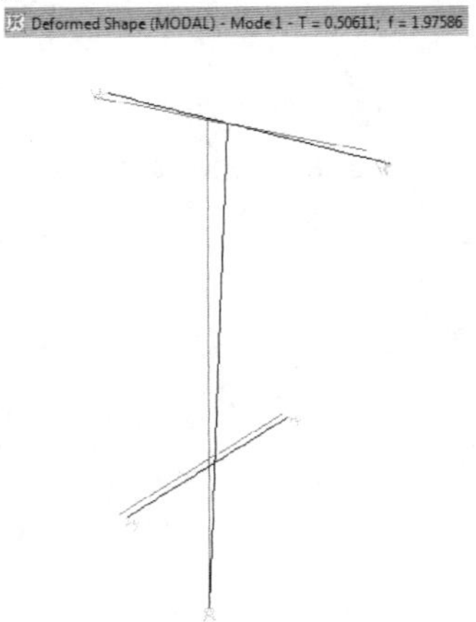

Figure 5. A finite element model of the considered body in SAP 2000. The deformed shape is result of modal analysis.

Five different load patterns concerning the point masses are considered. For each of them the results (natural frequencies and periods) are extracted and compared with the values, obtained from the MATLAB codes.

6. Loading patterns

The five loading patterns consider the definition of the point masses in the finite element analysis. While the point masses in the DE of motion are accounted for by the functions of kinetic and potential energies [10], the finite element solution requires the input of point masses as loads. The first load pattern (1st LP) assumes that no point masses are defined. The second one (2nd LP) defines the masses with the ability to oscillate at the direction of the spring. For example, mass at k_1 can only move parallel to y (see figure 6).

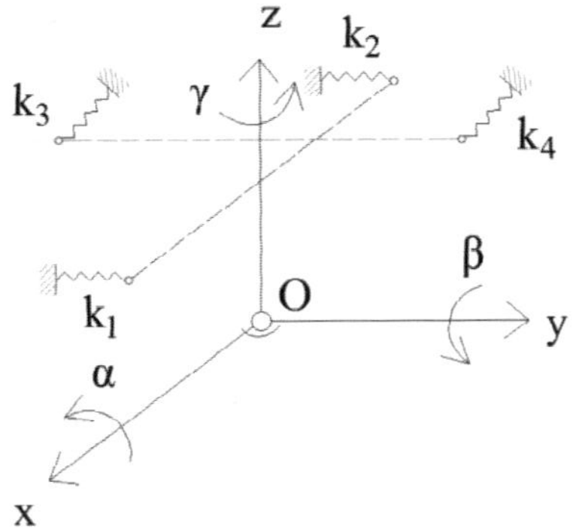

Figure 6. A model of the position and orientation of the springs that maintain the oscillations of the body. The point masses are assumed to be placed at the same position as the springs.

The third load pattern (3rd LP) assumes that masses can move at all three directions (x, y and z). The fourth pattern (4th LP) allows for the point masses to move at 2 directions (parallel to the considered spring and vertically) and also allows for the mass to move alongside the rod with half of its magnitude. This means, for example, that the mass of 1.25 kg at k_1 (see figure 6) can oscillate parallel to y and z, but it can only oscillate with magnitude of 0.625 kg parallel to x. The fifth pattern (5th LP) is quite similar to the fourth with the exception that it does not allow for the point mass to oscillate vertically. This means that point mass at the position of spring k_1 can move parallel to y as 1.25 kg mass, however, it can move as 0.625 kg mass parallel to x. No vibrations are allowed with respect to z.

The five patterns are modelled in SAP2000 FE program. The results are displayed at a table (see table 1). Two additional sets of results are presented as well – from the MATLAB codes when there are no point masses and when the point masses of 1.25 kg are applied.

7. Results and discussion

The results of the numerical experiments are listed in the table below (table 1). The first natural frequency (f_1) corresponds to the mode that is predominately translational with respect to the y axis. The second one (f_2) describes the mode that is predominately translational with respect to the x axis, while the third (f_3) is the natural frequency of the torsional mode.

Table 1. Results for the natural frequencies (f) and periods (T) of the numerical experiments with different loading patterns (LP) used in the finite element (FE) model and the solution with matrix equations based on MATLAB codes.

Natural frequencies (f) and periods (T)	MATLAB no PM	FEM no PM (1st LP)	FEM (2nd LP)	FEM (3rd LP)	FEM (4th LP)	FEM (5th LP)	MATLAB with PM
f_1 (s^{-1})	2.088	1.976	1.475	0.753	0.932	0.967	0.909
T_1 (s)	0.479	0.506	0.678	1.327	1.072	1.034	1.1
f_2 (s^{-1})	5.062	4.94	2.054	1.896	1.957	1.988	1.966
T_2 (s)	0.198	0.202	0.486	0.527	0.511	0.503	0.509
f_3 (s^{-1})	12.017	11.255	2.229	2.217	2.22	2.222	2.225
T_3 (s)	0.0832	0.0888	0.449	0.451	0.45	0.45	0.449

It is obvious (columns 2 and 3 of table 1) that when no point masses are applied the FE model gives approximately the same results as the MATLAB codes. Therefore, the deviations in the results for the case when point masses are applied to the body (columns 4 to 8 of table 1) can be explained with the definition of the degrees of freedom of the point masses (PM). If the solution with MATLAB codes is adopted as closest to reality (column 8 of table 1), then the FE models that approach its values are the models that allow oscillations in the direction of the given rod with a half of the mass (0.625kg). These are load patterns 4 and 5 (columns 6 and 7 of table 1) and the closest values are obtained by LP 4 (column 6 of table 1). This can be explained with the fact that the oscillations of the point masses occur harder at the direction of the rod (in a plain parallel to Oxy and perpendicular to the spring, see figures 1 and 6), while there is nothing to prevent them from oscillating vertically or along the direction of the spring. The contribution of the vertical oscillations to the overall performance is negligible.

The results at the first row of the fourth and fifth column of table 1 show a significant deviation in comparison with the corresponding value at the last column. The discrepancy decreases at the second row and practically diminishes at the third row, but the first mode is usually the most important one for the modal analysis of a structure [11].

8. Conclusion

The obtained results show that when 3D oscillations are considered it is important to properly define the degrees of freedom of the point masses (if such loads are used at the model) with respect to the geometry of the problem. For the studied body it is shown that the model which describes most accurately the oscillations via the determination of the natural frequencies and periods is the one that uses the 4th LP.

The problem of properly defining the degrees of freedom of the point masses can be avoided if objects with actual dimensions and density are modelled instead. That way a higher level of automation can be achieved. This is a direction for future studies.

References

[1] Chopra A 1995 *Dynamics of structures: Theory and Applications to Earthquake Engineering* ed. Scherwatzky K (Englewood Cliffs, New Jersy: Prentice Hall) p.35

[2] Pavlov P, Lilkova-Markova S and Ivanova G 2016 *Modular Approach in Creating the Matrix Equations, Describing the Free Vibrations of Descrete Plane Systems* ed. ICEDM Proceedings of the 11th International Congress on Engineering, Design and Manufacturing. Sustainability and resiliance approach, Barcelona, Spain, (Barcelona: IJTEE-ICEDM2016) Fullpapers E-Book pp 70-75

[3] Irgens F 2008 *Continuum Mechanics* (New York: Springer) chapter 3 p 37

[4] Vandiver J and Gossard D 2011 *2.003SC Engineering Dynamics.* [learning resources online] (Massachusetts: Massachusetts Institute of Technology MIT OpenCourseWare), [cited 2022 June 13]. Available from: https://ocw.mit.edu/courses/2-003sc-engineering-dynamics-fall-2011/

[5] Windall S, Deyst J and Greitzer E 2009 *16 07 Dynamics* [learnin resources online] (Massachusetts: Massachusetts Institute of Technology MIT OpenCourseWare), [cited 2022 June 13] Available from: https://ocw.mit.edu/courses/16-07-dynamics-fall-2009/

[6] Wilson E 2002 *Three-Dimensional Static and Dynamic Analysis of Structures* 3rd edittion (Berkeley, California: Computers and Structures Inc.) chapter 3 p 53

This reference has two entries but the second one is not numbered (it uses the 'Reference (no number)' style.

[7] Nagy G 2021 *Ordinary Differential Equations* (Michigan: Michigan State University) [cited 2022 June 13] Available from: https://users.math.msu.edu/users/gnagy/teaching/ode.pdf p 235

[8] Lewin W and Goldstein W 2011 *For the Love of Physics From the End of the Rainbow to the Edge of Time – A Journey Through the Wonders of Physics* (New York: Free Press) chapter 6 p 101

[9] Moore H 2013 *MATLAB for Engineers* ed Bhat S (Kendallville, USA: Pearson) chapter 4 p 137

[10] Tamrakar A and Harursampath D 2014 *Acing the GATE Mechanical Engineering* ed Wiley Editorial Office, India (New Delhi: Wiley) p 245

[11] Elnashai A and Di Sarno L 2008 *Fundamentals of Earthquake Engineering* (Chippenham, England: Wiley) chapter 4 p 185

Bearing capacity of surface-strengthened concrete beam elements. Parametric analysis.

A Yanakieva[1], G Nikolova[2] and A Baltov[3]

[1,2,3]Institute of Mechanics, Bulgarian Academy of Sciences, 1113 Sofia, Bulgaria

Abstract. The study treats diffusively strengthened concrete beam elements. The element bearing capacity is estimated within a Generalized space of events introduced by the authors. Safety zones of the element's safe operation under overloading are specified therein. The overloading is assumed to have a probabilistic character. A parametric analysis is performed based on specific criteria of loss of element's bearing capacity, i.e.: (1) a boundary conditions criterion; (2) a criterion considering the effect of the overloading probability of the Safety zone; (3) a criterion accounting for the effect of transition area strengthening on the element bearing capacity. Two cases of beam element deformation are assumed- pure elastic deformation (C1), and deformation accompanied by compression micro-damage in the compression area of the concrete core (C2). The results for the Safety zones are discussed and analysed according to each criterion. Assessment of the element bearing capacity is also proposed.

1. Introduction

Surface-strengthened concrete elements are increasingly used in construction practice. Construction workers usually apply technology of sticking high-strength foils to the element surface [1-3]. Yet, the diffusive deposition of strengthening coatings on the surface of concrete elements grows popular in recent years, [4,5]. The technology is known for the diffusive penetration of the strengthening material into the concrete forming so-called transition zones throughout the element, [6]. They provide a much stronger bond between concrete and coating as compared to that between concrete and a metal foil.

We consider in what follows diffusive deposition of strengthening coatings on concrete elements, as well as loss of element bearing capacity under extreme overloading. To assess it we use a previously designed approximate calculation model (ACM) [6] performing a quadratic approximation of the elastic moduli and the normal linear deformations through the transition zones. Furthermore, we expand and complete ACM, designing a detailed methodology for ACM practical application as shown in [7-8].

The task of the presented work is to perform a parametric analysis of the loss of bearing capacity of beam concrete elements undergoing overloading of variable probabilistic character. The parametric analysis is performed involving specific criteria of bearing capacity loss, i.e.: (1) a boundary conditions criterion - α; (2) a criterion considering the effect of overloading probability on the Safety zones - υ; (3) a criterion accounting for the effect of the transition zone strengthening on the element bearing capacity - λ. Two cases of beam elements are assumed - pure elastic deformation (C1), and deformation accompanied by compression micro-damage in the compression area of the concrete core (C2).

2. Geometry, boundary conditions and loading of structural elements

We study in particular beam concrete elements with a rectangular cross-section $(b \times h)$ and a reduced length l_r encompassing various beams - $l_{(i)}, i = 1,2,3,4$, and different boundary conditions - $\alpha_{(i)}, i = 1,2,3,4$. The element strengthening is bilateral, the thickness of the transition zones is h_R, and the height of half of the concrete core is $z_L = \frac{h}{2} - h_R$. Consider now the probability of applying loading exceeding the admissible one, i.e. element overloading. It may be due to technological factors, failure or natural impacts (hurricanes, strong earthquakes etc.). We assume that overloading occurs as a result of applying a distributed load $q = const$ $(Q = ql_r)$ and a normal compression force P, Fig. 1.

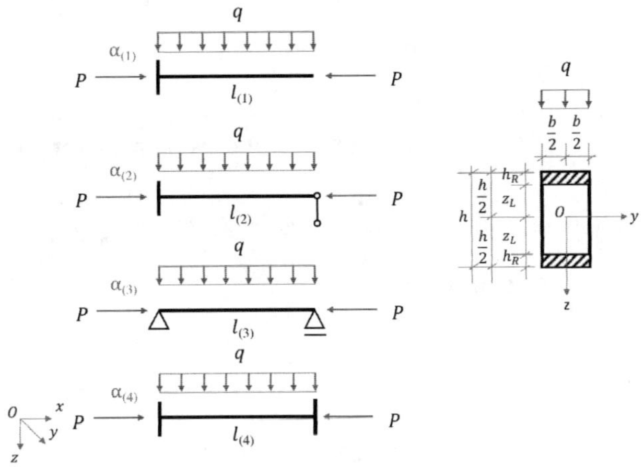

Figure 1. Geometry, boundary conditions and loading of the structural elements

Note that our assumptions are approximate for some types of beams in order to perform an express assessment of the beating capacity and avoid expensive strengthening procedures:

- Assume that loading is simple, and the probability of overloading occurrence depends on one and the same parameter ξ introduced as, [7,8]

$$P = P_0 \upsilon_p \xi, \; Q = Q_0 \upsilon_q \xi \tag{1}$$

where P_0 and Q_0 are the design normative loads, and υ_p and υ_q are overloading probabilities.

- Assume also that the weaker material (concrete) is the first one to lose its bearing capacity.

- Consider two cases: Case 1 (C1) – concrete elastic deformation in tension and under compression, and Case 2 (C2) – concrete elastic deformation in tension and under compression accompanied by micro-damage of the concrete core. The assumption yields more economical solutions.

- Make assumptions related to the maximal admissible normal stresses in the concrete core for C1, C2:

C1 in tension:

Concrete loses its bearing capacity in tension when the maximal normal stress in the core attains the admissible limit, i.e.:

$$\max \sigma_x^+ = \sigma_{B\,per}^+ = s_1 \tag{2}$$

where $\sigma_{B\,per}^+ = \frac{\sigma_{BP}^+}{C^+}$, σ_{BP}^+ is concrete elasticity limit in tension, and C^+ is a generalized safety factor.

C1 under compression:

$$\max \sigma_x^- = \sigma_{B\,per}^- = s_2 \tag{3}$$

where $\left|\sigma_{B\,per}^-\right| = \dfrac{\left|\sigma_{BP}^-\right|}{C^-}$, σ_{BP}^- is concrete elasticity limit under compression, and C^- is a generalized safety factor.

C2 in tension:

See condition (2).

C2 under compression, admitting micro-damage:

$$\left|\max \bar{\sigma}_x^-\right| = \left|\bar{\sigma}_{B\,per}^-\right| = s_3 \tag{4}$$

where $\left|\bar{\sigma}_{B\,per}^-\right| = \dfrac{\left|\sigma_{BC}^-\right|}{\bar{C}^-}$, σ_{BC}^- is concrete's limit strength under compression, and $\bar{C}^-$ is a safety factor.

- Assume superposition in both cases due to simple loading, i.e. $\sigma_x(M) + \sigma_x(N^-) = \sigma_x$.

3. A method of assessing the bearing capacity

3.1. Non-dimensional determining parameters

For overloading:

We find the loading parameter ξ in a non-dimensional form from eq. (1)

$$\xi = \frac{P}{P_0 \upsilon_p} = \frac{Q}{Q_0 \upsilon_q} \tag{5}$$

For probabilities:

Probabilities υ_p and υ_q belong to one and the same interval, i.e. $\upsilon_p \in [0,1]$ and $\upsilon_q \in [0,1]$. Assume $\upsilon_p \neq const$, while $\upsilon_q = const$. Note that they are independent as probability coefficients.

Introduce

$$\upsilon = \frac{\upsilon_p}{\upsilon_q} \tag{6}$$

For geometry:

$$\eta = \frac{b}{\tilde{b}} \frac{F^*}{J_y^*} |z_L| l_r \tag{7}$$

where "~" denotes the geometry characteristics of the selected basic element; $z_L = \dfrac{h}{2} - h_R$, h_R is the thickness of the transition zone emerging as a result of the coating diffusive deposition, Fig. 1; l_r is the reduced length of the beam concrete elements introduced for each of them by means of parameter $\alpha_{(i)}, i = 1,2,3,4$; F^* is the generalized area of the cross section of the strengthened concrete element (consisting of transition areas and a concrete core); J_y^* is the generalized inertial moment of the strengthened concrete element.

F^* and J_y^* participate in the equations of the generalized compression and bending stiffnesses - K_I^* and K_{II}^* [6]. The form of the generalized stiffness is:

$$K_I^* = E_B F^*, \quad F^* = \frac{2}{3} b z_L + 2bh_R\left[1 + 0.215\left(e_R - 1\right)\right]$$
$$K_{II}^* = E_B J_y^*, \quad J_y^* = \frac{2}{3} b z_L^3 + 2\sum_{i=1}^{3} K_i h_R^i$$
$$K_1 = \frac{bh^2}{12}\left(2 + e_R\right), K_2 = -\frac{bh}{12}\left(E_B + e_R\right), K_3 = \frac{b}{12}\left(3 + \frac{1}{3} e_R\right)$$

where $e_R = \frac{E_R}{E_B}$; E_R is the elasticity modulus of the strengthening material; E_B is the elasticity modulus of concrete.

For boundary conditions:

Introduce a non-dimensional parameter

$$\alpha_{(i)} = \frac{l_r}{l_{(i)}} \tag{8}$$

For strengthening:

Introduce a non-dimensional parameter λ of the form

$$\lambda = e_R \frac{h_R}{\tilde{h}} \tag{9}$$

3.2. Generalized space of events

Based on the non-dimensional quantities in eqs. (5-9), introduce a multi-dimensional Generalized space of events. Its general form in both cases C1 and C2 reads:

$$\Pi_1 = \left\{\xi_J, \upsilon, \eta, \alpha_{(i)}, \lambda\right\}, \quad J = I, II, \quad (i) = 1 \ldots n$$
$$\Pi_2 = \left\{\xi_J, \upsilon, \eta, \alpha_{(i)}, \lambda\right\}, \quad J = I, III, \quad (i) = 1 \ldots n \tag{10}$$

Indexes $J = I, II$ are for C1, where I is for tension and II for compression, respectively, without micro-damage. Indexes $J = I, III$ are for C1, where I is for tension and III for compression, accounting for micro-damage.

Introduce a corresponding coordinate system and consider the quadrant of positive parameters in eqs. (5-9), i.e., their physical meaning. If some parameters are assumed constant, the space order decreases. We look for Safety zone in such a transformed space where the structural elements operate safely under overloading. Safety zones emerge between the minimal values of ξ_I and ξ_{II} for C1 or between ξ_I and ξ_{III} for C2, and the admissible values of $\eta \in \left[\eta_{\min}, \eta_{\max}\right]$. They are found at the expert level such as to account for tension and compression.

The task of the presented parametric analysis is to follow the variation of the Safety zones with respect to three criteria: (1) a boundary conditions criterion; (2) a criterion considering the effect of overloading probability on the Safety zones; (3) a criterion taking into account the effect of strengthening of the transition area on the element bearing capacity. These criteria are in relation with three of the parameters belonging to the multi-dimensional space - α, υ and λ in eq. (10).

To find the Safety zones, we proceed as follows:

For each criterion, and in both cases C1 and C2, draw the limit curves $\xi_J = f_J(\eta)$ considering different values of α, υ and λ according to formulas (11). Find the interval of values of η, where curves cross each other, i.e., the structural element undergoes tension and compression simultaneously.

$$\xi_I = \frac{C_I}{\eta - A}, \quad \xi_{II} = \frac{C_{II}}{\eta + A}, \quad \xi_{III} = \frac{C_{III}}{\eta + A} \tag{11}$$

where $C_J = \dfrac{s_J \tilde{F}^*}{\upsilon_q \left|\tilde{Q}_0\right|}$, $(J = I, II, III)$; Calculate η according to formula (7); $A = \upsilon \dfrac{\tilde{P}_0}{\left|\tilde{Q}_0\right|}$.

Find the coordinates of the cross point of the limit curves using formulas (12) and (13). Find it in C1 when $\xi_I = \xi_{II}$ for $\eta = \eta^*$, and in C2 – when $\xi_I = \xi_{III}$ for $\eta = \bar{\eta}^*$.

$$\eta^* = s\,A, \quad s = \frac{s_I + s_{II}}{s_{II} - s_I} \tag{12}$$

$$\bar{\eta}^* = \bar{s}\,A, \quad \bar{s} = \frac{s_I + s_{III}}{s_{III} - s_I} \tag{13}$$

where $s > 0$ and $\bar{s} > 0$, since normal stresses are always $s_{II} > s_I$ and $s_{III} > s_I$ for concrete.

Find the areas under the curves using formulas (14) and (16). They are the Safety zones sought in C1 and C2 respectively, without/with the consideration of the compression micro-damage.

$$S_1 = \int_{\min \eta}^{\eta^*} \xi_{II}(\eta)d\eta + \int_{\eta^*}^{\max \eta} \xi_I(\eta)d\eta \tag{14}$$

$$S_2 = \int_{\min \eta}^{\bar{\eta}^*} \xi_{III}(\eta)d\eta + \int_{\bar{\eta}^*}^{\max \eta} \xi_I(\eta)d\eta \tag{15}$$

Introduce parameter ΔS according to formula (16), characterizing the tribute of micro-damage within the compression zone of the concrete core.

$$\Delta S = \frac{S_2 - S_1}{S_2} \% \tag{16}$$

4. A parametric analysis in different spaces. Results

Perform a parametric analysis for specific values of the three parameters α, υ and λ belonging to the space of events, which change the Safety zone discussed above. We have fixed and varying quantities for each criterion. The fixed values for the three criteria are: $s_I = 300 N/cm^2$, $s_{II} = 1031 N/cm^2$, $s_{III} = 1200 N/cm^2$, $K_I^* = 2{,}06 \times 10^9 N$, $K_{II}^* = 16.4 \times 10^{10} N/cm^3$, $\upsilon_q = 0.01$, $\tilde{P}_0 = 100N$.

Select six cross sections $(b \times h)cm$ - (20×30); (25×35); (30×40); (35×45); (40×50); (45×55) of structural elements with a reduced length l_r regarding four types of beams, Fig. 1. Assume the smallest cross section $(\tilde{b} \times \tilde{h}) = (20\times30)$ to be the basic one.

4.1. Boundary conditions criterion - α

Consider four typical boundary conditions for concrete beam elements. The different boundary conditions are presented by the reduced length l_r thus comprising the four beams with lengths $l_{(i)}$ via parameter $\alpha_{(i)}$, Fig. 1, eq. (8). Hence, we find $l_r = 50cm$ for four lengths: $l_{(1)} = 100cm$, $l_{(2)} = 300cm$, $l_{(3)} = 400cm$, $l_{(4)} = 600cm$ and four values of $\alpha_{(i)}$: $\alpha_{(1)} = \frac{1}{2}$, $\alpha_{(2)} = \frac{1}{6}$, $\alpha_{(3)} = \frac{1}{8}$, $\alpha_{(4)} = \frac{1}{12}$. Those values of α correspond to the maximal beam moment and the respective boundary condition.

The normative basic loading is $|\tilde{Q}_0| = 1000N$, $3000N$, $4000N$ and $5000N$ regarding the different values of α. The probability ratio υ in eq. (6) is $\upsilon = 19,58,76,94$.

The results plotted in Fig. 2(a) - Fig. 5(a) present the limit curves $\xi_I(\eta)$ and $\xi_{II}(\eta)$ in C1, while those in Fig. 2(b) - Fig. 5(b) – the limit curves $\xi_I(\eta)$ and $\xi_{III}(\eta)$ in C2. Note that the results are valid only for those cross sections and interval of η, where the curves cross each other. Here, the only cross sections are (20×30), (25×35), (30×40), (35×45) for $\eta\in[3.0,3.6]$ and fixed $h_R = 1cm$. The cross points between the limit curves in C1 and C2 are specified in the figure captions.

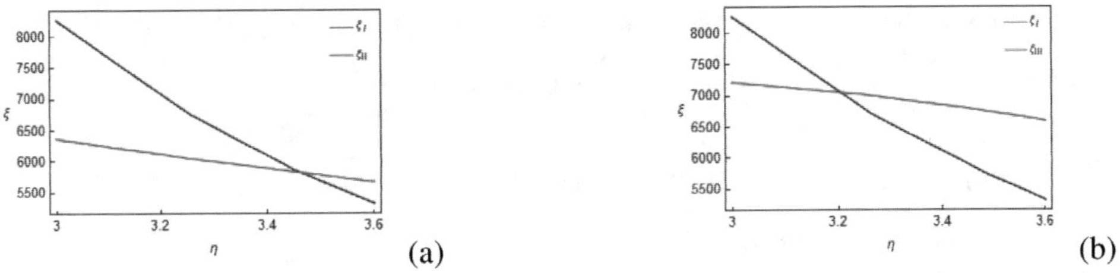

Figure 2. For $\alpha = 1/2$: $\eta^* = 3.46$ and $\bar{\eta}^* = 3.17$

Figure 3. For $\alpha = 1/6$: $\eta^* = 3.52$ and $\bar{\eta}^* = 3.22$

Figure 4. For $\alpha = 1/8$: $\eta^* = 3.46$ and $\bar{\eta}^* = 3.17$

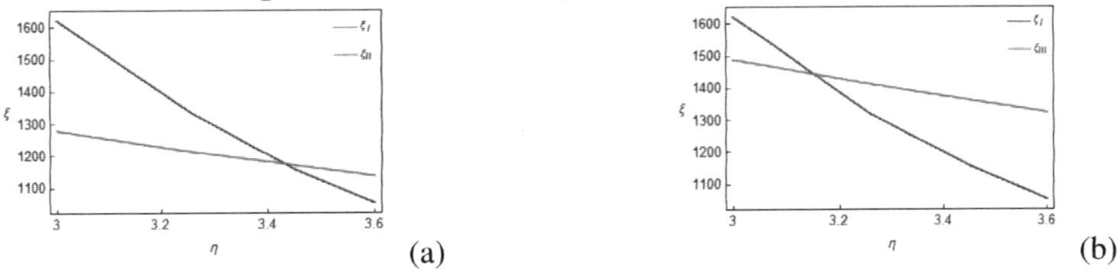

Figure 5. For $\alpha = 1/12$: $\eta^* = 3.42$ and $\bar{\eta}^* = 3.13$

The results calculated by using eqs. (14), (15) in both cases, employing the current criterion α, are plotted in Fig. 6(a). The results for ΔS found using formula (16), presenting the tribute of the micro-damage within the compression zone with respect to the current criterion α, are plotted in Fig. 6(b).

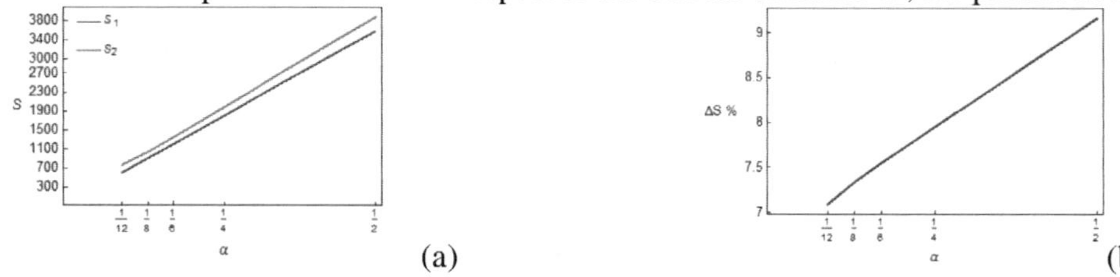

(a) (b)

Figure 6. Safety zones S_1 and S_2, and ΔS for each α

The results in Fig. 6(a) show that **there is no** essential difference between the areas of the Safety zones in C1 and C2, using the current criterion. Hence, the element bearing capacity is almost one and the same in both cases at a specific value of α. Yet, it increases with the increase of α. Fig. 6(b) proves that the parameter accounting for the tribute of micro-damage depends on the boundary conditions. For $\alpha = \frac{1}{2}$, i.e. when treating a beam, the value of ΔS is the largest one.

4.2. Criterion considering the effect of overloading probability on the Safety zones - υ

Employing this criterion, we perform a parametric analysis at fixed $\alpha = \frac{1}{2}$ and beam length $l_{(1)} = 200cm$, Fig. 1. Thus, we find $l_r = 100cm$. The overloading basic values are fixed as $\left|\tilde{Q}_0\right| = 2000N$ and $\tilde{P}_0 = 100N$ for the assumed value of α. We perform a number of calculations herein to find the interval of υ, where $\xi_I(\eta)$ and $\xi_{II}(\eta)$, as well as $\xi_I(\eta)$ and $\xi_{III}(\eta)$ cross each other. The interval found is $\upsilon \in [74,84]$.

The results plotted in Fig. 7(a)- Fig. 11(a) present $\xi_I(\eta)$ and $\xi_{II}(\eta)$ in C1, while those in Fig. 7(b) - Fig. 11(b) – $\xi_I(\eta)$ and $\xi_{III}(\eta)$ in C2. They concern all the initially assumed 6 cross sections for $\eta \in [6.0, 7.7]$ and fixed $h_R = 1cm$. Figure captions specify the cross points between the limit curves in C1 and C2.

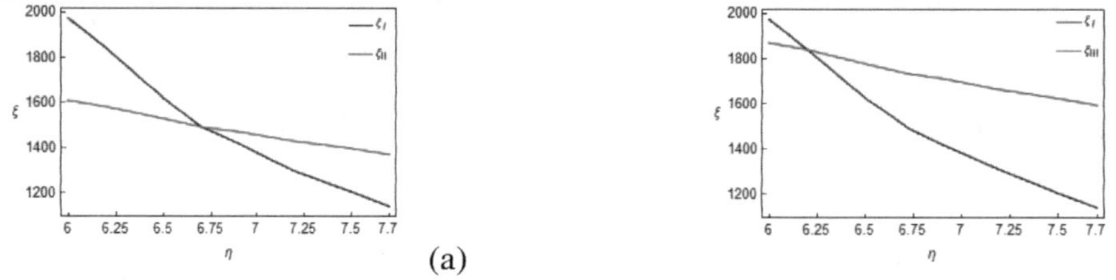

(a) (b)

Figure 7. For $\upsilon = 74$: $\eta^* = 6.7$ and $\overline{\eta}^* = 6.2$

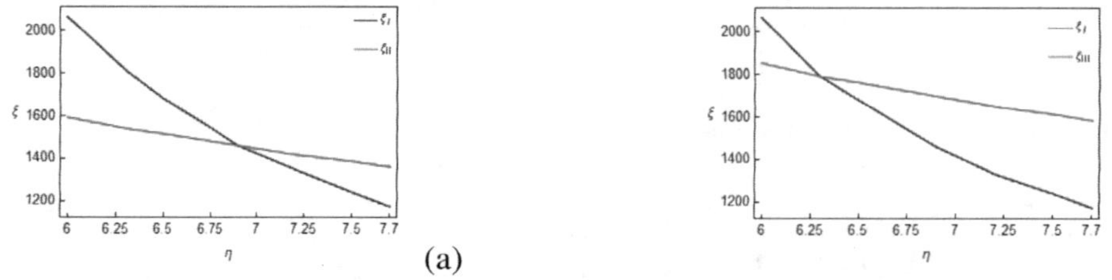

(a) (b)

Figure 8. For $\upsilon = 76$: $\eta^* = 6.9$ and $\overline{\eta}^* = 6.3$

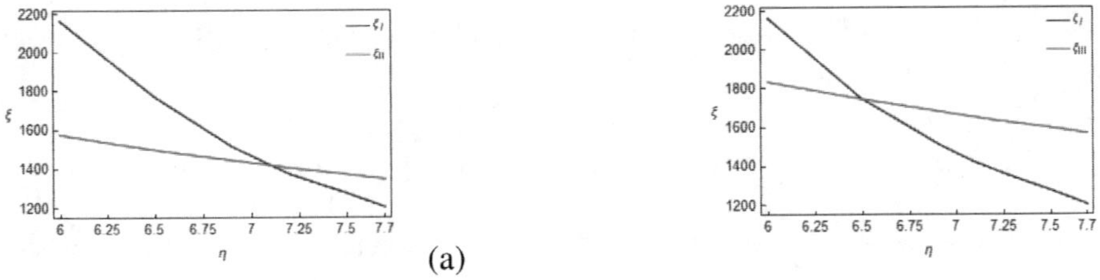

Figure 9. For $\upsilon = 78$: $\eta^* = 7.1$ and $\bar{\eta}^* = 6.5$

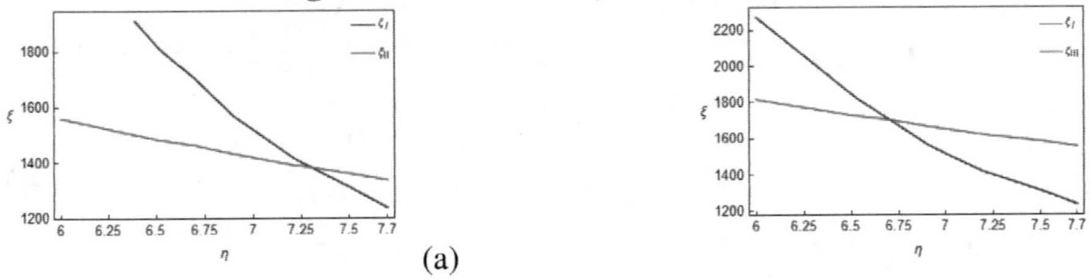

Figure 10. For $\upsilon = 80$: $\eta^* = 7.3$ and $\bar{\eta}^* = 6.7$

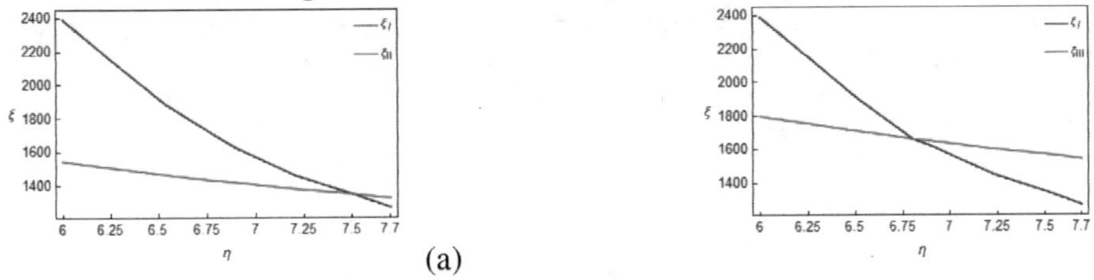

Figure 11. For $\upsilon = 82$: $\eta^* = 7.5$ and $\bar{\eta}^* = 6.8$

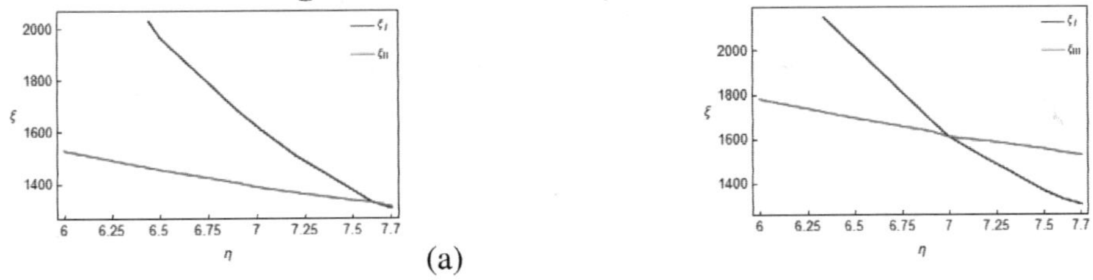

Figure 12. For $\upsilon = 84$: $\eta^* = 7.6$ and $\bar{\eta}^* = 7.0$

The calculation results according to eqns. (14) and (15) for the Safety ones in C1 and C2, pursuant to the current criterion υ, are plotted in Fig. 13(a). The results for ΔS according to formula (16), presenting the tribute of micro-damage in the compression zone considering the current criterion υ are plotted in Fig. 13(b).

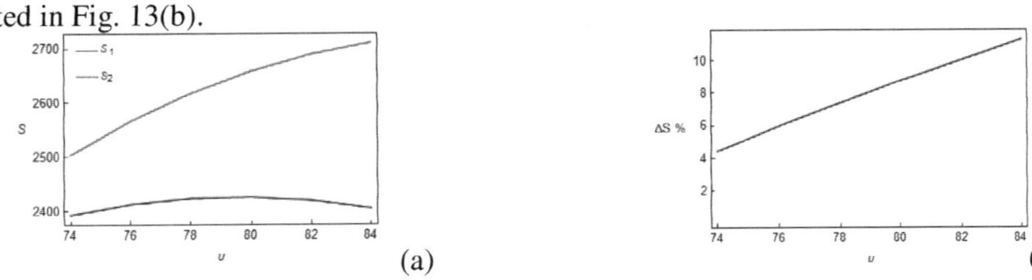

Figure 13. Safety zones S_1 and S_2, and ΔS for each value of υ

The results in Fig. 13(a) show that **there is an essential difference** between the areas of the Safety zones in C1 and C2 according to the current criterion. The effect of parameter υ on the element bearing capacity is much stronger when assuming the emergence of compression micro-damage.

Fig. 13(b) shows that the increase of υ within the specified interfal $\eta \in [6.0, 7.7]$ yilds increase of the effect of ΔS .

4.3. A criterion taking into account the effect of strengthening of the transition zone on the bearing capacity - λ

According to the current criterion, we perform a parametric analysis specifying three values of parameter $\lambda = 0.40, 0.67, 1.00$ for $h_R = 0.6, 0.8, 1.0$ and $e_R = 20, 25, 30$. We fix $\alpha = \frac{1}{2}$, beam length $l_{(1)} = 200cm$, $\upsilon = 70$, $l_r = 100cm$, and the basic values of overloiading $|\tilde{Q}_0| = 2000N$ and $\tilde{P}_0 = 100N$, Fig. 1. We select the three values of λ on the basis of several solutions where the limit curves cross each other in C1 and C2.

The results in Fig. 14(a) - Fig. 16(a) present the limit curves $\xi_I(\eta)$ and $\xi_{II}(\eta)$ in C1, while those in Fig. 14(b) - Fig. 17(b) – thre limit curves $\xi_I(\eta)$ and $\xi_{III}(\eta)$ in C2. The limit curves cross each other for cross sections (20×30), (25×35), (30×40), (35×45) and respective intervals of $\eta \in [5.76, 6.97]$, $\eta \in [5.43, 6.62]$ and $\eta \in [5.16, 6.37]$. Figure captions specify the cross points of the limit curves in C1 and C2.

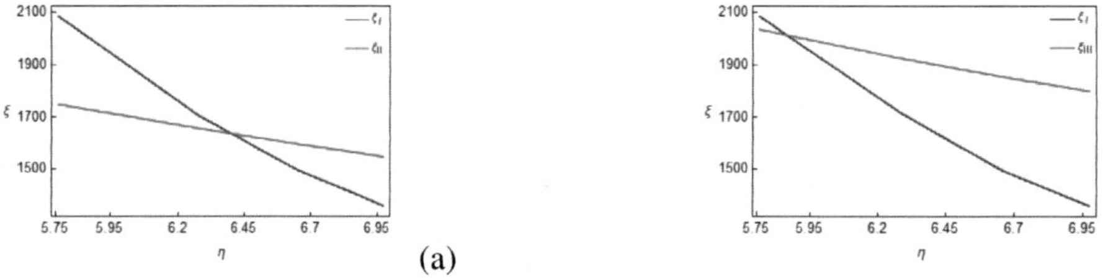
(a) (b)

Figure 14. For $\lambda = 0.4$: $\eta^* = 6.37$ and $\bar{\eta}^* = 5.83$

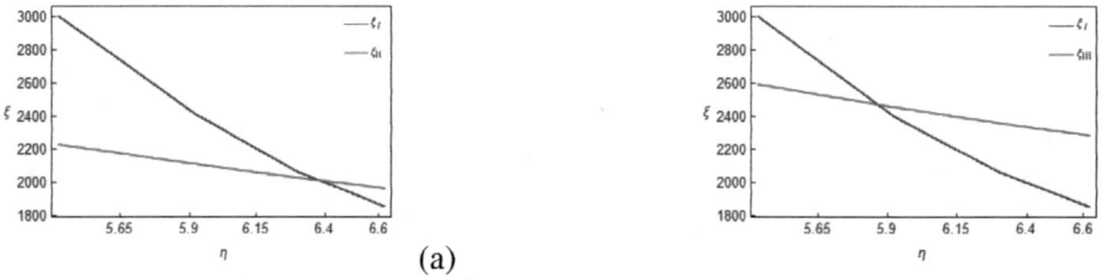
(a) (b)

Figure 15. For $\lambda = 0.67$: $\eta^* = 6.37$ and $\bar{\eta}^* = 5.83$

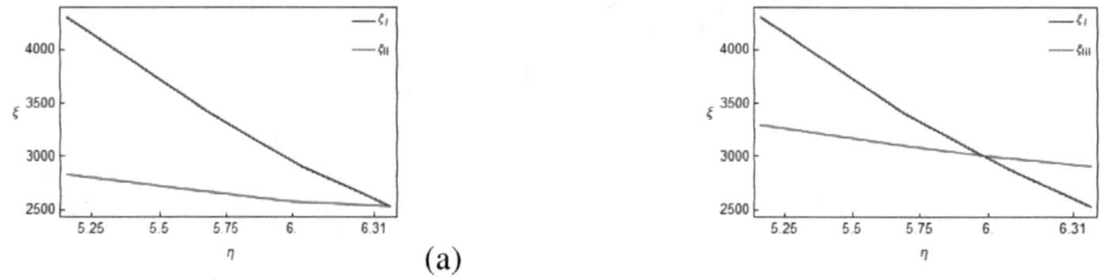
(a) (b)

Figure 16. For $\lambda = 1$: $\eta^* = 6.37$ and $\bar{\eta}^* = 5.83$

The calculation results according to eqs. (14) and (15) for the Safety zones in C1 and C2 according to the current criterion λ are shown in Fig. 17(a). Fig. 17(b) shows the results for ΔS according to formula (16), illustrating the tribute of compression micro-damage (criterion λ).

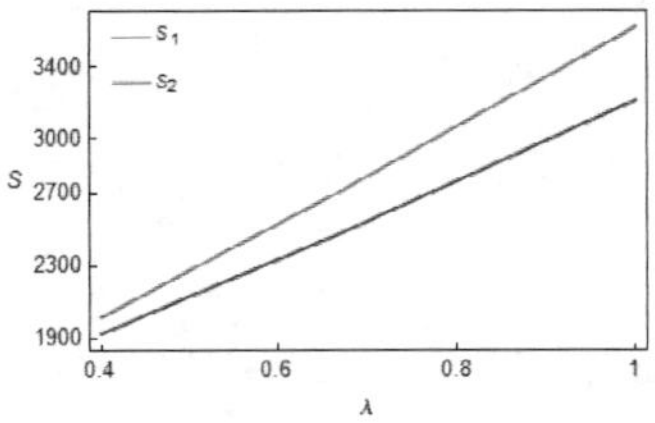

(a)

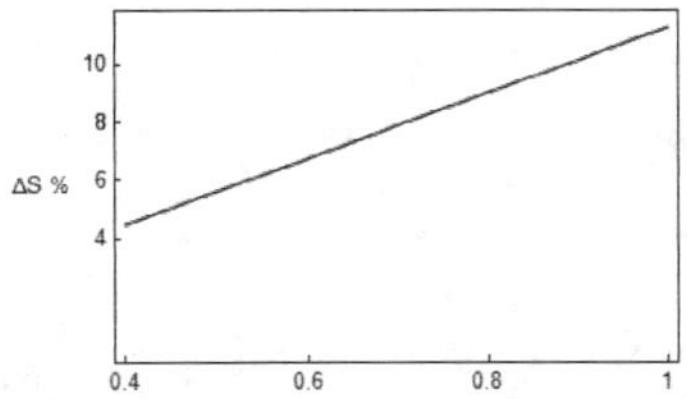

(b)

Figure 17. Safety zones S_1 and S_2, and ΔS for each value of parameter λ

The results in Fig. 17(a) prove that **there is** a difference between the areas of the Safety zones in C1 and C2 considering the current criterion. Fig. 17(b) shows that λ affects ΔS - the parameter accounting for the tribute of micro-damage, yet in different intervals of η depending on the current boundary conditioins.

5. Conclusions

The paper presents an assessment of the bearing capacity of concrete beam elements, diffusively strengthened and undergoing overloading. The bearing capacity is assessed by specifying Safety zones in different spaces of events introduced by the authors. The elements operate safely in those zones under overloading, and two cases of element operation are considered. It is found that the largest difference between the Safety zones occurs when micro-damage is/isn't accounted for under criterion υ. The element bearing capacity increases in both cases with the increase of each criterion parameter. Note that this holds true only for a specific interfal of the criterion parameter, i.e. when limit curves $\xi_I(\eta)$ and $\xi_{II}(\eta)$ cross each other in C1, and limit curves $\xi_I(\eta)$ and $\xi_{III}(\eta)$ cross each other in C2.

The present study can be of practical use in the design of civil structures.

6. References

[1] Zhelyazov T, 2007 *Renforcement des structures en béton arme par collage de CFRP*, (PAF, Paris)

[2] Rabinovitch O, Frosting Y 2001 Delamination failure of RC Beams strengthening with FRP strips *ASCE Journal of Eng. Mechanics, 127 (8)*

[3] Ivanova I, Assih J and Dontchev D 2020 Influence of anchorage length of composite fabrics and bonded surface on the strengthened short reinforced concrete corbel by bonding CFRF *European Journal of Environmental and Civil Engineering, 2018 - Taylor & Francis, 24 (12)*

[4] Jakob A, Sarott A, and Spieler P 1999, Diffusion and sorption on hardened cement pastes - experiments and modelling results *PSI Bericht 99-05, 30 (49)*

[5] Cherneva S, Iankov R, and Stoychev D 2010 Characterization of mechanical properties of electrochemically deposited thin silver layers, *Transactions of the Institute of Metal Finishing, Birmingham, England, 88 (4)*

[6] Baltov A and Yanakieva A. 2018 *Energy saving and surface-reinforced structural elements with transition areas* (Series Appl. Math. and Mech., Vol. 10, Inst. of Mech.- BAS, Sofia)

[7] Baltov A, Yanakieva A and Nikolova G 2019 Residual Resource of Emergency-loaded Concrete Columns under Pressure and Micro-damage, *Comptes rendus de l'Académie bulgare des Sciences, 72 (8)*

[8] Baltov A, Yanakieva A and Nikolova G 2022 Safety Zones within the Load Space of Diffusively Surface-reinforced Structural Components *Journal of Theoretical and Applied Mechanics, 52 (4) (in print)*

Seismic Response of Embedded Retaining Wall for Excavation Support

L Mihova[1,2] and M Topalska[1]

[1]University of Architecture, Civil Engineering and Geodesy (UACG), Sofia, Bulgaria

[2]L_mihova@yahoo.com

Abstract. A comparative analysis of the solutions for the seismic impact on a flexible retaining wall designed to strengthen the construction excavation for a multi-storey building in Sofia has been performed. For this purpose, the following FEM calculation models of the interaction of the wall with the surrounding soil body were compiled: (1) a 1D beam-column-spring model of the wall, including spring support in its embedded length and pseudostatic seismic loading of the wall with active earth pressure, determined by the Mononobe-Okabe method; (2) a 2D model of the wall and soil body, with pseudostatic seismic loading in the form of inertial forces applied at the finite element nodes; (3) a time history FE analysis conducted with the accelerograms of two landmark earthquakes from the Balkan Peninsula - the 1976 Vrancea (Romania) earthquake and the 1986 Kalamata (Greece) earthquake. The pseudostatic seismic loading in analyzes (1) and (2) was determined on the basis of the reference pseudostatic acceleration of 2.3 m/s^2, determined for the city of Sofia according to the seismic hazard map for Bulgaria. The accelerograms in analysis (3) were scaled to a maximum value of acceleration of PGA = 2.3 m/s^2. The physical modelling of the soil body uses the HSM and HSsmall advanced constitutive laws, implemented in the Plaxis software.

1. Introduction

The design of geotechnical structures is performed based on the solutions of computational modelling of soil-structure interactions. Historically, these models have evolved in terms of a number of aspects, such as improving the constitutive laws for materials, accurately describing the geometry of structures and their response. This process is directly related to the implementation of numerical modelling and computer technology in mechanics.

The Mononobe-Okabe (M-O) method for determining the pseudostatic earth pressure in the event of seismic impact on retaining walls is a well-established concept in geomechanics, derived from the theoretical and experimental studies of the authors Okabe [1] and Mononobe & Matsuo [2]. The method is an extension of Coulomb's classical limit state theory, with pseudostatic inertial forces applied at the centre of gravity of active and passive soil wedges and their values determined by the force equilibrium of the wedge. Okabe [1] provide formulae for the coefficients of pseudostatic active and passive earth pressure. The angles of active and passive soil wedges that produce the total thrust on the retaining structure are given by Zarrabi-Kashani [3]. The M-O method is described in Eurocode 8 [4], and with the help of coefficients the stiffness of the wall and the amplification of the seismic impact depending on the type of soil base are determined. The limitations of the method [5], [6], most of which have been "inherited" from Coulomb's theory, restrict its application. However, due to the possibility of simplified engineering solutions for the seismic response of retaining walls, this method is the most commonly used in design practice. In the present study, the M-O method was applied in combination with a beam-column-spring model of a retaining wall.

The development of constitutive modelling in geomechanics has led to the creation of a number of physical models describing the specific features of the mechanical behaviour of soils. The application of FEM together with discretization of structures and the soil massifs, and the use of appropriate constitutive models is a modern approach that allows for precise modelling of the soil-structure interaction.

In the present study, a comparative analysis of computational models for the behaviour of embedded retaining wall in the event of a horizontal seismic impact was performed. The models were implemented using the SAP2000 and Plaxis2D software. The wall structure, made of steel profiles, was modeled as linear elastic. The advanced constitutive models Hardening Soil (HS) [7] and HSsmall [8] were used for the soil, which use more than 10 material patameters. The HSsmall model is an extension of the HS model and has been designed to describe dynamic soil behavior characterized by stiffness degradation when increasing shear strains [9]. A study on determining the values of the material parameters of the HS and HSsmall soil models in Bulgaria has been made in the sources [10], [11], [12].

2. An embedded retaining walls in Sofia

For the strengthening of the construction excavation of a multi-storey building in Sofia, along the contour of the excavation, which has a total length of 116.3 m, retaining walls of the Soldier Pile Wall type, also known as a Berlin Wall (Fig. 1, 2) were built. The vertical load-bearing elements are steel 2T profiles (St235), with an axial distance of 1.0 m between the profiles. The 2T profiles are 6.0 m to 8.0 m long and are incased in concrete with a diameter of 0.35 m below the level of the excavation. Walls braced by strut and cantilever walls were erected.

Figure 1. View of the support structure.

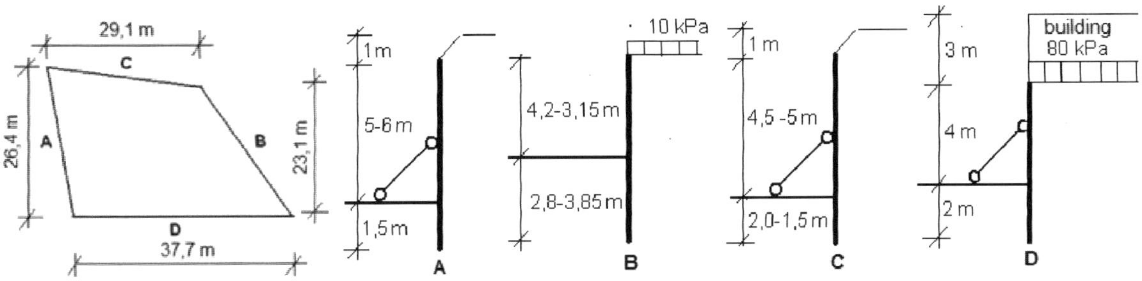

Figure 2. Layout of the excavation and cross sections of the retaining walls along sides A, B, C and D.

The ground to the depth of the walls is made up of soil layers with basic parameters listed in Table 1.

For the purposes of the study, a type A wall from Figure 2 was examined, with a depth of 5 m.

Table 1. Soil parameters

Layer	γ (kN/m^3)	(deg)	c (kPa)
1 – fill material	18.0	18	24
2 – silty sandy clay	19.5	21	30.0
3 – clayey sand	20.0	34	31.8
4 – clay	18.5	19	26.0

3. FE analyses of the seismic response of the type A retaining wall

3.1. Beam-column-spring model

The computational model of the wall shown in Figure 3a is a vertical beam-column with stiffness of 2T steel profile IPE220. The soil above the upper end of the wall is simulated with surcharge q = 19 kPa. The wall is supported with discrete horizontal springs along the embedded length. The stiffness of the springs, k_i, is determined by the soil subgrade modulus and it increases linearly in depth [13]. The load on the wall is created by pseudostatic active earth pressure p_{ae}, which is determined by the M-O method, taking into account the contribution of the surcharge q and the cohesion of the soil c. For this purpose, the coefficient of pseudo-static earth pressure K_{ae} is calculated according to the following formula

$$K_{ae} = \frac{\sin(\alpha-\varphi+\theta)\cos(\alpha-\beta)[\cos(\beta-\varepsilon)+\dfrac{2q}{\gamma H(1-k_v)}\cos\beta}{\cos^2\beta\cos\theta\sin(\alpha-\varepsilon)\cos(\alpha-\beta-\varphi-\delta)} - \frac{2c}{\gamma H(1-k_v)}\frac{\cos(\beta-\varepsilon)\cos\varphi}{\cos\beta\sin(\alpha-\varepsilon)\cos(\alpha-\beta-\varphi-\delta)} \tag{1}$$

where:

$$\theta = tg^{-1}\left[\frac{k_h}{1-k_v}\right] \tag{2}$$

$$\alpha = \varphi-\theta+tg^{-1}\left[\frac{tg(\varphi-\theta-\varepsilon)+C_{1E}}{C_{2E}}\right] \tag{3}$$

$$C_{1E} = \sqrt{tg(\varphi-\theta-\varepsilon)\left[tg(\varphi-\theta-\varepsilon)+cotg(\varphi-\theta-\beta)\right]}\sqrt{\left[1+tg(\delta+\theta+\beta)cotg(\varphi-\theta-\beta)\right]}$$

$$C_{2E} = 1+\left\{tg(\delta+\theta+\beta)\left[tg(\varphi-\theta-\varepsilon)+cotg(\varphi-\theta-\beta)\right]\right\}$$

The parameters included in formulae (2) and (3) are shown in Figure 3b: W is the weight of the soil wedge; k_h, k_v – the earthquake acceleration coefficients; $k_h g$, $k_v g$ – the pseudostatic accelerations; g = 9.81 m/s^2 – the gravitational acceleration; φ – the soil friction angle; c – the soil cohesion; δ – the wall friction angle; ε – the terrain slope angle; β – the wall slope angle; α – the failure surface angle; P_{ae} – the total active thrust. The present study assumes that k_v = 0 and δ = 0.

In accordance with [4], the coefficient k_h is determined by the expression

$$k_h = \frac{PGA}{g}\frac{S}{r} \tag{4}$$

where: PGA is the peak ground acceleration, $PGA = 2.3$ m/s$^2 = 0.23g$ after the present Bulgarian map of seismic risk; $S = 1.2$ – amplification factor; $r = 1$ – wall stiffness factor.

The following values of the parameters were obtained: $k_h = 0.276$, $\theta = 15.4°$, $\alpha = 46.4°$ and for Layers 3 and 4 from Figure 3, the following values of the coefficients of pseudostatic active earth pressure were determined: $K_{ae,3} = 0.368$ and $K_{ae,4} = 0.329$.

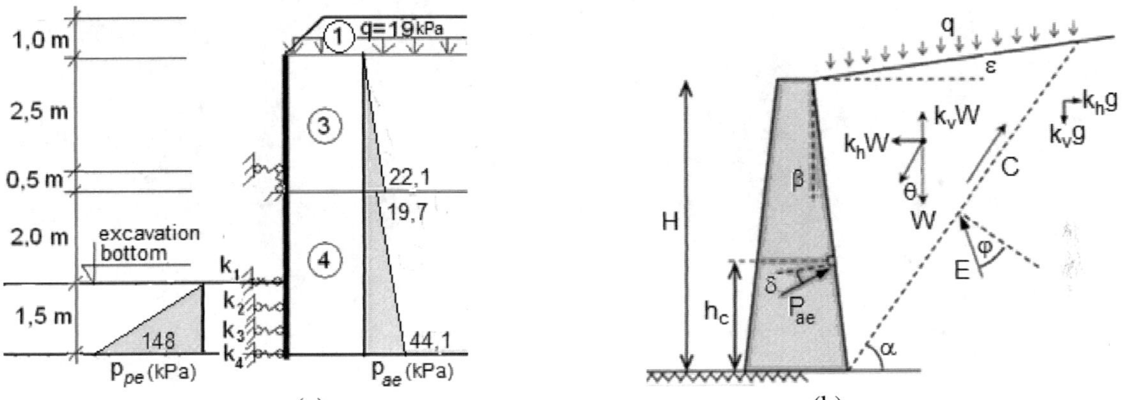

Figure 3. Beam-column spring model: (a) Static scheme and earth pressure diagrams; (b) Forces acting on the soil wedge.

The coefficient of pseudostatic passive earth pressure for Layer 4, where the wall is embedded, was determined from the charts in Figure 4 based on the Log-Spiral-Rankine Limit Equilibrium Model [14]. Its value is $K_{pe,4} = 4.0$.

The diagrams of active and passive earth pressure p_{ae} and p_{pe} are shown in Figure 3a. The Diagram p_{pe} is a limit one in terms of soil resistance. Performing the check that the sum of spring forces does not exceed the total passive thrust, proved that the chosen embedded length of the wall was sufficient.

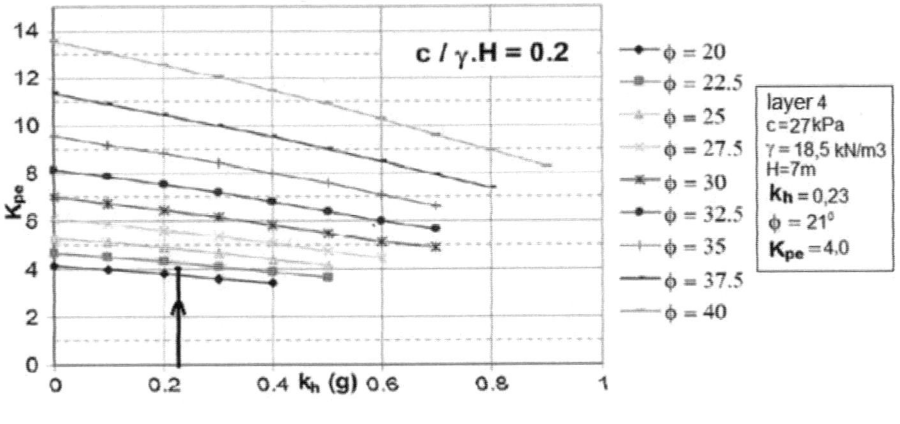

Figure 4. Charts for passive earth coefficient K_{pe} [14].

3.2. *Analyses of the soil-structure interaction*

A plane strain FE model was created (Figure 5) using the Plaxis 2D software. 15-node triangular elements with node parameters displacements U_x and U_y in the *xy* plane were used to discretize the soil body, and for the wall are used 5-node frame elements which work for bending moment, axial and shear forces. The soil-wall interaction was modeled by interface elements. The construction period was simulated by means of five phases of excavation works (Figure 6).

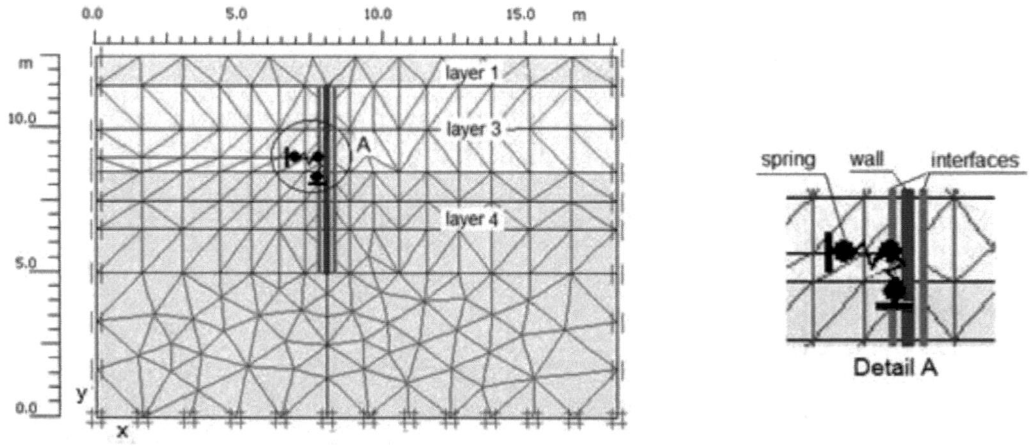

Figure 5. Plane strain FE model.

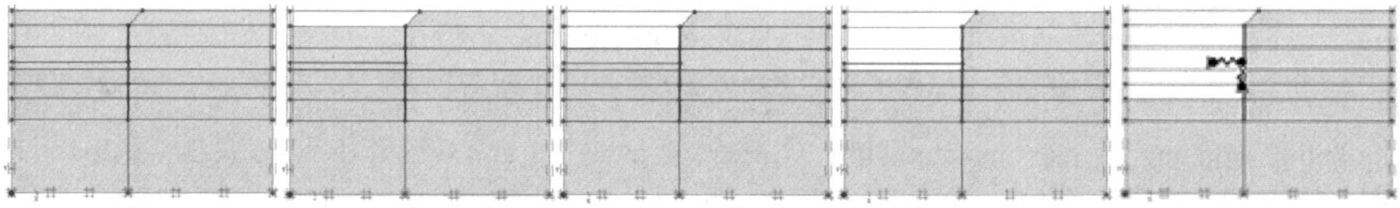

Figure 6. Construction period phases.

3.2.1. *Pseudostatic FE analysis*

For the mathematical description of the soil, the constitutive model HS was adopted. The material parameters of the soil layers are presented in Table 2. The boundary conditions of the assumed soil body (Figure 5) are the following: free top boundary; restrained horizontal and vertical displacements at the bottom boundary, i.e. $U_x = 0$ and $U_y = 0$; restrained horizontal displacements $U_x = 0$ at the vertical boundaries. The seismic load is pseudostatic, where for each node *i* of the finite element mesh, a pseudostatic horizontal inertial force $F_{h,i}$ was applied, determined by the following formula [4]

$$F_{h,i} = 0.5 \times PGA \times S \times W_i / g \tag{5}$$

where: $PGA = 2.3$ m/s^2; $S = 1.2$; $g = 9.81$ m/s^2; W_i – weight of the soil in node *i*.

3.2.2. *Time history analysis*

The FE seismic time history analysis was performed in accordance with the following prerequisites:

- The geometry of the model and the solution phases were adopted in accordance with Figure 5 and Figure 6.
- The boundary conditions include dampers along the boundaries of the assumed soil body; the dampers absorbs seismic energy and thus simulates infinite soil space.

Table 2. Parameters of the HS and HSsmall models

Parameter	Symbol	Measure	Layer 1	Layer 3	Layer 4
Reference stress for stiffness	p_{ref}	kPa	100	100	100
Tangent stiffness for primary oedometer loading	$E_{oed,ref}$	kPa	12 000	26 000	20 000
Secant stiffness in the standard drained triaxial test	$E_{50,ref}$	kPa	12 000	26 000	20 000
Unloading/Reloading stiffness	$E_{ur,ref}$	kPa	36 000	80000	60 000
Initial shear modulus*	$G_{0,ref}$	kPa	80 000	150 000	120 000
Poisson's ratio	v'	-	0.3	0.3	0.3
Poisson's ratio for unloading/reloading	v_{ur}	-	0.2	0.2	0.2
Shear strain for the 0.722 of the reduction of the shear modulus*	$\gamma_{0.7}$	%	0.06	0.015	0.05
Power factor for the stress-level dependency of stiffness	m	-	0.6	0.6	0.6
Failure ratio	R_f	-	0.9	0.9	0.9
Angle of dilatancy	ψ	°	0	2	0
Coefficient of lateral earth pressure	k_0	-	0.700	0.470	0.642

* Additional parameters of the HSsmall model

- The seismic impact is an accelerogram. The solutions used the accelerograms of the 1976 Vrancea (Romania) earthquake and the 1986 Kalamata (Greece) earthquake scaled to PGA = 2.3 m/s^2 The accelerogram of these earthquakes scaled to PGA = 1.0g = 9.81 m/s^2 are shown in Figure 7 and Figure 8 and the response spectra are shown in Figure 9 and Figure 10.
- HSsmall is the constitutive model of the soil. In this model, the parameters $G_{0,ref}$ and $\gamma_{0.7}$ (Table 2) defining the stiffness reduction of the soil during the seismic response were added to the HS model.
- To account for the hysteretic behavior of the soil, the viscous damping coefficients α_0 and β_0 according to the Rayleigh's theory [15] were introduced. In accordance with [16], [17], the coefficients α_0 and β_0 were determined from the predominant frequency of the ground motion f_p and from the natural frequency of the ground f. The α_0 and β_0 values are given in Table 3.
- The natural frequency of the ground f was determined by the following formula [5]

$$f = V_s \,/\, (4H) \quad (6)$$

where V_s is the shear wave velocity and H is the thickness of the soil base. The obtained value is f = 1.0 Hz.

Table 3

Earthquake	Predominant period, T_p (s)	Predominant frequency, f_p (Hz)	α_0 (-)	β_0 (-)
Vrancea (1976)	1.4	0.7	0.102	2.586×10^{-3}
Kalamata (1986)	0.3	3.3	0.153	1.290×10^{-3}

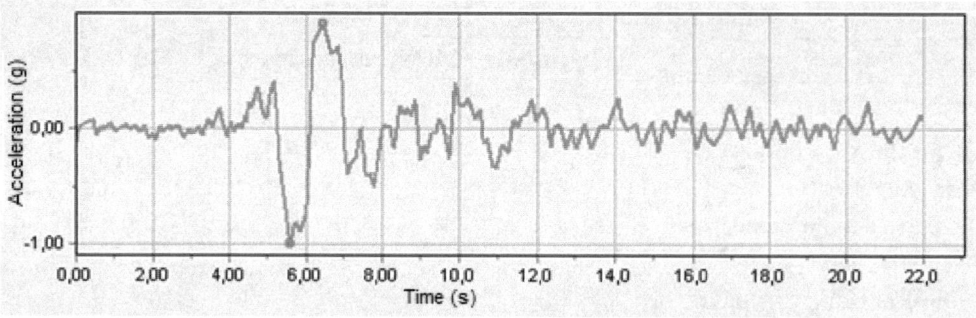

Fig. 7. Accelerogram of the 1976 Vrancea earthquake.

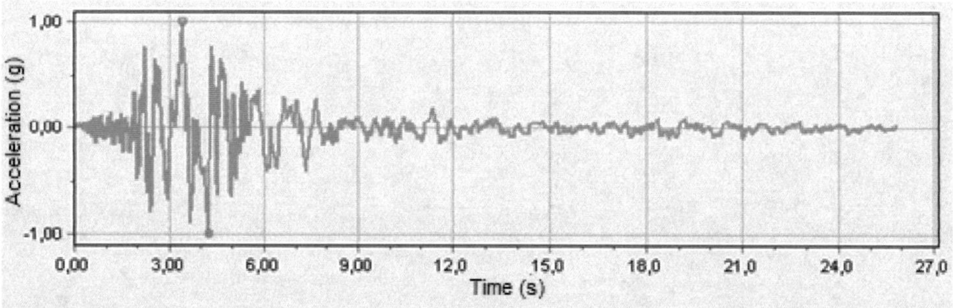

Fig. 8. Accelerogram of the earthquake 1986 Kalamata earthquake.

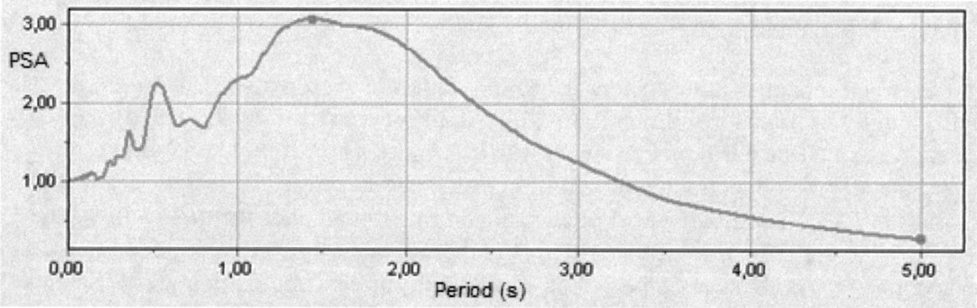

Fig. 9. Response spectrum of the 1976 Vrancea earthquake.

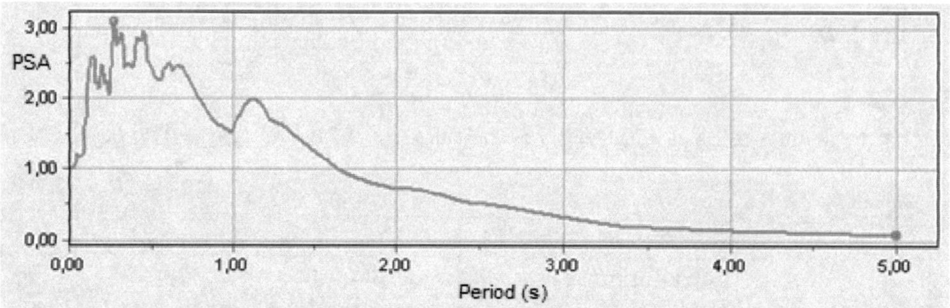

Fig. 10. Response spectrum of the 1986 Kalamata earthquake.

4. Analysis of results for the seismic response of the wall

Table 4 shows the seismic response of the wall using the alternative numerical models.

IOP Conf. Series: Materials Science and Engineering 1276 (2023) 012015 doi:10.1088/1757-899X/1276/1/012015

Table 4. Results for the seismic response of the wall

Model	Earth pressure (kPa)	Shear force envelope (kN)	Bending moment envelope (kNm)	Displace-ment (cm)	Strut force (kN)
Pseudostatic beam-column-spring	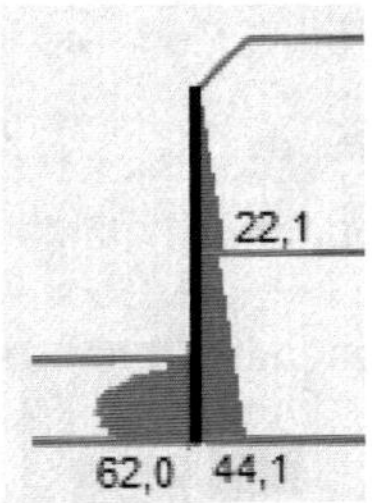	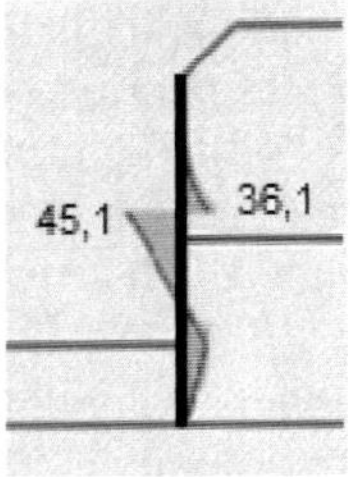	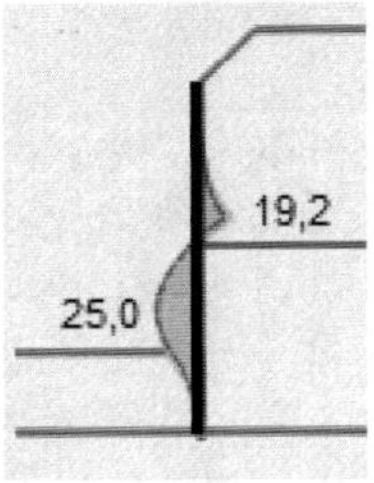	1.0	68.1
Pseudostatic FE	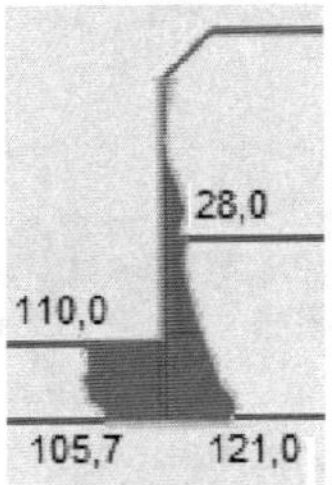	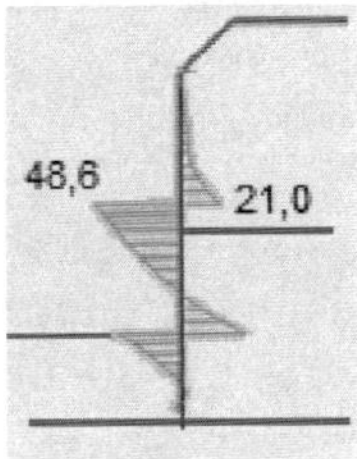	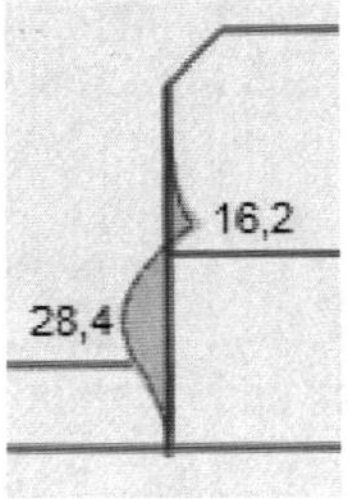	2.6	71.8
Time history Vrancea	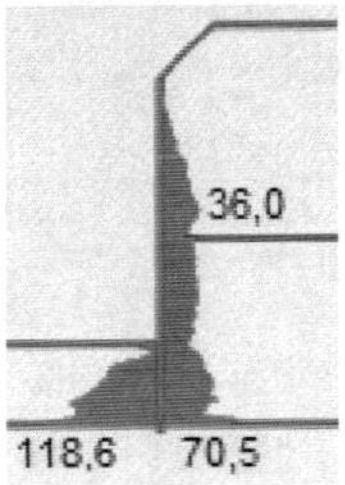	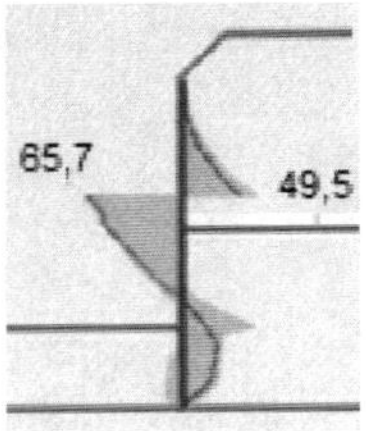	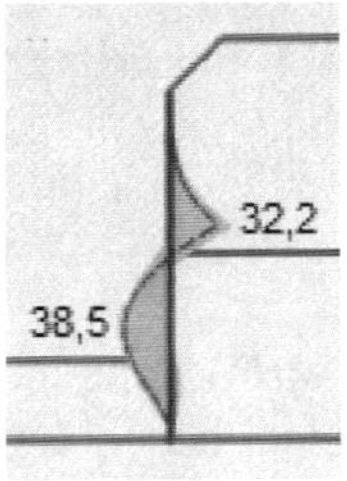	9.7	76.1
Time history Kalamata	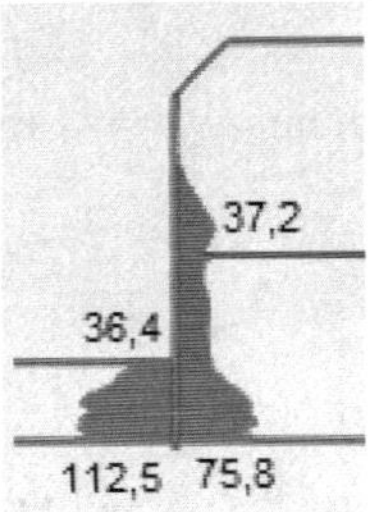	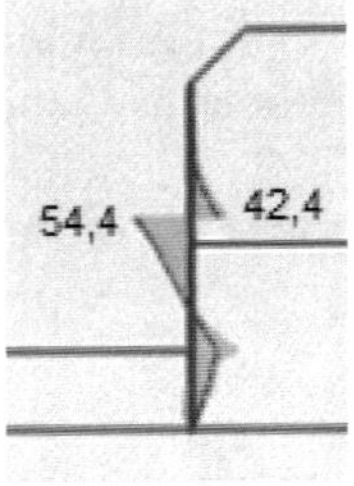	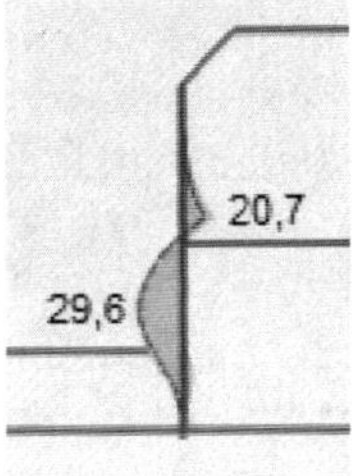	2.4	99.3

Based on the comparative analysis the following conclusions were made:

- The diagram of active earth pressure is characterized by a gradual increase of the values with the wall, which is determined by the one-way motion of the wall towards the excavation. The Mononobe-Okabe method used in the Pseudostatic beam-column-spring model yielded the lowest value of total active thrust. The value is twice lower than the value in the Pseudostatic FE model and 1.5 times lower than the value in the Time history models.
- The diagram of passive earth pressure develops from the side of the excavation in the embedded part of the wall. It is different for the different models: parabolic in the Pseudostatic beam-column-spring model, uniform in the Pseudostatic FE model, triangular in the Time history Vrancea model and trapezoid in the Time history Kalamata model. The Pseudostatic FE model yielded the highest value of total passive earth thrust of 162 kN. The lowest value of total passive earth thrust of 71 kN was obtained by the Pseudostatic beam-column-spring model.

- All models yielded identical diagrams in terms the of internal forces acting within the wall - shear force and bending moment.
- The most unfavourable seismic response of the wall was obtained by the Vrancea Time history model. This is indicated by the highest displacement, shear force and bending moment values obtained. The maximum horizontal motion of the wall is 9.7 cm and this is approximately 4 times higher than when using the Pseudostatic FE and the Kalamata Time history models and 10 higher than the motion obtained when using the Pseudostatic beam-column-spring model. The maximum shear force and the bending moment are respectively 1.2 to 1.5 times higher than the corresponding values in the other models. The reason for this is the fact that the Vrancea earthquake had a frequency of $f_p =$ 0.7 Hz, which is similar to the natural frequency of the ground f = 1.0 Hz. Although the Kalamata earthquake had the same amplitude as that of the Vrancea earthquake, the more favourable seismic response of the wall during the latter was due to its comparatively higher predominant frequency $f_p =$ 3.3 Hz compared to the natural frequency of the ground.
- The Time history models yielded 1.1 to 1.5 higher values of the internal forces acting within the wall and the strut as compared to the Pseudostatic models.

5. Conclusion

The dynamic behaviour of geotechnical structures is a complex process whose modelling should be based on established concepts in modern theoretical and experimental geomechanics. The present study shows that the simplified classical mathematical solutions are not always viable when trying to ensure structural safety. That is why precise modelling of all the elements of the structure, i.e. its geometry, materials and responses is needed implementing the possibilities of numerical modelling and computer technologies. Some advanced trends in this approach have been studied in [18].

References

[1] Okabe S 1926 General Theory of Earth Pressures *Journal of Japan Society of Civil Engineers* **12** 1 pp 123-134

[2] Mononobe N and Matsuo M 1929 On the Determination of Earth Pressures during Earthquakes *Proc. World Engineering Congress* (Tokyo) **9** 388 pp 179-187

[3] Zarrabi-Kashani K 1979 *Sliding of Gravity Retaining Wall during Earthquakes ConsideringVvertical Accelerations and Changing Inclinations of Failure Surface* S. M. Thesis (Department of Civil Engineering, Massachussetts Institute of Technology, Cambridge, Massachussetts)

[4] EN 1998 2003 Eurocode 8: Design of Structures for Earthquake Resistance, Part 5: Foundations, Retaining Structures and Geotechnical Aspects (CEN European Committee for Standardization, Bruxelles, Belgium)

[5] Kramer S 1996 *Geotechnical Earthquake Engineering* (Prentice Hall, Upper Saddle River) 653p

[6] Yazdani M, Azad A, Farshi A and Talatahari S 2013 Extended Mononobe-Okabe Method for Seismic Design of Retaining Walls *Journal of Applied Mathematics* ID 136132 doi.org/10.1155/2013/136132

[7] Schanz T, Vermeer P and Bonnier P 1999 The Hardening-Soil Model: Formulation and Verification (In Brinkgreve R B J *Beyond 2000 in Computational Geotechnics*) (Balkema Rotterdam)

[8] Benz T 2006 *Small-Strain Stiffness of Soil and its Numerical Consequences* PhD thesis (Universitat Stuttgart)

[9] Atkinson H and Sallfors G 1991 Experimental Determination of Soil Properties *Proc. of the 10th ECSMFE* **3** (Florence) pp 915-956

[10] Kerenchev N 2015 *Analysis of Seismic Slope Stability and Deformations* PhD Thesis (University of Architecture Civil Engineering and Geodesy, Sofia) (in Bulgarian)

[11] Kerenchev N 2019 On the E50 Modulus of Tailings Dam Materials *Proc. 19th International Multidisciplinary Scientific GeoConference SGEM 2019* (ISBN: 978-619-7408-77-5)

[12] Kerenchev N and Milev N 2019 Assessment of the Dynamic Soil Properties for the FEM Model of the Liuliakovitsa Tailings Dam *Proc. of the 8th Conference Geotechnics in Civil Engineering* (Vrnjačka Banja, Serbia)
[13] Bulgarian Pile Design Standard 1992 (in Bulgarian)
[14] Shamsabadi A, Xu S -Y and Taciroglu E 2013 A Generalized Log-Spiral-Rankine Limit Equilibrium Model for Seismic Earth Pressure Analysis *Soil Dynamics and Earthquake Engineering* **49** pp 197–209
[15] Rayleigh J and Lindsay R 1945 *The Theory of Sound* (Dover Publications, New York)
[16] Hashash Y and Park D 2002 Viscous Damping Formulation and High Frequency Motion Propagation in Non-Linear Site Response Analysis *Soil dynamic earthquake engineering* **22** 7 pp 611-624
[17] Hudson M, Idriss I and Beikkae M 2003 *QUAD4M : A computer program to evaluate the seismic response of soil structures using finite element procedures and incorporating a compliant base* (Center for Geotechnical Modeling, Dept. of Civil and Environmental Engineering, University of California)
[18] Tsvetkova T and Kerenchev N 2020 Review of Open Source Software for Modelling and Analysis of Structures *IOP Conference Series: Materials Science and Engineering* **951** 012003 pp 1-7

IOP Conf. Series: Materials Science and Engineering 1276 (2023) 012016 doi:10.1088/1757-899X/1276/1/012016

Comparison analysis of bearing capacity of arch bridges, where the influence of the stability is taken into account

A K Jiponov

Road and Transport Structural Facilities Department, FTE, University of Architecture, Civil Engineering and Geodesy, 1 Hr. Smirnenski Blvd, Sofia 1046, Republic of Bulgaria
ajiponov_fte@uacg.bg

Abstract. The structural stability failure is a critical factor for arch bridges which are not transversally braced and that must be considered when the bearing capacity has to be determined. One of the main parameters dictating the behaviour of arched structures in case of loss of stability is their slenderness. Accordingly, structures with high stiffness and low flexibility are expected to have significantly increased bearing capacity. The current paper consists analysis of complex 3-dimensional FE models, where the non – linear behaviour of geometry and materials is considered. The methodology is based on typical for bridge construction geometry, where are used typical cross – sections. The critical forces in the analysis are limited by the bending bearing capacity of the cross – sections. The results provide information about the advantages and disadvantages of the different structural solutions.

1. Introduction

The arch bridges are elegant and economical solution for mid to large span structures where the terrain is not accessible. The increasing of production technologies leads to the implementation of arch bridges which are non – braced in transversal direction and that cause problems with the structural stability in outer plane.

The appropriate design of cross – section and relevant usage of different materials should be considered in order to solve the mentioned problems. The reinforced concrete and its behaviour is favourable against stability failure for one side, from another side the structural steel could be used for scaffolding and for achieving of confined reinforced concrete. Moreover, the structural steel contributes for the bearing capacity, bending stiffness and torsional stiffness. This is essential for arch elements, which are outer – plane loaded.

Then, the composite (structural steel and reinforced concrete) elements unite the advantages and reduce the disadvantages of both materials. They ease the construction processes like transportation, scaffolding, assembling, detailing and their implementation could behave positively of the economy of one infrastructure project.

This paper presents a series of analyzes based on the Finite Element Method (FEM) to study the behavior in the presence of a load outside the plane of the arch element. Geometric and material nonlinearity is set in the conducted analyzes in order to obtain the most accurate results. The main goal is to compare the bending bearing capacity of all cross sections depending on the angle of the inclination of the arches and then to show the graphical relation with the curve of the stability failure.

2. Description of prerequisites and definition of cross - sections

Within the presented topic, a series of cross – sections of the main arch structure are considered with identical geometry. Each of the presented variants differs only in the parameters of the cross – section. The geometry is selected based on experience and practice applied on the construction of arch bridges in the recent decades. The main geometrical parameter of the main arch structure, which is selected for the comparative analysis, is circumscribed circle. All sections selected for analysis have different parameters of inertial characteristics, bearing capacity and own weight. They are limited to a particular contour which they can fit.

The geometry of the cross – section is inspired by an existing arch bridge. The Jiubao bridge (China) was chosen to determine the main geometry of the FE models. Alternative composite cross – sections are presented in the different analysis, and they meet the structural requirements of EN1994.

The supporting conditions of the arch are considered as "fixed" in order to reduce the parameters of the analysis.

2.1. Basic geometry of the model

The shape of the circle arch is with radius of 30m, which defines the span length distance – 60m and the rise – 30m. The contour of the rectangular cross – section is with dimensions 3.2m x 2.2m – **Figure 1**

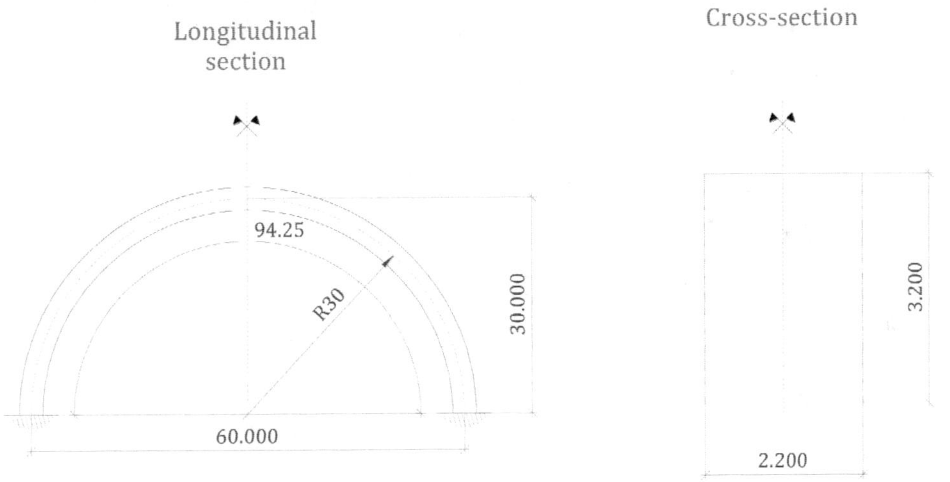

Figure 1. Geometry of the FE model.

2.2. Defined cross – sections for analysis

In that study, there are 8 cross – sections taken into account:

- Reinforced concrete box – (1 type, see **Figure 2**a) – Reinforced concrete section is considered as homogeneous. The presence of reinforcement in the section is defined indirectly through a specific idealization of the material "reinforced concrete". This allows performance of non-linear analysis for the stability failure investigation.
- Steel box – (3 types, see **Figure 2**b) – The steel section is presented directly taking into account the full geometric shape, including the stiffening longitudinal ribs. In the analysis, 3 similar sections were considered, with a difference in the wall thicknesses of the box, respectively 22, 32 and 42 mm.
- Steel truss – (1 type, see **Figure 2**c) – A truss section is considered, which consists of a group of 4 steel pipe sections type CHS508F50. The premise is that truss diagonals have only a bracing function. This ensures a planar section.

- Composite truss – (1 type, see **Figure 2**c) – This is an additional variant of the steel truss type with pipes filled with reinforced concrete.
- Composite box – (2 types, see **Figure 2**d and **Figure 2**e) – the two composite types have similar concept. The solid reinforced concrete core is enclosed in thin 11mm thick steel frame. The difference between the 2 types is inside the concrete core where in **Figure 2**d the core forms a concrete box like the one in **Figure 2**a. In **Figure 2**e, the reinforced concrete core has 2 circular openings with diameter of 800mm.

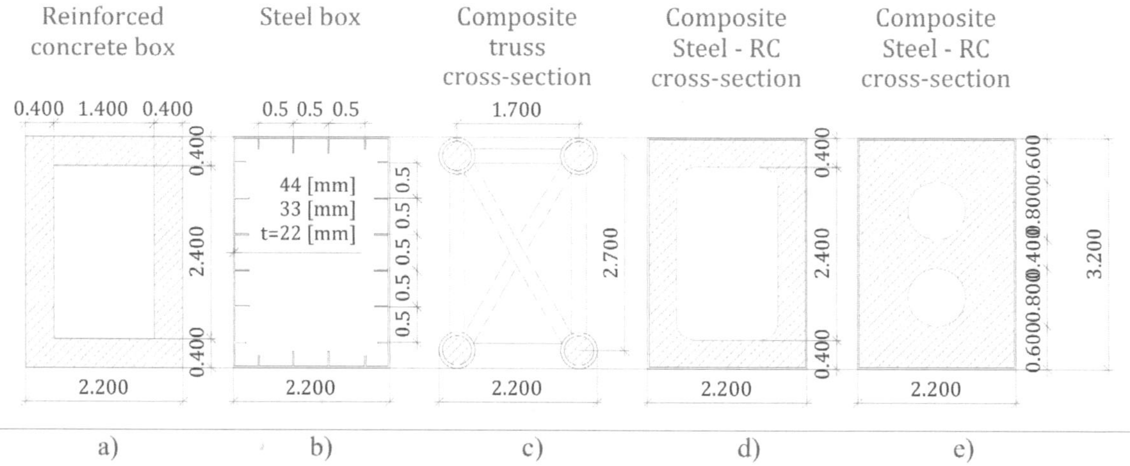

Figure 2. Cross – sections.

3. Analysis of the results

3.1. Bending bearing capacity of the cross – sections

A number of calculations have been made to find the bending capacity of the cross-sections presented in Chapter 2.2. It is calculated the outer plane bearing capacity of the cross – sections where the angle of inclination varies **Figure 3**.

Due to the comparation aim, the bending capacity is presented as a function of evenly distributed load. The results are presented graphically as the ratio between the maximum value of the uniformly distributed load leading to rupture by bending moments in the cross sections **Figure 4, Figure 5**.

To determine the load-bearing capacity of a reinforced concrete section, it is assumed a reinforcement coefficient of 2.75% for bending and 5.5% for the entire section. Plastic bearing capacity is calculated based on considering continuous deformation diagrams of the cross section [1].

3.2. Interrelation between criteria of the bending bearing capacity rupture and stability failure in the section

The results presented in Chapter 3.1. are worked up and related with a static solution of the statements for the presented angles. The critical bending moment is given nearby the fixed support of the arch. The bending bearing capacity is expressed as a function of the uniformly distributed load on the base of the bearing capacity of the section and the static solution for each angle. The correlation is interpreted as envelope from the bending rupture criteria along the two perpendicular axes of the section.

The graphical solution presented in this way is comparable to the results from the stability failure analysis.

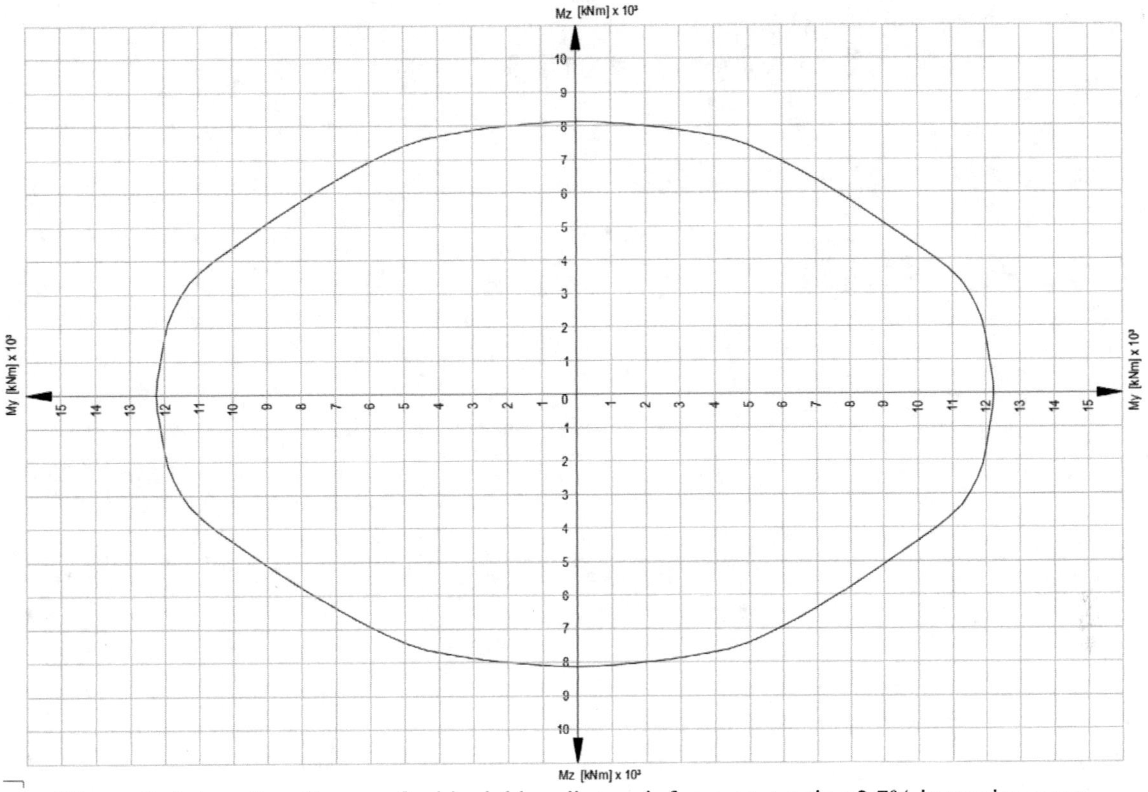

Figure 3. Interaction diagram for biaxial bending, reinforcement ration 2,7% in tension zone.

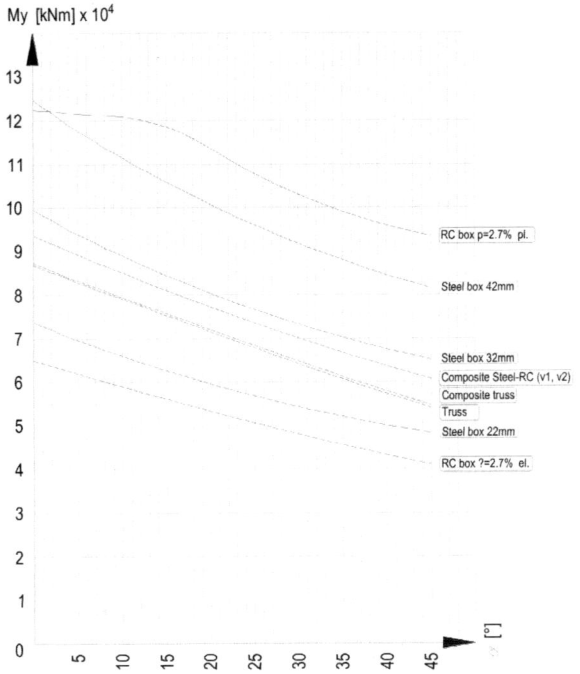

Figure 4. Graphic of the variation of the bending capacity of the section on direction M_y.

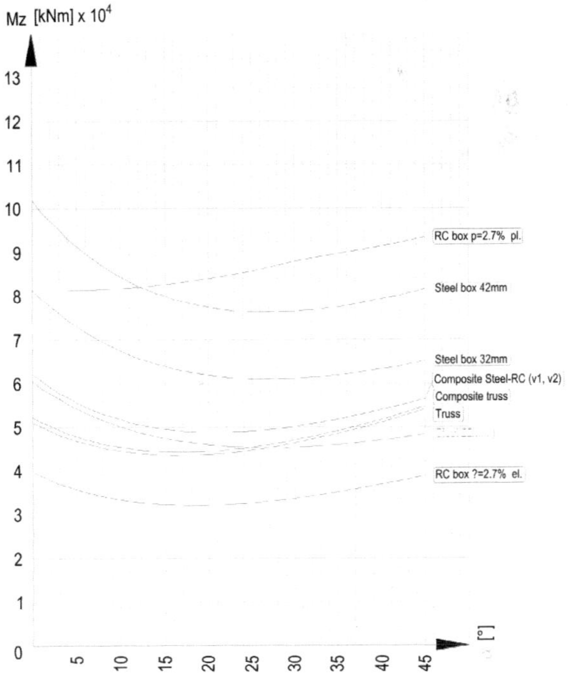

Figure 5. Graphic of the variation of the bending capacity of the section on direction M_x.

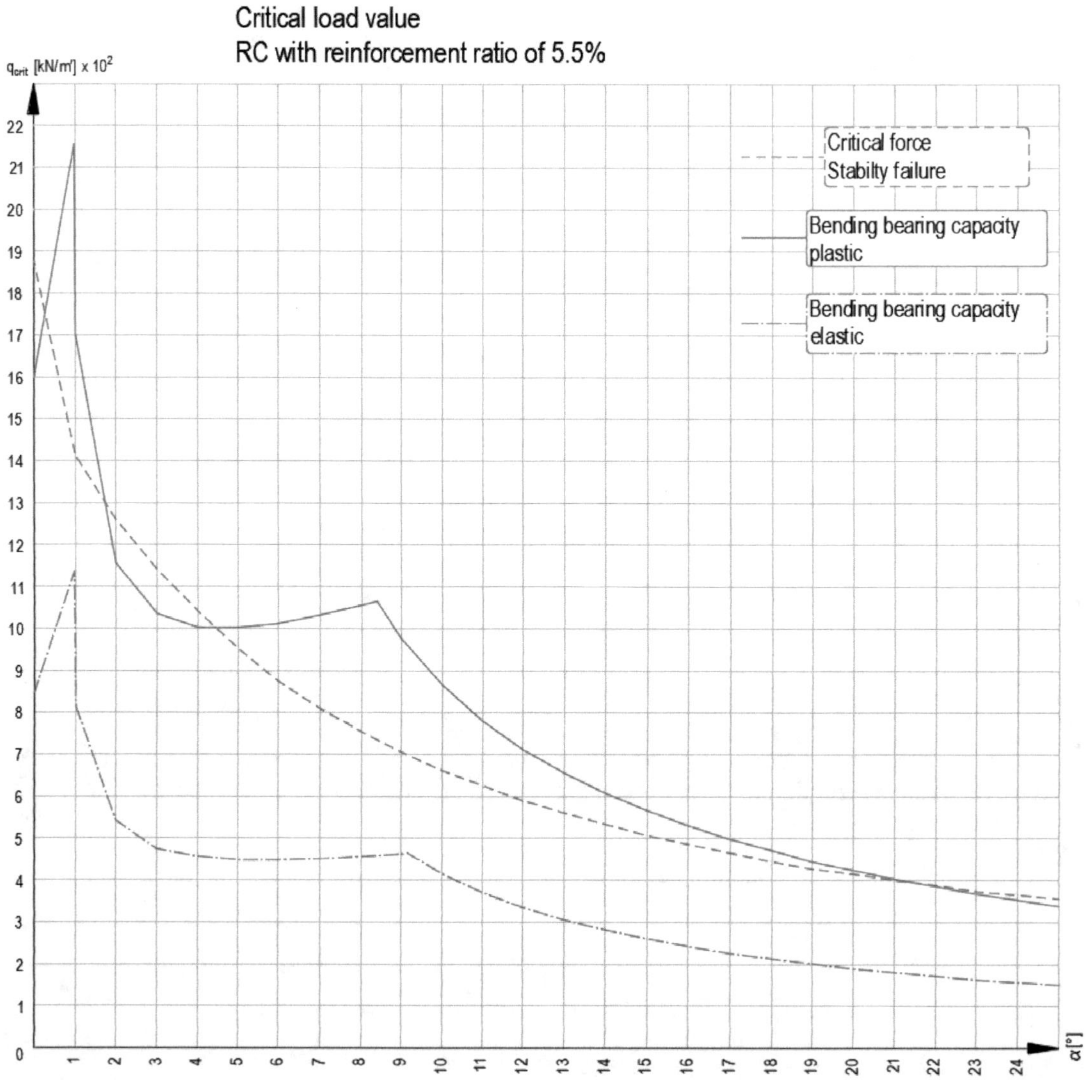

Figure 6. Interrelation between criteria of the bending bearing capacity rupture and stability failure in Reinforced concrete section.

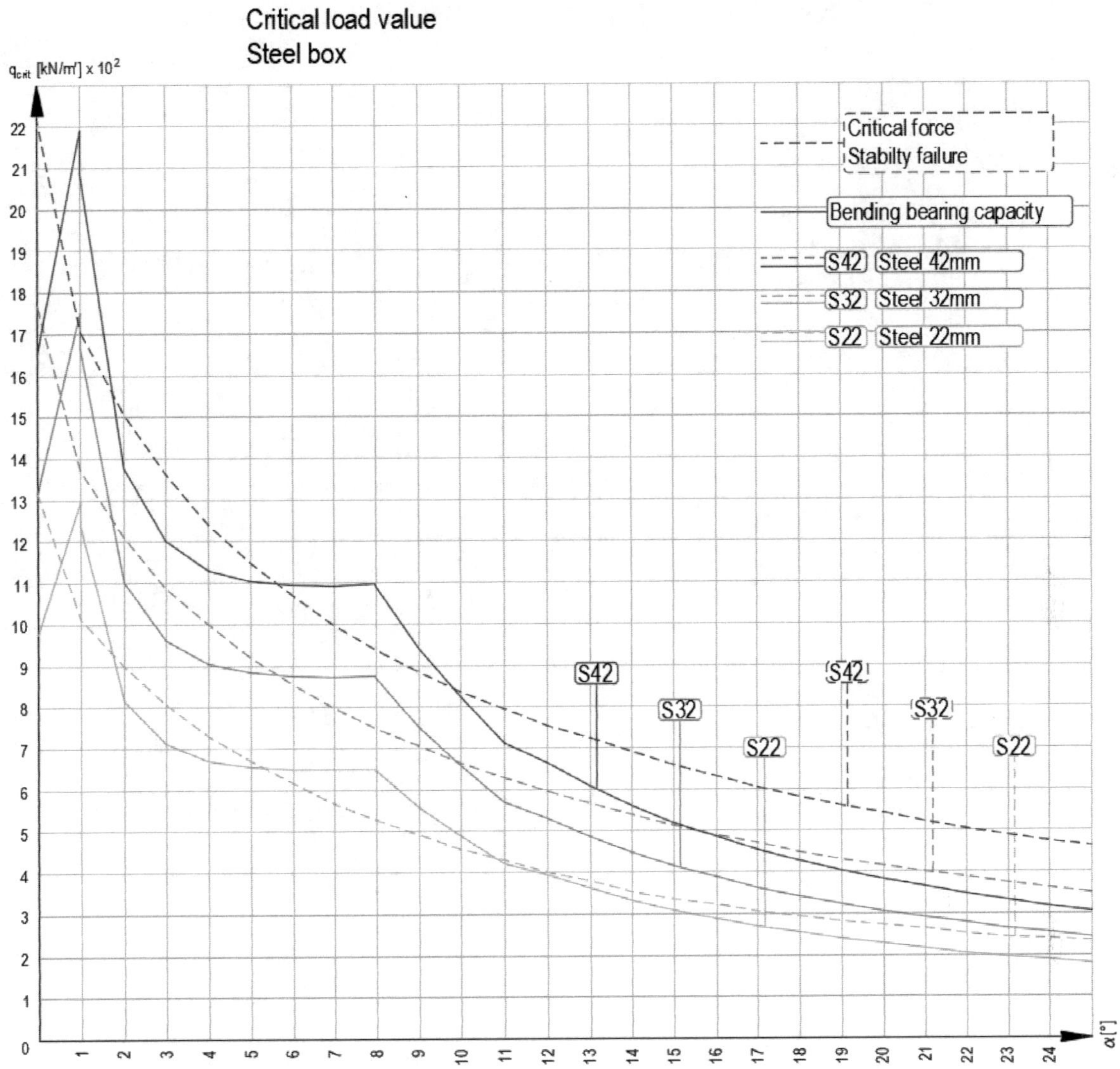

Figure 7. Interrelation between criteria of the bending bearing capacity rupture and stability failure in Steel box sections – 42mm, 32mm, 22mm.

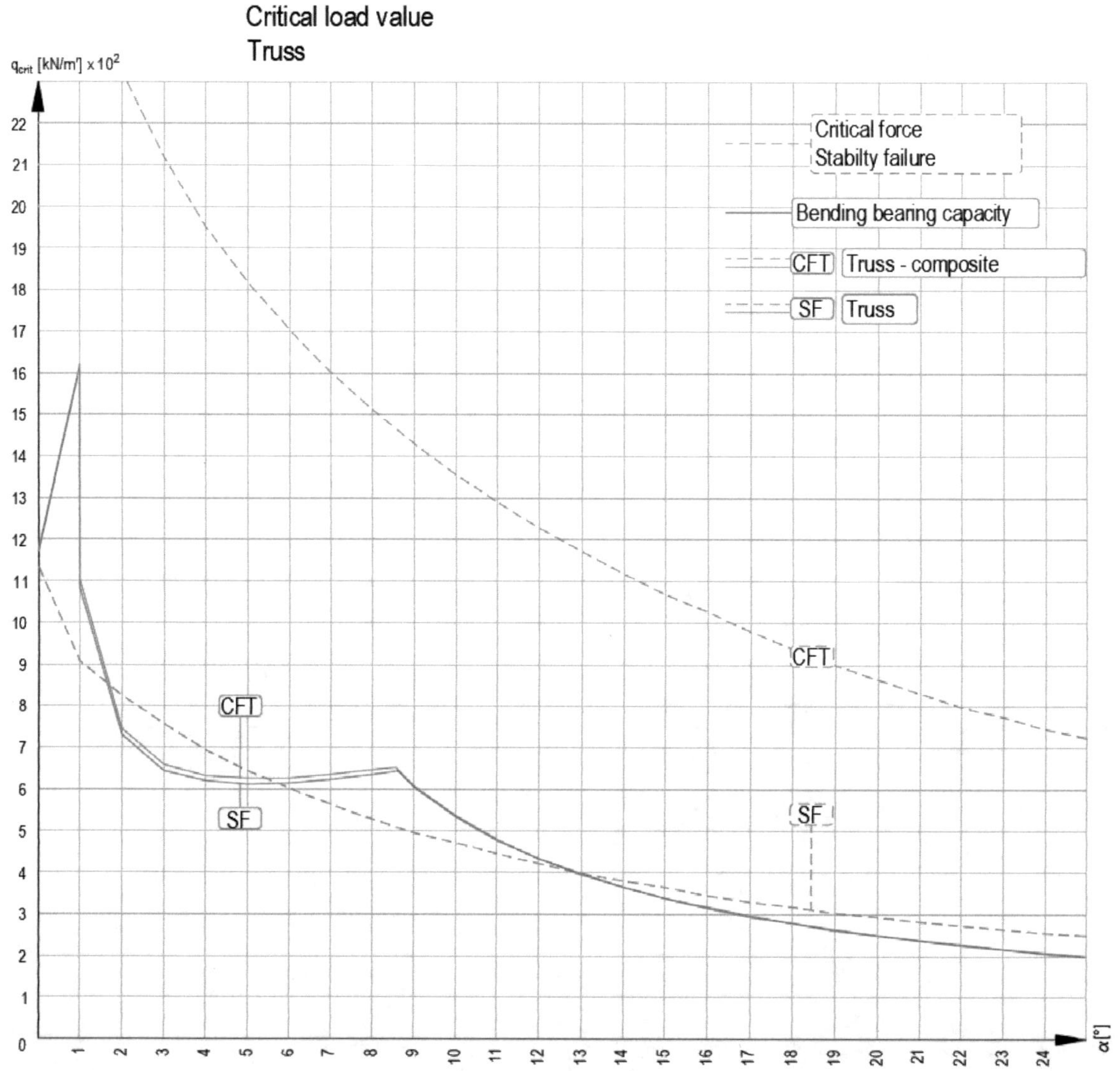

Figure 8. Interrelation between criteria of the bending bearing capacity rupture and stability failure in Truss and Truss - composite sections.

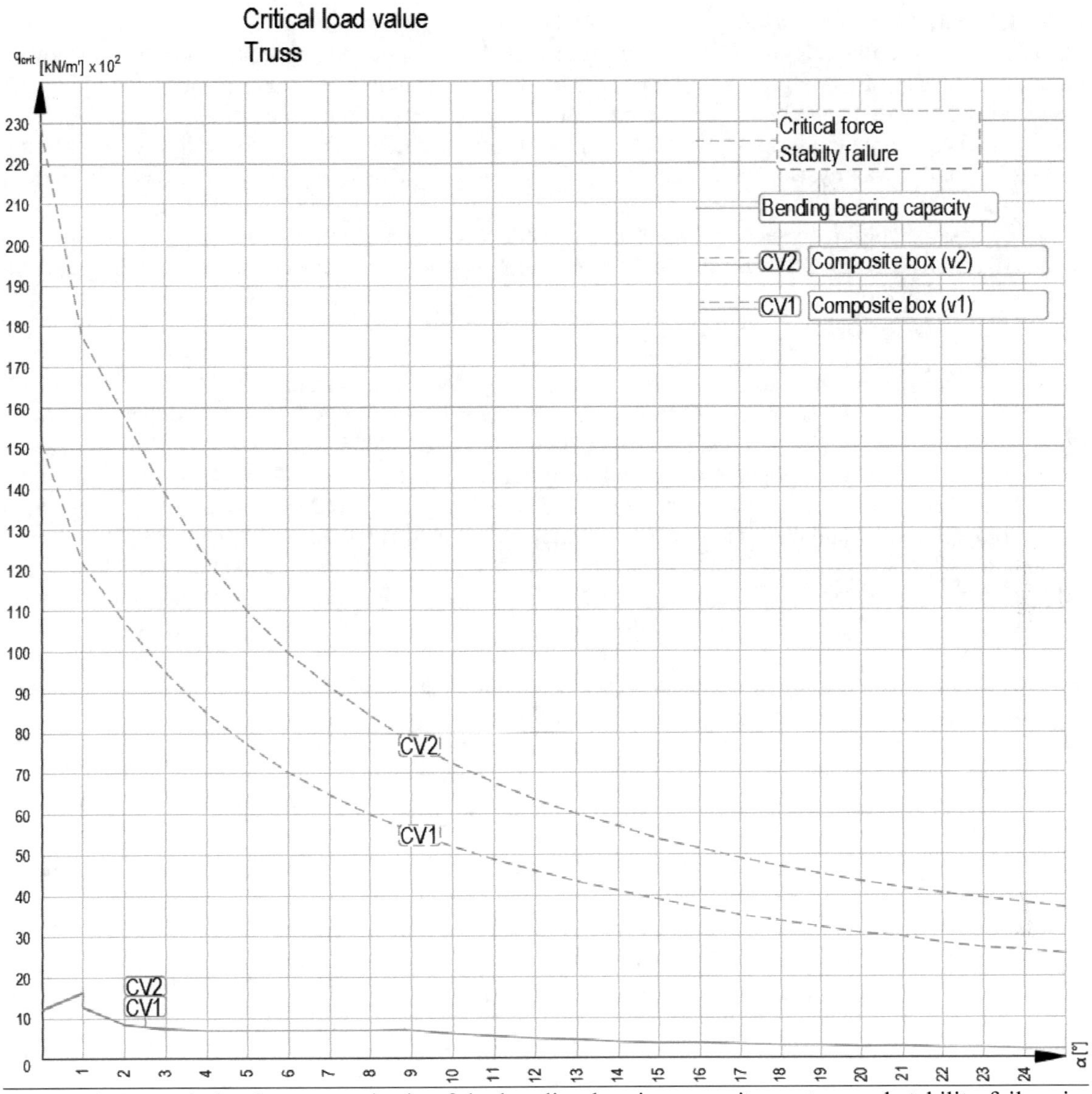

Figure 9. Interrelation between criteria of the bending bearing capacity rupture and stability failure in Composite box sections – V1 and V2.

4. Conclusion

The criteria represented in graphics from 3.2. where the bending bearing capacity rupture and stability failure curves are compared, leads to the conclusion that the bending bearing capacity out of plane of the arch is a key parameter concerning the global bearing capacity of the structure.

The intersection of the graphs proves the impossibility of determining uniform failure criteria and the bearing capacity analysis need to include both bending-bearing capacity analysis and stability failure analysis.

In the case of combined sections, there is a clear tendency for significant increase of the stability capacity. That leads to the conclusion where the determining criterion in that cases is the bending bearing capacity of the section.

References

[1] European Committee for Standardization, 2015, EN 1992-1-1:2004+A1:2014, Eurocode 2: Design of concrete structures. General rules and rules for buildings, ISBN: 978-0-580-83726-5
[2] European Committee for Standardization, 2010, EN 1992-2:2005, Eurocode 2. Design of concrete structures Concrete bridges. Design and detailing rules, ISBN: 978-0-580-64127-5
[3] European Committee for Standardization, 2009, EN 1994-1-1:2004, Eurocode 4. Design of composite steel and concrete structures General rules and rules for building, ISBN: 978-0-580-67675-8
[4] European Committee for Standardization, 2010, EN 1994-2:2005, Eurocode 4. Design of composite steel and concrete structures General rules and rules for bridges, ISBN: 978-0-580-64129-9
[5] Hendy C R and Smith D A 2007 Designers` guide to EN 1992-2. Eurocode 2: Design of concrete structures. Part 2: Concrete bridges, ISBN: 978-0-7277-3159-3 Thomas Telford
[6] Chen W., Duan L., 2014 Handbook of International Bridge Engineering, ISBN: 978-1-4398-1030-9, CRC Press
[7] ©LUSAS authors collective, 2021 Theory Manual Volume 1 Version 19.1 - Issue 1, Distributed worldwide by LUSAS - Finite Element Analysis Ltd
[8] ©LUSAS authors collective, 2021 Theory Manual Volume 2 Version 19.1 - Issue 1, Distributed worldwide by LUSAS - Finite Element Analysis Ltd
[9] ©LUSAS authors collective, 2021 Element Reference Manual Version 19.1 - Issue 1, Distributed worldwide by LUSAS - Finite Element Analysis Ltd
[10] Karnovsky I., 2009 Advanced Methods of Structural Analysis, ISBN: 978-1489983350, Springer
[11] Petel V., 2018 Concrete-Filled Stainless Steel Tubular Columns, ISBN: 978-0367656836, CRC Press
[12] Lacarbonara W., 2013 Nonlinear Structural Mechanics: Theory, Dynamical Phenomena and Modelling, ISBN: 978-1441912756, Springer
[13] Narayanan R., 2014 Steel-Concrete Composite Structures: Stability and Design (Stability and strength), ISBN: 978-1851661343, CRC Press
[14] Karnovsky I., 2011 Theory of Arched Structures: Strength, Stability, Vibration, ISBN: 978-1461404682, Springer

Inspection, structural analysis, and assessment of the existing condition of the tunnel beneath the National Palace of Culture, Sofia, Bulgaria

A K Jiponov

Road and Transport Structural Facilities Department, FTE, University of Architecture, Civil Engineering and Geodesy, 1 Hr. Smirnenski Blvd, Sofia 1046, Republic of Bulgaria
ajiponov_fte@uacg.bg

Abstract. The current paper considers and summarize the elaborated inspection of the existing condition and operational capacity of the structure of a tunnel beneath the National Palace of Culture and together with the collected data and conclusion concerning the capacity of the top slab of the structure to undertake an additional load from a secondary track from a light rail load. It also includes an assessment concerning the durability of the structure and the compliance with the regulation requirements. The report includes – review and analysis of the available design and other documentation for the site; Surveying and scanning of the main structural elements, incl. localization and registration of visible defects of the structure; performing static and dynamic field testing; structural analysis and calculation of the actual bearing capacity of the reinforced concrete slab of the tunnel according to the regulatory requirements.

1. Subject, scope of works and main activities

The present report is related to the current technical condition and assessment of the existing structural elements of the Tunnel beneath the National Palace of Culture, Sofia, Bulgaria.

The main activities included in the scope of works are described as follows:

- Review and analysis of the available archive documentation – identification of the structural system, results of past inspections, description of implemented measures for rehabilitation of the structure, definition of deficiencies and non-compliances with current regulations;
- Surveying of the geometric parameters of the tunnel structures, survey points, numbering and marking of the surveyed objects, laying of a reference network.
- Visual inspection and register of the key elements of the tunnel structure – cracks, delamination, deformations, lack of concrete cover, leaks, corrosion of reinforcement steel bars, changes in concrete structure, the presence of unacceptable deformations, degree of neutralization, adhesion between concrete and steel bars;
- Non- Destructive Testing (NDT) methods – Ferroscan, Rebar detector, Ultrasonic device, Rebound hammer;
- Core drilling – Concrete Core Sample Testing, Insulation Sample Testing;
- Perform Static and Dynamic on-situ tests - Inductive Standard Displacement Transducers, Strain gauge bridge amplifier;
- Calculation of actual bearing capacity of the superstructure of the tunnel based on the FEM analysis and comparation of the results from the static and dynamic tests [5], [6].

2. Visual inspection and scanning of the tunnel via Lidar

2.1. Visual inspection

The visual inspection of the tunnel is carried out in order to make a preliminary assessment of the existing condition of the structure, based on external signs and to establish the need (or lack thereof) for a detailed inspection with NDT and Core drilling.

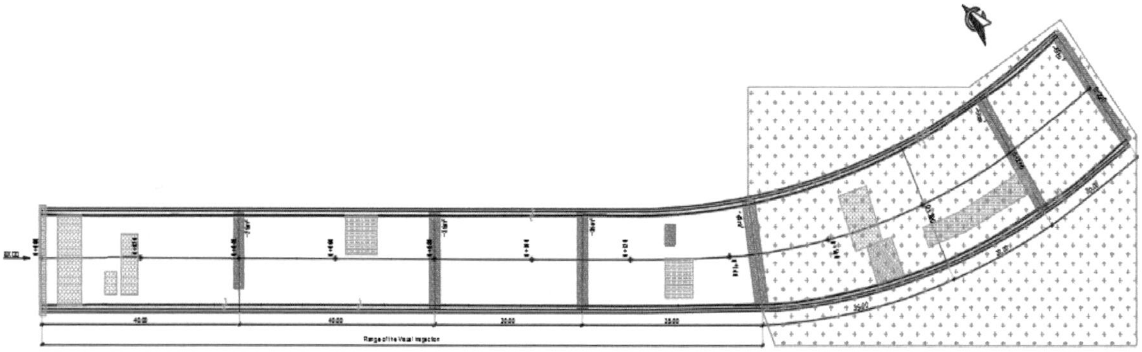

Figure 1. Range of the visual inspection.

The basis of the visual inspection is an identification of characteristic defects and damages on the structure. The following indicators are considered during the technical visit:

- Visible deformations of the structural elements - significant (visible) deviation of the horizontal elements (slabs, beams) and the vertical (columns, walls);
- Establishment of visible defects on the concrete surface from the structural ones – collapse, shrinkage signs, presence of areas with uncompacted or layered concrete, presence of joints, etc. similar;
- Establishing the presence, location (orientation relative to the sides of the fields) and type of cracks in the structure (if any), measuring the width of the cracks in individual sections along them;
- Identification of sections with exposed reinforcement and / or unprotected reinforcement (if any), and condition of the reinforcement bars regarding corrosion processes;
- Identification of visible changes in the structure of concrete caused by atmospheric, exploitation or other types of impacts (if any);
- Presence of leaks, filtration of water through the volume of concrete and extraction of free lime.

From the following activities were carried out, some findings could be stated:

- on the lower surface of the ceiling slab is not found the presence of visible deformations, neither local nor of the structure as a whole;
- the surface of the concrete is smooth, without variations in colour (diversity) except for local areas, which is an indirect sign of consistency of the composition of concrete and the technology of concrete work;

Figure 2. Lower surface of the slab - general view.

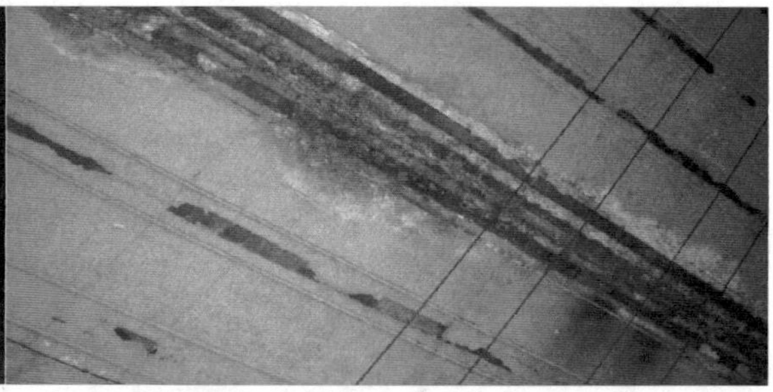

Figure 3. Expansion joint area.

- the presence of visible cracks on the surfaces of the slab from shrinkage ore creeping has not been established;
- there are no areas with uncompacted and / or layered concrete on all inspected surfaces;
- areas with corroded concrete which has changed its structure under the atmosphere effects or t gases from vehicles are not visible;
- the slab area, located between the expansion joints, no traces of water filtration are found through the volume of the concrete, except for isolated local areas;
- the slab area, located between the expansion joints, the presence of sections with exposed or unprotected reinforcement has not been established, except for isolated local areas;
- the slab area, located in the zones of the expansion joints, massive leaks were found, which are active during the inspection;
- in the zones of the expansion joints, there is a filtration of rainwater through the volume of the concrete, combined with the extraction of free lime;
- in the areas of the expansion joints, there is a corrosion on the surface of the reinforcement;

Figure 4. Corrosion of reinforcement in the area of expansion joint.

Figure 5. Active leakage and extraction of free lime from the concrete nearby the expansion joint.

- The colours of the concrete follow the tram route. This implies the presence of compromised waterproofing, a consequence of the dynamic load of tram traffic;
- The registered cracks do not have structural origin and do not assume compromised load-bearing capacity of the structure. Given the static scheme of the structure, the cracks are in a lightly loaded area. Leaks in cracks area have led to their expansion and deepening.
- The concrete cover of the slab has been compromised in several places due to leaks. Depending on the degree of water saturation, the insulation system of the structure is compromised, the concrete is carbonized, the concrete cover layer collapses or it is completely lost.

2.2. Laser scanning of the structure

With the help of a laser scanner, a database was prepared to generate a three-dimensional "point cloud" model. The data from the model were used as a control parameter of the surveying and extension of its scope.

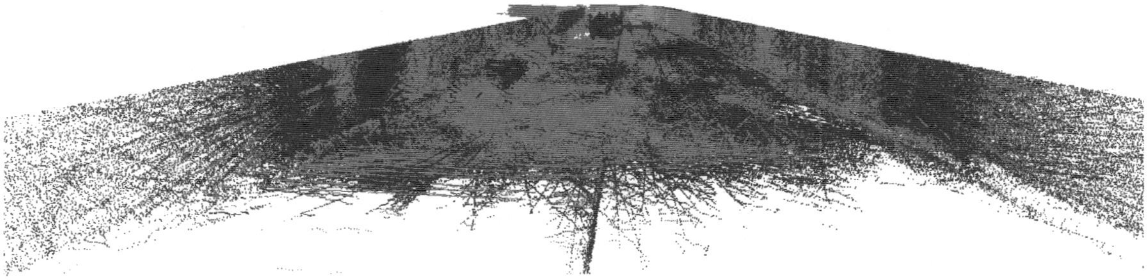

Figure 6. Laser scan data visualization.

The results were used to validate geometric data and positions of temporary structures and facilities serving the needs of the field test.

IOP Conf. Series: Materials Science and Engineering 1276 (2023) 012017 doi:10.1088/1757-899X/1276/1/012017

3. Results from NDT methods and laboratory tests

The purpose of the NDT methods and Laboratory tests is to evaluate the used materials, to determine the location and the quantity of the reinforcement and to establish the condition of the waterproofing.

3.1. Determining the degree of corrosion of the reinforcement in the concrete and adhesion in the concrete in a non-destructive way

In order to determine the condition of the reinforcement in the areas where the concrete cover is not disgraced and the condition of the reinforcement cannot be determined visually, an instrumental inspection with a corrosion analysis device was performed. The condition of the reinforcement in the reinforced concrete elements is determined with the help of a specialized device "CANIN" (Corrosion Analysing Instrument). The obtained results show that the reinforcement in the tunnel slab is in a passivated state, except for the areas around the expansion joints. In these neuralgic parts of the slab, the reinforcement is covered by corrosion processes, which are visible on visual inspection, as the entire surface of the reinforcing steel is covered with corrosion

3.2. Determination of the diameter and distance between the tension reinforcement bars of the tunnel – longitudinally and transversally

Information for the thickness of the concrete surface, the location and the diameter / number (N) of the reinforcement was sought using a specialized device called – Ferroscan.

The accuracy of the device, according to the information from the data manufacturer: for diameter of the reinforcement - ± 1 mm, for the location of the reinforcement - ± 3 mm.

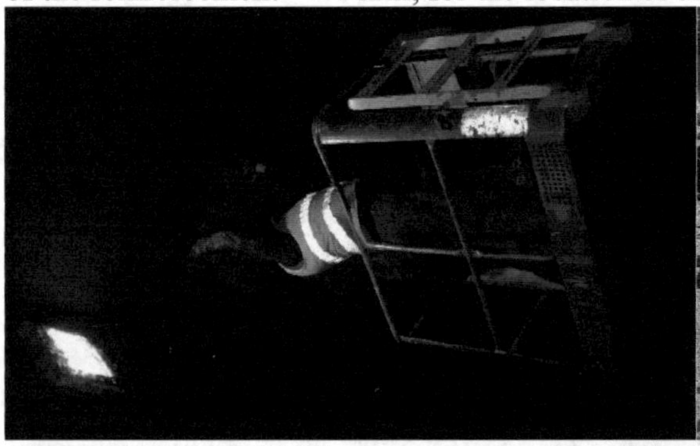

Figure 7. Scan of the bottom surface of the slab.

Figure 8. Scan of the top surface of the slab.

In the scanned sections with dimensions of 600/600 mm, reinforcing bars with a nominal diameter of 32 mm, class A-III are found, which are located in about 10 cm, in the direction of the supporting distance of the top slab. Above them there are second and third rows of reinforcement. The second row of reinforcement is located perpendicular to the supporting distance of the slab, located in about 15cm. The third row is in the same direction as the first, with a nominal diameter of 32 mm, class A-III, located in the same step.

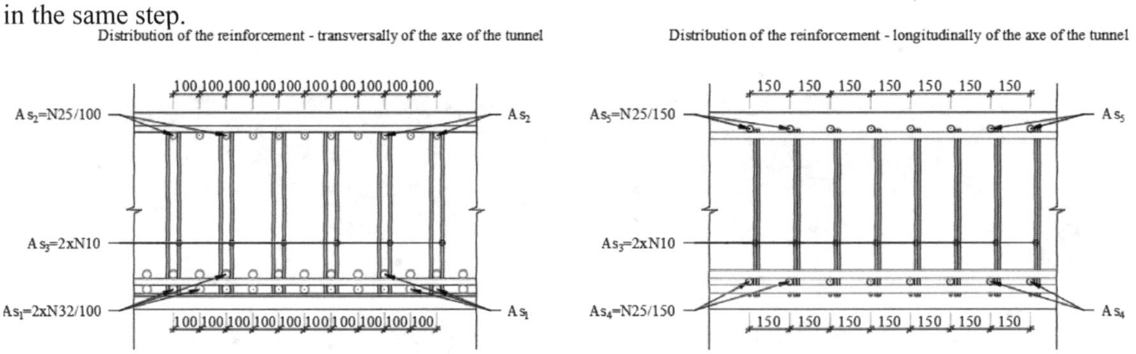

Figure 9. Distribution of the reinforcement in the slab.

3.3. Determining the diameter of the stirrups, their shear length along the axis of the tunnel and the distance between them transverse to the axis of the tunnel

In the scanned area with dimensions of 600/600 mm, the presence of reinforcement against shear - type of stirrup, located in about 10 cm. Based on the analysis of the samples taken, the presence of paired stirrups with a nominal diameter of 10 mm, made of steel grade A III, was determined.

3.4. Determining the compressive strength and condition of concrete by Core drilling

The preparation and testing of the samples, and the evaluation of the results are done in accordance with the requirements of the normative documents.

The drilling cores are cut from the structure with the help of a machine. Drilling is performed perpendicular to the concrete surface so as not to damage the cores (see the figures below):

Figure 10. Taking a sample from the bottom surface of the tunnel slab.

Figure 11. Taking a sample from the top surface of the tunnel slab.

The following measurements shall be made before the test – the diameter, length, mass of the samples

Figure 12. General view of the aligned samples.

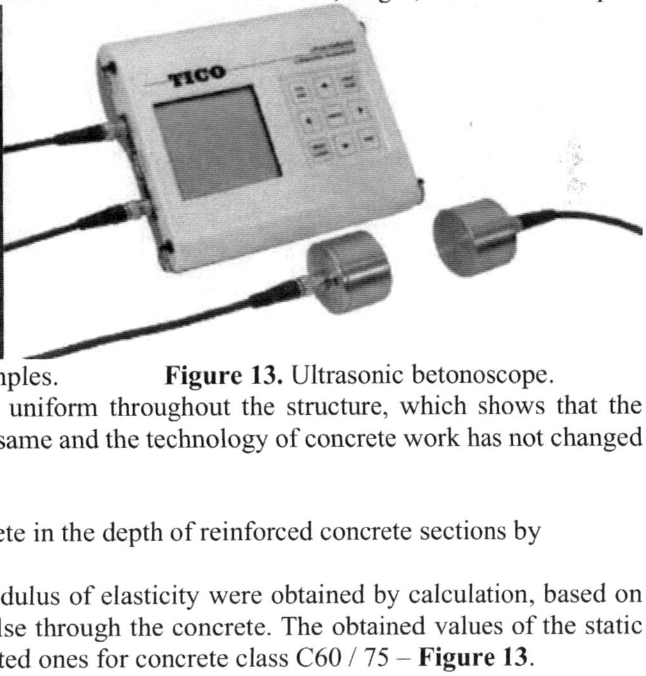

Figure 13. Ultrasonic betonoscope.

The density of the concrete is high and uniform throughout the structure, which shows that the content of the concrete is approximately the same and the technology of concrete work has not changed during construction.

3.5. Determination of the structure of concrete in the depth of reinforced concrete sections by ultrasonic testing

The values of the dynamic and static modulus of elasticity were obtained by calculation, based on the measured velocities of the ultrasonic pulse through the concrete. The obtained values of the static modulus of elasticity are close to the calculated ones for concrete class C60 / 75 – **Figure 13**.

3.6. Assessment of the condition of the waterproofing and protective layers on the structure

The condition of the waterproofing can be defined as satisfactory, excluding the areas of expansion joints where repair works are needed to ensure their water tightness.

4. Determination of the bearing capacity of the top slab of the top slab of the tunnel structure based on Finite Element Method

The calculation checks aim to prove the bearing capacity and serviceability of the top slab of the inspected tunnel. The used models and standards for loading exceed the current ones and those envisaged in the future operation, due to additional load from the displacement of the light rail route from the local lane of "Gen. MD Skobelev".

The tunnel is about 460m long. It is built via "cut and cover" method, using standard construction technologies. From the ramps inside to strengthen the excavation in front of the existing high-rise buildings, reinforced concrete piles of the "Benoto" type ∅1000mm were made. In the middle zone was used a common foundation slab together with the construction of the National Palace of Culture and the metro tunnel.

The geometry modelled for the analysis is based on the information collected from archive data and a visual inspection:

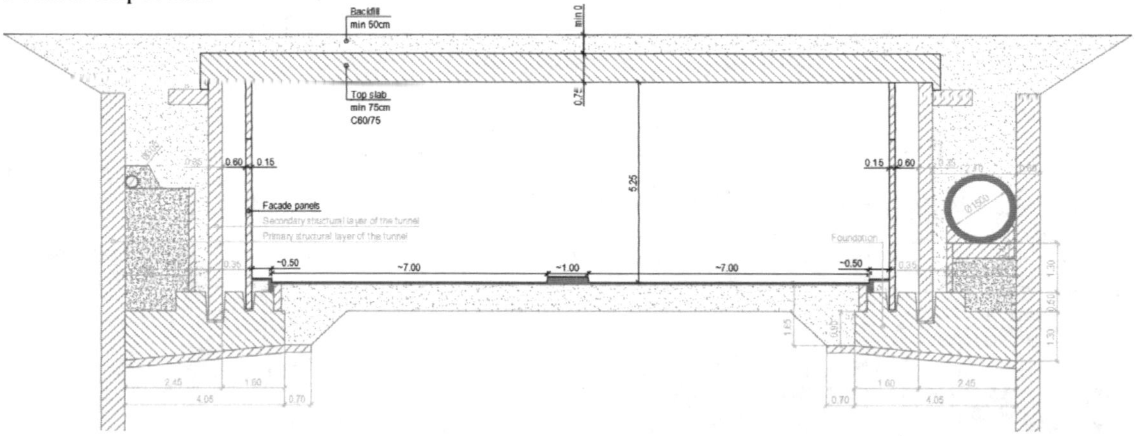

Figure 14. Typical cross – section.

The LM1, is applicable for a traffic load [7]. This load model generates results for the calculation checks, which exceed those of the special wheel NK-80 according to [4] BDS 1050-72 "Loads movable for calculating road bridges". This is in favor of safety and is due to the difference in axial loads.

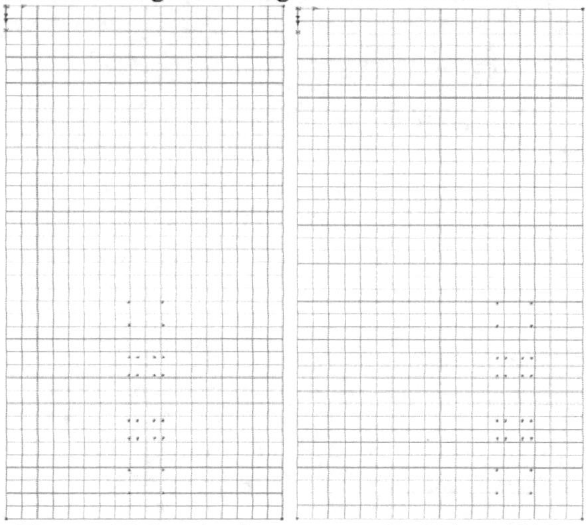

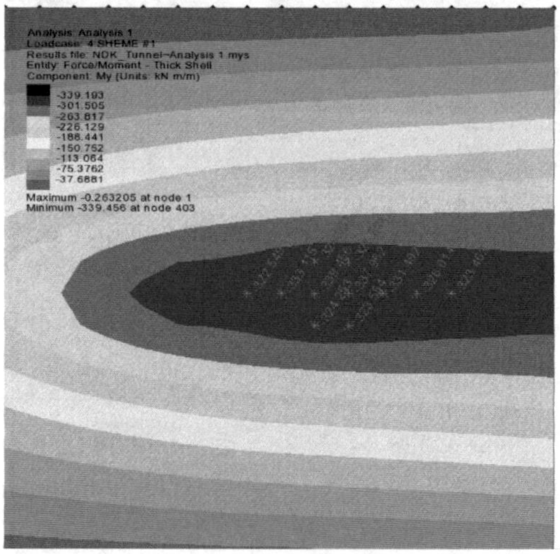

Figure 15. Load scheme scenarios №1 and №2. **Figure 16.** Ultrasonic betonoscope.

The results of the performed calculation checks prove that the top slab of the tunnel meet the serviceability requirements and has a bearing capacity to withstand an additional load from a second light railway track from the local lane on "Gen. MD Skobelev" [8].

IOP Conf. Series: Materials Science and Engineering 1276 (2023) 012017 doi:10.1088/1757-899X/1276/1/012017

5. Static and dynamic in – situ tests

The test is carried out on site, providing working conditions, scaffolding, installation of sensors and load on the structure. According to the STATIC AND DYNAMIC TEST PROGRAM [3], the test of the superstructure is performed, which includes:

- Static field test in two sections;
- Dynamic tests of the top slab.

The field test of the construction was conducted during the night in the period from 00:00 to 4:00. Given the conditions of the site, the following test environment was adopted with monitoring of deformations at 4 points, located in two key sections of the structure. Two dynamic studies with dynamic load passing at different speeds through the middle of the structure have been prepared. The static test was performed in two sections (**Figure 17**) with three load scheme scenarios – **Figure 18**, **Figure 19**, **Figure 20**

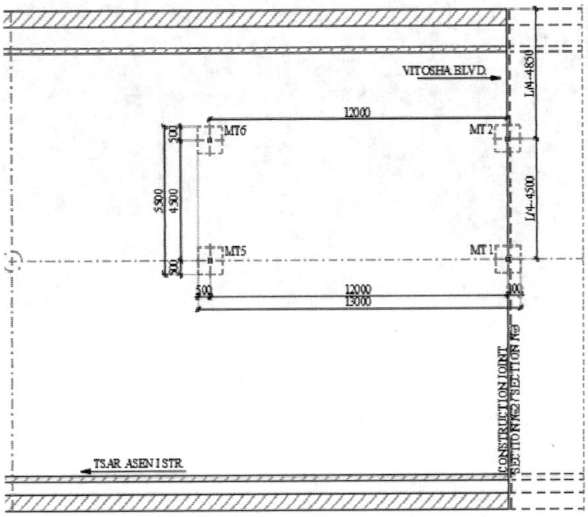

Figure 17. Position of measuring points.

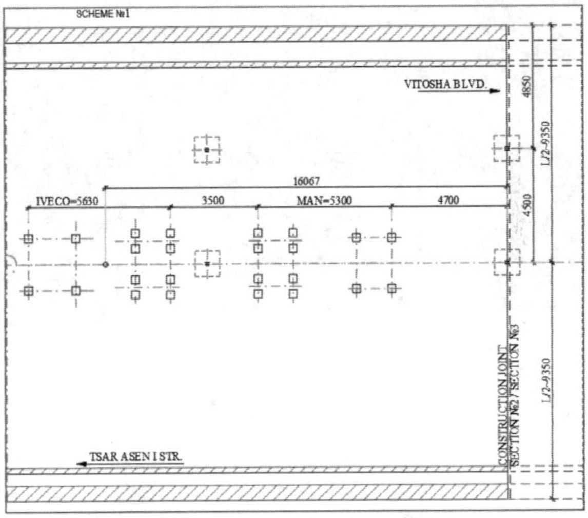

Figure 18. Load scheme №1.

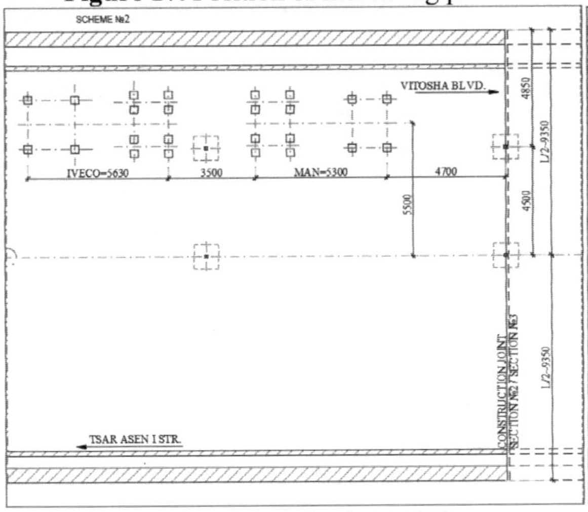

Figure 19. Load scheme №2.

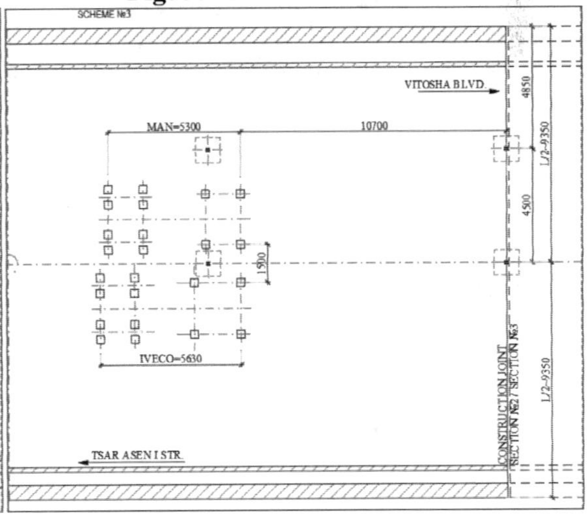

Figure 20. Load scheme №3.

Figure 21. Photos from the Static and Dynamic tests.

5.1. Static test

The relative deformations in the sensors previously placed at the measuring points were measured. Numerous reports have been recorded in order to mark the deformation of the structure over time. The records are presented from the state of rest of the structure to the attenuation of deformations in the presence of load.

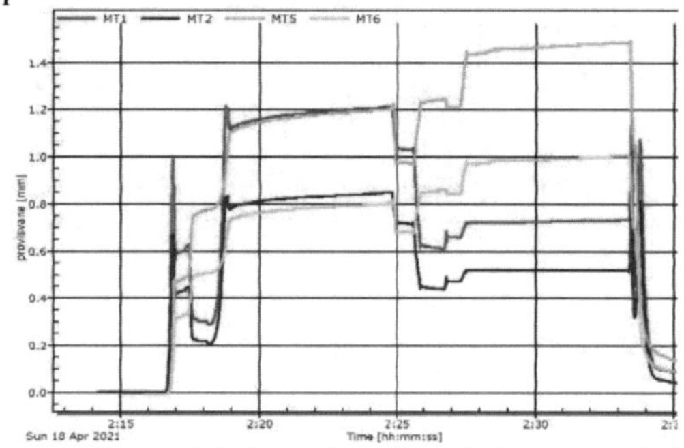

Figure 22. Results of static load test №1.

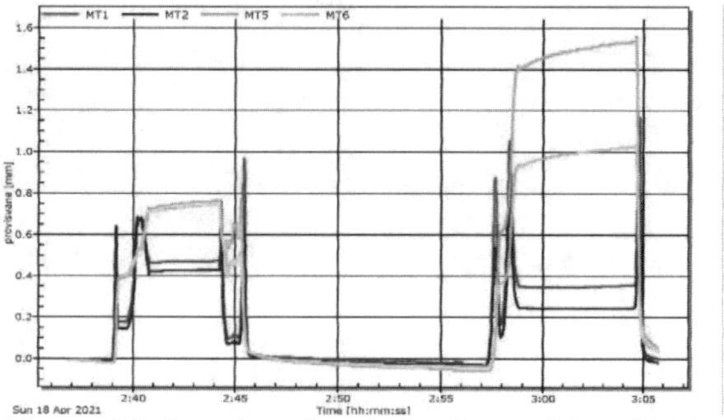

Figure 23. Results of static load test №2 and №3.

Maximum displacements presented for the respective load schemes are as follows:

Load Scheme №1	MT6 = 1.01mm,	MT5 = 1.49mm,	MT2 = 0.52mm,	MT1 = 0.74mm
Load Scheme №2	MT6 = 0.74mm,	MT5 = 0.76mm,	MT2 = 0.43mm,	MT1 = 0.47mm
Load Scheme №3	MT6 = 1.02mm,	MT5 = 1.54mm,	MT2 = 0.24mm,	MT1 = 0.36mm

5.2. Dynamic test

The dynamic test is performed by passing a dynamic load through a pre-set obstacle in order to induce forced oscillations in the structure. In order to control the results, the test was performed at two different speeds of the dynamic load - 15 and 30 km / h.

According to the reports made, the natural frequency of the construction is approximately 4.5 ÷ 5.0 [Hz] with a period of 0.25 ÷ 0.20 [s].

6. Conclusions

As a result of the performed researches the following conclusions can be made:

- Significant leaks are present in the connection areas. Leaks are typical vulnerable zones for structures with separate structural sections. The presence of leaks indicates for compromised waterproofing in the areas of connections. The observed water quantities assume for a compromised drainage system or drainage / sewerage system. The damages from the leaks are in an advanced stage, which implies prolonged exposure to aggressive agents (water leaks). In some places, water catchment and drainage measures have been observed. The implemented measures are ineffective and have compromised functionality. The damages are significant for the durability and serviceability of the tunnel;
- Up to the date of the performed inspections, there are no prerequisites suggesting reduced load-bearing capacity of structural elements. To rehabilitate the serviceability of the structure, the causes of leaks must be removed, and the water tightness of the waterproofing system restored;
- Damages to the structure are present in the areas of connection between the structural sections, as a result of continuous leakage. The integrity of the reinforced concrete elements is failed, the reinforcing bars are visible, there are marks of advanced corrosion of both concrete and steel. Impaired integrity of the load-bearing reinforcement has not been established. The damages are significant for the durability and serviceability of the tunnel;
- Up to the date of the performed inspections, there are no prerequisites suggesting reduced load-bearing capacity of structural elements. To recover the durability of the structure, the affected elements must be treated, their integrity restored and detailed to ensure the necessary parameters for the operating conditions of the working environment;
- Concrete discolorations due to corrosion of the concrete due to infiltrated water are present in local areas of the top slab. The zones are not widespread in the mass part of the roof structure. Their location is in the range of the tram route passing from above. The parameters indicated in this way suggest a compromised base of the track, which led to the appearance of local defects in the waterproofing layer. The distribution of the zones is wide, without the presence of advanced corrosion of concrete. This implies long-term exposure to aggressive environments, but in minimal quantities. Given the above, it can be assumed that the defects in the waterproofing are minimal, but the time range of action is long and there is a significant retention of water masses in the area of the roof slab, signalling a compromised drainage system;
- Up to the date of the performed inspections, there are no prerequisites suggesting reduced load-bearing capacity of structural elements. To restore the durability of the structure, the tram route must be revised, and its structure rehabilitated. Restoration of the functionality of the drainage system and the water tightness of the waterproofing system must be considered;
- In local areas of the top slab there is a lack of concrete cover. The defects are local and are not characteristic of the overall condition of the structure. There are no marks of leaks in the area up the date of the inspection and the damaged cover is not presumed to be due to corrosion of the concrete. Exposed reinforcement bars show no significant loss of material. The connection between concrete and steel is not broken;
- Up to the date of the performed inspections, there are no prerequisites suggesting reduced load-bearing capacity of structural elements. To restore the durability of the structure, the concrete cover and the surface protective layer in the compromised area must be restored;
- Cracks are present in local areas of the roof structure. The cracks are in an area with typically low stress values. There are signs of leaks and corrosion of the concrete in the crack areas. The observed corrosion is not at an advanced stage and does not compromise the bearing capacity of the element. According to archival data, there is a utilities / ventilation opening near the mentioned section, confirmed by the presence of vents / inspection openings. Given the observed damage, the exposure to aggressive agents is significant and the corrosion of the concrete is greatly accelerated;

7. Proposal for refurbishment of the tunnel

Repair and refurbishment activities are recommended to be carried out in the following form and order:

7.1. Overall measures

- Full disclosure of the top slab in the range from the entrance portal of the structure to the intersection with Vitosha Blvd.;
- Revision, repair and, if necessary, reconstruction of the water supply /sewerage systems;
- Demolition of the existing waterproofing insulation;
- Reconstruction of the connection areas between the individual structural blocks (joints). Restoration of the surface of the reinforcing bars with the initial stages of corrosion. Treatment of compromised areas to ensure the necessary adhesion and connection between old and new concrete. Restoration of the concrete cover to ensure the serviceability of the structure;
- Application of a functioning solution guaranteeing the necessary range of dilatation and water tightness of the facility. Connections between sections with and without repair works are carried out according to adequate detailing;
- Construction of a new waterproofing system to ensure the water tightness of the structure;
- Construction of a drainage system for catching infiltrated waters and reconstruction of the ROAD drainage in the area of the tunnel;
- Implementation of a protective layer of concrete to protect the waterproofing or an alternative method of protection;

7.2. Refurbishment measures for the top slab

- The lower surface of the structure is cleaned and treated to ensure a suitable basis for taking measures to preserve the slab structure.
- Areas with leaks, cracks, damaged concrete surface are additionally treated with appropriate measures – cleaning and refurbishment of concrete surface, sealing of cracks.
- Treatment of the entire lower surface of the slab with a system for preservation of the structure, protecting it from the negative impact of the environment.

References

[1] European Committee for Standardization, 2015, EN 1992-1-1:2004+A1:2014, Eurocode 2: Design of concrete structures. General rules and rules for buildings, ISBN: 978-0-580-83726-5
[2] European Committee for Standardization, 2010, EN 1992-2:2005, Eurocode 2. Design of concrete structures Concrete bridges. Design and detailing rules, ISBN: 978-0-580-64127-5
[3] Instruction for load testing of Road Bridges, 1980г. – ИИПМ-80 (in Bulgarian)
[4] BDS 1050-89 "Loads movable vertical for calculating road bridges" (in Bulgarian), 1989, TK-56
[5] Narayanan R S and Beeby A 2005 Designers` guide to EN 1992-1-1 and EN 1992-1-2. Eurocode 2: Design of concrete structures. General rules and rules for buildings and structural fire design ISBN: 07277 3105 X Thomas Telford
[6] Hendy C R and Smith D A 2007 Designers` guide to EN 1992-2. Eurocode 2: Design of concrete structures. Part 2: Concrete bridges ISBN: 978-0-7277-3159-3 Thomas Telford
[7] Topurov K., 2012 Reinforced concrete bridges, ISBN: 978-954-8873-93-2, ABC Technic
[8] Nikolov P 2012 Modelling of structural and non-structural elements of bridges International Jubilee Conference UACEG2012: Science & Practice

IOP Conf. Series: Materials Science and Engineering 1276 (2023) 012018 doi:10.1088/1757-899X/1276/1/012018

Helicopter service of hospital building

S Tsvetkov

University of Structural Engineering and Architecture "Lyuben Karavelov" Bulgaria, Faculty of Construction, Department "Building Structures", 1373 Sofia, 175 Suhodolska str.

E-mail: st.cvetkov@vsu.bg

Abstract. The report considers two variants of a building with a special purpose of the roof slab. Option 1 is a model of the building - without a helicopter pad, and Option 2 has a helicopter pad. For basic structural elements, the values of: displacements, shear forces and load-bearing reinforcements are shown. Specialized software for computational models is used. Combinations of impacts are proposed, including dynamic loading. The solutions of the two options are compared and a conclusion is made about the efficiency of the solutions with the helicopter service. The maximum values of the internal forces are obtained by taking into account together the horizontal and vertical components of the seismic impact.

1. Introduction

Nowadays, in Bulgaria, the operation of helicopters and their application in various fields (private transport, tourism, transport of goods and goods, etc.), including ambulances, is especially important. The type of "ground" parking with reinforced terrain is often used, given the reduced construction costs. "Raised" podiums are used for building constructions, with developed strong steel or combined constructions, under heavy impacts. Sometimes such projects reach significant economic values and make design difficulties.

In addition for up-ground parking, the roofs of buildings are also used for helicopters - and this is also used as a symbol - the emblem of the building. Therefore, the structures of the buildings on which the helicopters land, it is necessary to perform and design with quality. The roof sites of: Grand Hotel Millennium - Sofia (the first hotel with such a site) and Mall Varna - Towers are widely known in the country. The conceptual designs are, for example: the one for the construction of a landing site for modern helicopters on roof the parking lot of the University Hospital "St. Georgi "- Plovdiv, with a direct corridor (overpass) and a connection with the Emergency. Another solution is the extravagant project for a business center on the Bulgaria boulevard and the bypass of the capital (see figure 1).

Figure 1. a). [1]

Figure 1. b). [2]

c). [3]

d). [4]

Figure 1. Examples from Bulgaria - realized: a). and b).; conceptual: c). and d).

2. General information about the object of the study

A 9-storey hospital building in the city of Pleven (Bulgaria) is considered [5]. The total built-up area is 11094,30 m^2. The structure is monolithic, skeletally beamless (with a plate thickness of 26 *cm*), and the structural system - with shear walls (with 3 independent walls and 3 cores, with 25 *cm* wide) – in figure 2. Beams have been constructed on the periphery of the building. The materials are accepted for the structural elements: concrete grade C30/37 and reinforcing steel grade B500B, by Bulgarian State Standart BDS-EN (European norms).

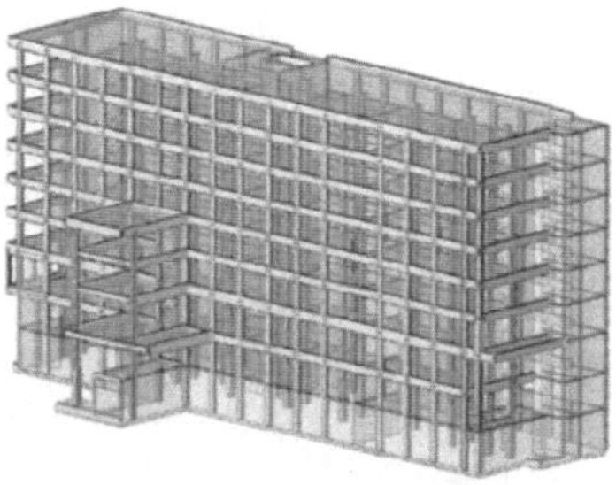

Figure 2. 3D model – visualization.

3. Basic analysis [6]

Model 1: The analysis was performed with the software product Radimpex Tower 7, based on work on the FEM. Impacts - constant and variable, as well as seismic, are accepted on the basis of architectural plans and zoning of the site. The seismic analysis (with Modal – see table 1) was performed for medium ductility class - design level - DCM (with two critical levels in the walls, above the rigid basement), preceded by Modal analysis (with 24 forms of own oscillations reported - with activated 90,07% and 92,11% modal masses - percentage participation in directions X and Y).

Table 1. Load factors for mass calculation – Model 1

№	Load case	φ.ψ	№	Load case	φ.ψ
1	Own weight construction	1	7	Permanent impacts - floors	1
2	Permanent impacts - for ground floor and basement	1	8	Permanent impacts - roof	1
3	Variable impacts - for ground floor	0,48	9	Variable impacts - floors	0,3
4	Technological equipment	0,50	10	Variable impacts - roof	0,3
5	Snow	0	11	Reverse embankment	0
6	Walls	1			

A period T_1 = 1,41 *s* was received. According to the performed checks: the construction is regular in plan and is not a torsion-deformable system. The coefficient of behavior q=3. For the city of Pleven the reference acceleration a_{gR}=0,11.g. In addition, the area is affected by vibrations from Vrancea (Romania), so the building has been studied for Specters Kind 1 and Kind 3. The foundation is solved with a common slab (80 *cm* - thick), in a layer of clay (classified as type C), with conditional calculated soil resistance 300 kN/m^2 and Winkler constant k_W=28000 $kN/m^2/m'$- for static (vertical) impacts and three times increased value - for seismic impacts, according to the engineering-geological assignment. The coefficient of significance is accepted γ_I=1,4. The attenuation is accepted 5% of the critical (coefficient ξ=0,05 - for reinforced concrete structures). Accidental eccentrics e_a=±0,05.L_i were taken into account in the analysis. The checks for the interfloor displacements and deformations have been performed, they are below the admissible maximum values.

Other solutions of constructions, for: types analysis, decisions and technology, there are in: [7], [8], [9] and [10].

4. Checks

Model 2: If (at the new design) helicopter service of the building is required, new computer modeling is required. The study was performed for impact from a helicopter with a maximum weight of 3700 *kg* and a power of 2 x 894 horsepower, at a speed of 230 *km/h*. The capacity of the machine is for patients: 1 bed and 1 seat. For this check: extended in height are the walls for the elevator, which will connect the roof - storeys, as corrected and the construction of the roof slab - from beamless to beam with the construction of a beam grid (on the right side of the plan) in front of the elevator. The thickness of the slab is 25 cm, the beam grid (with overall dimensions - *L/B* =18/12,50 *m*) is accepted with main beams *b/h* =35/75 *cm* and secondary beams - *b/h*=25/65 *cm* (see figures: 3, 4, 5 and 6).

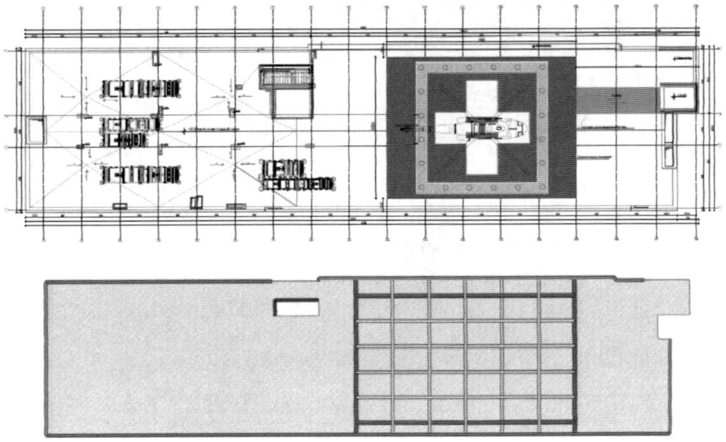

Figure 3. Site location – plans.

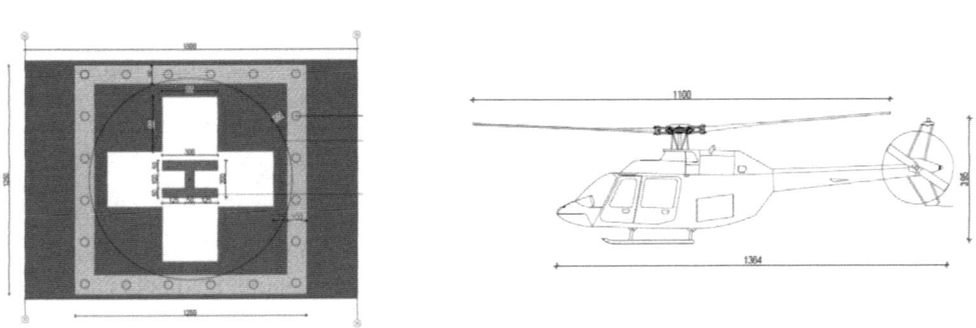

Figure 4.: a). Helicopter pad - dimensions and shaping.; b). Helicopter – dimensions.

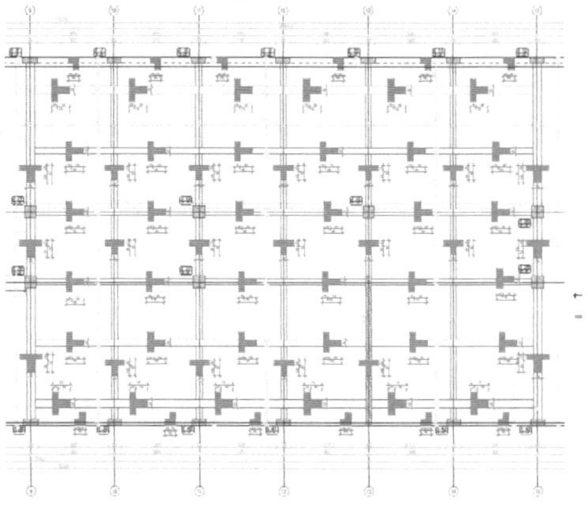

c). Formwork scheme (plan).

Figure 4. Main architectural-technological and constructive features and characteristics.

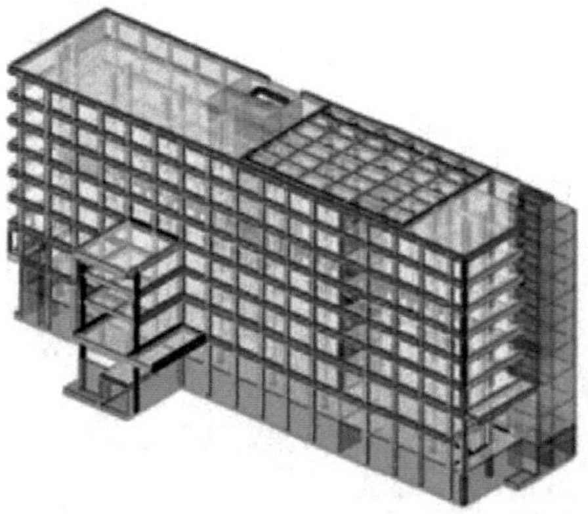

Figure 5. 3D model with visualization of the roof structure.

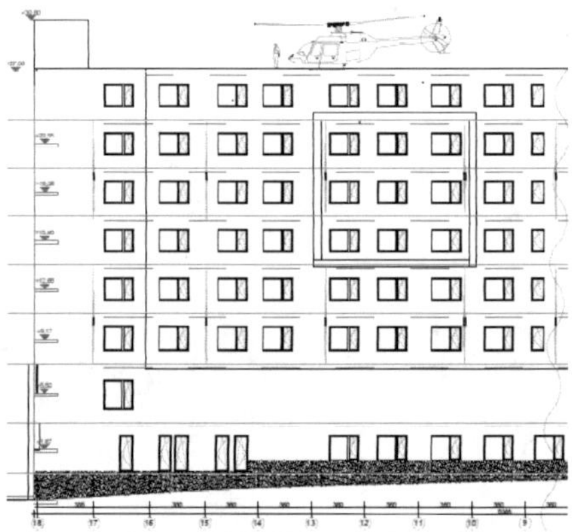

Figure 6. Facade North.

For the analysis of the model, almost the same characteristics are valid as for Model 1.
Modal analysis (see table 2): with 24 forms of own oscillations reported - with activated respectively 90,13% and 92,77% modal masses - percentage participation in directions X and Y.
The period $T_1 = 1{,}24\ s$.

Table 2. Load factors for mass calculation – Model 2

№	Load case	φ.ψ	№	Load case	φ.ψ
1	Own weight construction	1	8	Permanent impacts - roof	1
2	Permanent impacts - for ground floor and basement	1	9	Variable impacts - floors	0,3
3	Variable impacts - for ground floor	0,48	10	Variable impacts - roof	0,3
4	Technological equipment	0,50	11	Reverse embankment	0
5	Snow	0	12	Helicopter: landed position	0,333
6	Walls	1	13	Helicopter: lift-off ↑	0,333
7	Permanent impacts - floors	1	14	Helicopter: in the process of landing ↓	0,333

For **ULS combination**, the impact is accepted: Evenly distributed impact:
$p_k = Q / A = 37 / 42 = 0,88 \approx 1\ kN / m^2$; $p_d = p_k \cdot (\gamma_{dyn} = 1,5)$ - landed helicopter with engine running.

For an **Extraordinary combination** (distributed load over the entire beam grid structure) with a dynamic coefficient $\gamma_{dyn} = 1,6$: a). lift-off: + $p_k \uparrow = 15 kN / m^2$; b). landing: - $p_k \downarrow = 15 kN / m^2$

For **Seismic combination**: $p_k = 1 kN / m^2$ - at landed position. Impacts from: snow, pavements and equipment/maintenance (2 and 0,75 *kN/m²* - inside and outside the helicopter pad) are in accordance with the accepted design.The model also takes into account the vertical component of seismic impact for both kinds of spectrums, with a coefficient of behavior q=1,5 and a_{vg}/a_g=0,85.

Analysed are all possible and logical combinations of impacts, with coefficients, as well as with the directions - the components of the seismic impact, depending on the response spectrum and their combinatorics of interaction. The maximum values of the internal forces are obtained by taking into account together the horizontal and vertical components of the seismic impact.

5. Results

Below on the figures are shown schematically: the vertical displacement of floor and roof slabs (see figure 7 and figure 9); the resulting reinforcement for foundation slab (figure 11); internal forces for selected: shear walls, columns and beams (see figures: 12, 13, 14 and 15). The acepted reinforcement for: shear walls, columns and beams, are shown in tables: 3a)., 3b)., 4a)., 4b)., 5a)., 5b). and figure 16. Figure 8 and figure 10 shown comparative analysis for the displacements. All this are for Model 1 and Model 2.

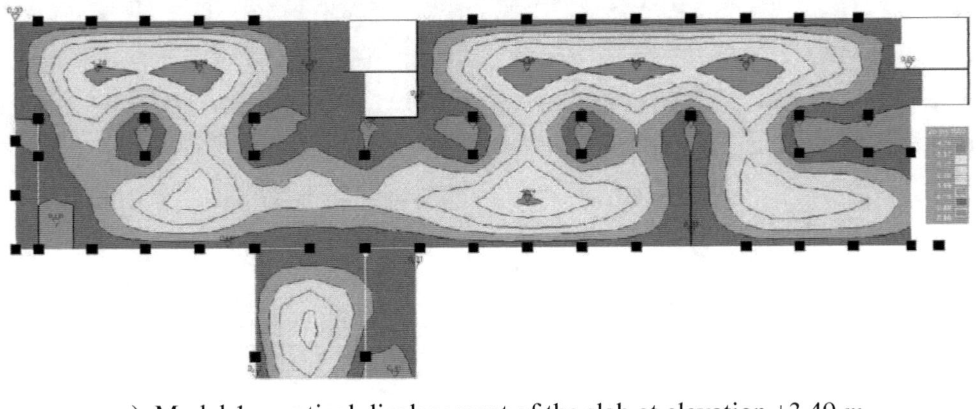

a). Model 1 - vertical displacement of the slab at elevation +3,49 *m*.

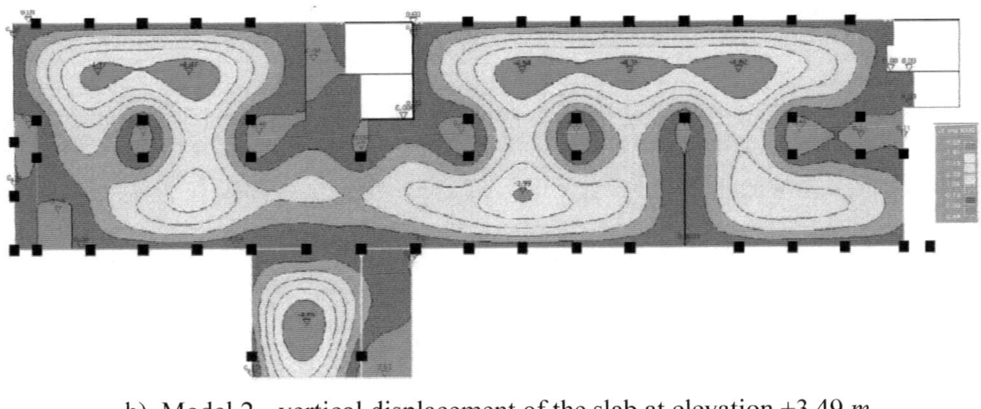

b). Model 2 - vertical displacement of the slab at elevation +3,49 *m*.

Figure 7. Displacement on Z-axis in the floor slabs – isolines.

$$\boxed{w_{el.-pl.} = 1,43\,cm < w_{\max} = 2,48\,cm}$$

$$w_{el.-pl.} = w_{el.}.k_{reol.} = 0,475.3 = 1,43\,cm$$

$$w_{\max} = \frac{L_{\max}}{250} = \frac{620}{250} = 2,48\,cm$$

$$\boxed{w_{el.-pl.} = 1,41\,cm < w_{\max} = 2,86\,cm}$$

$$w_{el.-pl.} = w_{el.}.k_{reol.} = 0,469.3 = 1,41\,cm$$

$$w_{\max} = \frac{L_{\max}}{250} = \frac{715}{250} = 2,86\,cm$$

Figure 8. Comparative analysis – from figure 7: left - Model 1; right - Model 2.

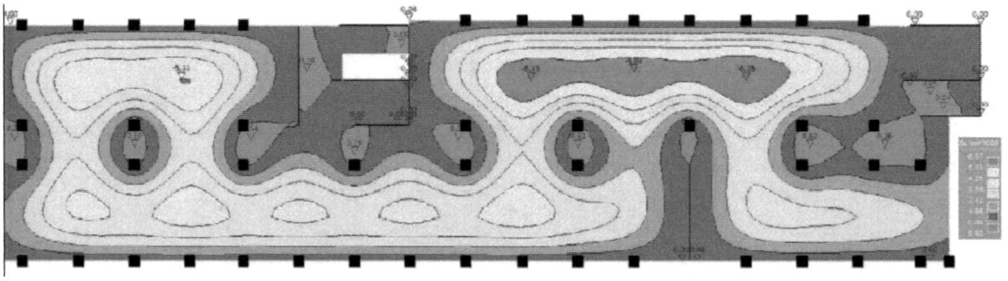

a). Model 1 - vertical displacement of the slab at elevation +27,50 *m*.

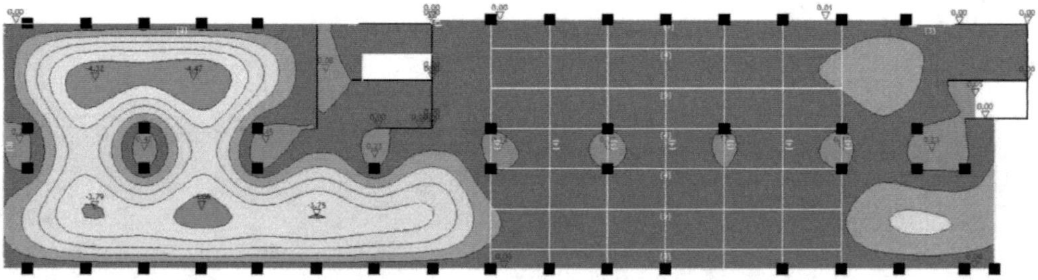

b). Model 2 - vertical displacement of the slab at elevation +27,50 *m*.

Figure 9. Displacement on Z-axis in the roof slab – isolines.

$$\boxed{w_{el.-pl.} = 1,91\,cm < w_{\max} = 2,48\,cm}$$

$$w_{el.-pl.} = w_{el.}.k_{reol.} = 0,637.3 = 1,91\,cm$$

$$w_{\max} = \frac{L_{\max}}{250} = \frac{620}{250} = 2,48\,cm$$

$$\boxed{w_{el.-pl.} = 1,34\,cm < w_{\max} = 2,48\,cm}$$

$$w_{el.-pl.} = w_{el.}.k_{reol.} = 0,447.3 = 1,34\,cm$$

$$w_{\max} = \frac{L_{\max}}{250} = \frac{620}{250} = 2,48\,cm$$

Figure 10. Comparative analysis – from figure 9: left - Model 1; right - Model 2.

Figure 11. a). Model 1.

b). Model 2.

Figure 11. Lower reinforcement of the foundation slab – isolines.

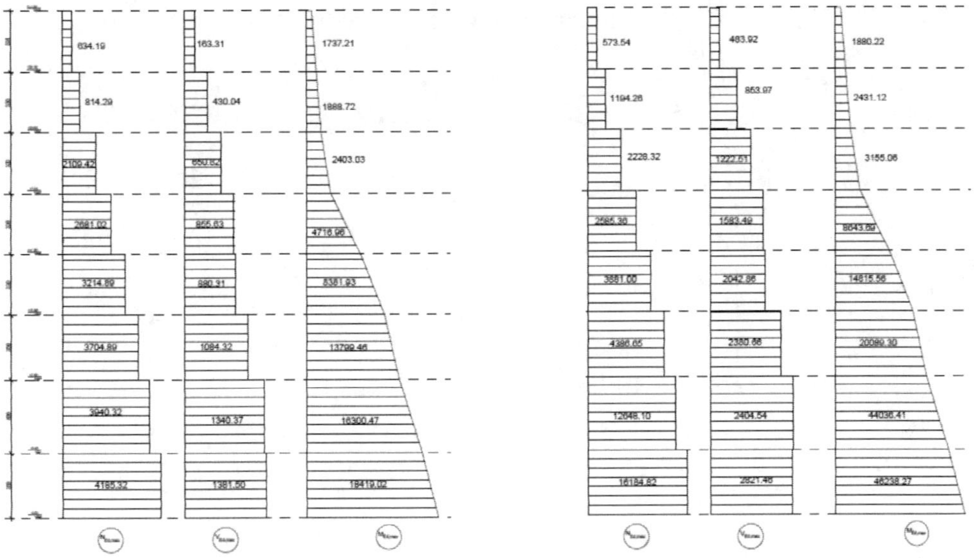

Figure 12. Results for Shear Wall (SW) 9 (650/25 *cm* – top left in plan), internal forces: N, V, M – diagrams: left - Model 1.; right - Model 2.

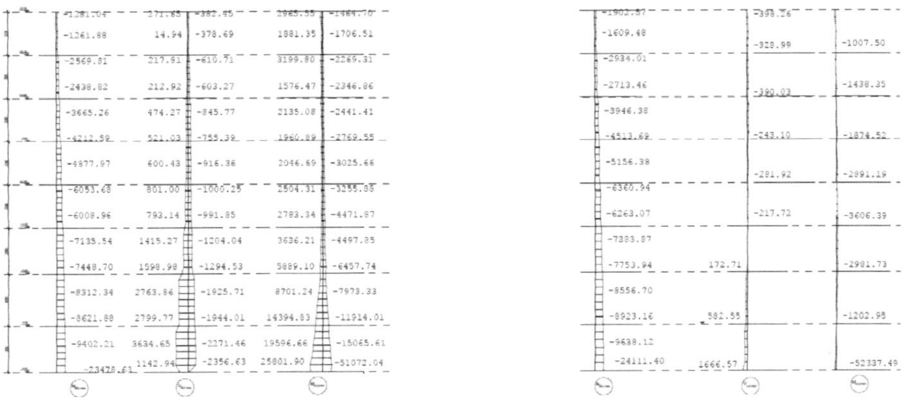

Figure 13. Results for Shear Wall (SW) 13 (640/25 *cm* – between the cores), internal forces: N, V, M – diagrams: left - Model 1.; right - Model 2.

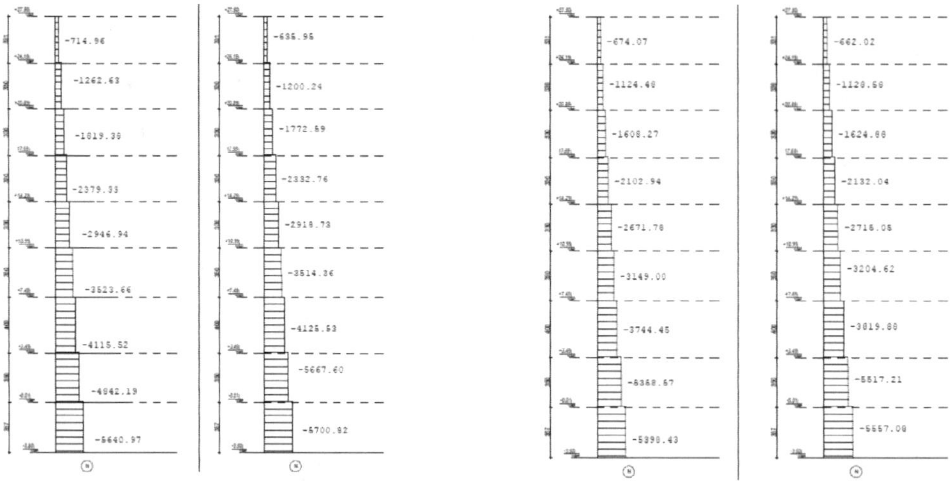

a). Column 15 (50/50 *cm* – left in plan). b). Column 19 (50/60 *cm* – in right, between the cores).

Figure 14. Results for columns (C), internal forces: N – diagrams *(on the left - Model 1, on the right - Model 2).*

Table 3a. Acepted RB for SW 9

Level, [m]	Model 1			
	End zone	Web		End zone
		Vertical RF bars	Horizontal RF bars	
found.	18N22	N8/20	N8/15	18N22
-3,50	18N22	N8/20	N8/15	18N22
-0,01	18N22	N8/20	N8/15	18N22
+3,49	18N20	N8/20	N8/15	18N20
+7,49	18N16	N8/20	N8/15	18N16

Table 3b. Acepted RB for SW 9

Level, [m]	Model 2			
	End zone	Web		End zone
		Vertical RF bars	Horizontal RF bars	
found.	18N25	N8/20	N10/20	18N25
-3,50	18N25	N8/20	N8/20	18N25
-0,01	18N25	N8/20	N8/20	18N25
+3,49	18N25	N8/20	N8/20	18N25
+7,49	18N20	N8/20	N8/20	18N20

Table 4a. Acepted RB for SW 13

Level, [m]	Model 1			
	End zone	Web		End zone
		Vertical RF bars	Horizontal RF bars	
found.	16N22	N8/20	N12/15	16N22
-3,50	16N22	N8/20	N12/15	16N22
-0,01	16N22	N8/20	N12/15	16N22
+3,49	16N20	N8/20	N12/15	16N20
+7,49	16N16	N8/20	N12/20	16N16

Table 4b. Acepted RB for SW 13

Level, [m]	Model 2			
	End zone	Web		End zone
		Vertical RF bars	Horizontal RF bars	
found.	16N18	N8/20	N12/15	16N18
-3,50	16N18	N8/20	N12/15	16N18
-0,01	16N18	N8/20	N10/15	16N18
+3,49	16N16	N8/20	N8/20	16N16
+7,49	16N14	N8/20	N8/20	16N16

Table 5a. Acepted RB for C 15 and C 19

Level, [m]	Model 1	
	Longitudinal load-bearing reinforcement	Transverse reinforcement (stirrups)
found.	12N22	N10/20
-3,50	12N22	N10/20
-0,01	12N22	N10/20
+3,49	12N18	N10/20
+7,49	12N14	N10/20

Table 5b. Acepted RB for C 15 and C 19

Level, [m]	Model 2	
	Longitudinal load-bearing reinforcement	Transverse reinforcement (stirrups)
found.	12N22	N10/20
-3,50	12N22	N10/20
-0,01	12N22	N10/20
+3,49	12N18	N10/20
+7,49	12N14	N10/20

Note: In the tables: 3a., 3b., 4a., 4b., 5a. and 5b., "RB" are the reinforcement bars!

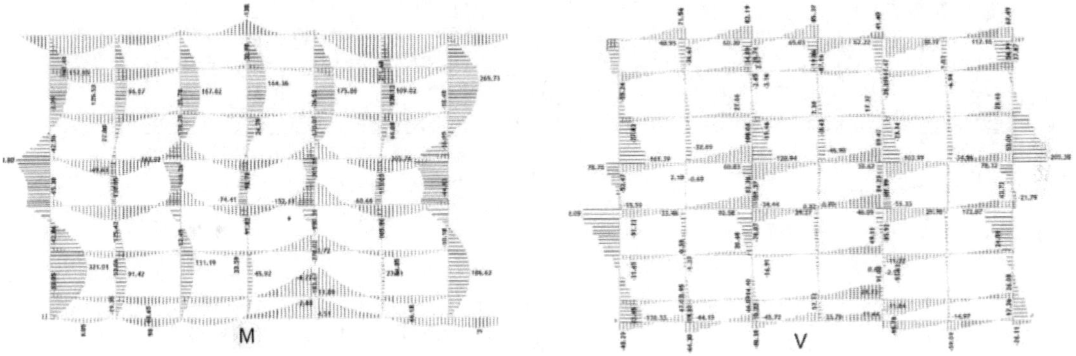

Figure 15. Internal forces for roof beams: M and V – diagrams.

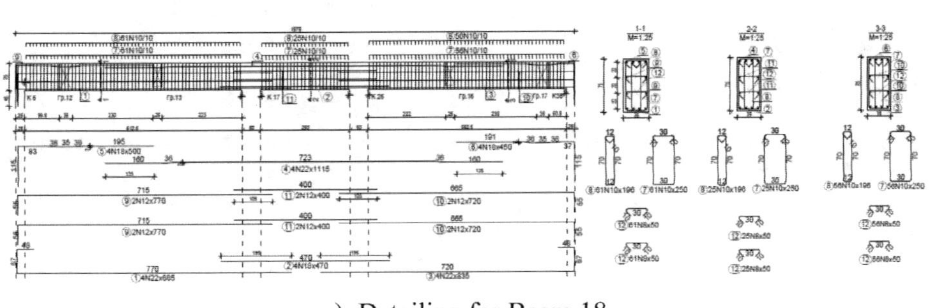

a). Detailing for Beam 18.

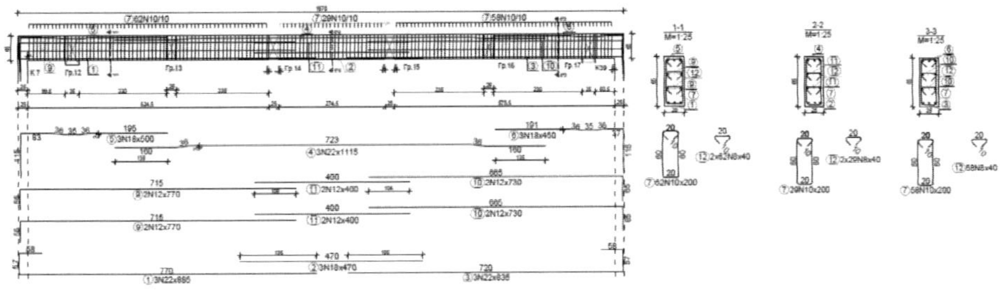

b). Detailing for Beam 19.

Figure 16. Fragments of reinforcement plans at elevation + 27,50 m (roof).

6. Conclusions

Examination of the calculated example shows that even a relatively light helicopter on the roof of the building, leads to an increase in the internal forces in all vertical and horizontal elements, as well as the construction of a reinforced roof structure (which is a local grid with beams) connected directly to the roof slab. And its needs significant reinforcement. Seismic analysis shows that the decision of the vertical load-bearing elements - shear walls, is correct.

The simplified solution of the roof construction is suitable, from options: coffered/waffle slabs, slabs with arched capitals, etc. types of constructions.

In phase of detailed design, the accident condition can be checked in case of possible passage of the machine near the board, through the possibility of falling. It is also possible to check the temperature field caused by fire in the event of a machine accident.

The future of air transport will definitely belong to drones, given technical progress. The constructions of the buildings must correspond to their device, dimensions and functions.

References

[1] https://opoznai.bg/hotels/view/grand-hotel-millennium-sofia
[2] http://wikimapia.org/592074/bg/%E2%80%9E%D0%92%D0%B0%D1%80%D0%BD%D0%B0-%D0%A2%D0%B0%D1%83%D1%8A%D1%80%D1%81%E2%80%9C%D0%94%D0%B2%D0%B5%D1%82%D0%B5-%D0%BA%D1%83%D0%BB%D0%B8#/photo/4895890
[3] https://www.informiran.net/%D1%81%D0%BD%D0%B8%D0%BC%D0%BA%D0%B8-%D0%BD%D0%B0%D0%B9
%D0%B3%D0%BE%D0%BB%D1%8F%D0%BC%D0%B0%D1%82%D0%B0-%D0%B1%D0%BE%D0%BB%D0%BD%D0%B8%D1%86%D0%B0-%D1%83-%D0%BD%D0%B0%D1%81-%D1%81-%D0%BF%D0%BB%D0%B0/
[4] https://www.24chasa.bg/Article/85043 (*Last visited June 10, 2022 for Internet - links*)
[5] *Diploma thesis* of Petkova Kr 2022 Master's degree on the topic: "Hospital building in Pleven" VSU "L. Karavelov "- Sofia diploma supervisor: Assoc. Prof. Eng. Stanislav Tsvetkov, PhD
[6] Georgiev G and Tsvetkov St 2013 Methodological guide for development of course project Reinforced Concrete Structures Part 1 Chapter 2 – Wall constructive systems & G.G. Ts.St. 2014 …Part 2 – Frame constructive systems – *VSU Publishing house Sofia*
[7] Stoynova Il 2021 Dynamic response of a reinforced concrete cantilever beam subject to harmonic loads, *XXI International Scientific Conference on Construction and Architecture VSU'2021, 14-16 October Sofia Proceedings* ***vol. II*** pp 74-83
[8] Handruleva An and Atanasov S 2021 Analysis of the stress and strain of discrete spherical dome subjected to asymmetric vertical load, *XXI International Scientific Conference on Construction and Architecture VSU'2021, 14-16 October Sofia Proceedings* ***vol. II*** pp 84-96
[9] Georgiev T 1986 New construction decision for the reconstruction of the electrostatic precipitators of a thermal power plant *Bulletin "Construction activity" published by GDCT* year XVIII issue **9** pp 31-40
[10] Hrischev L, Todorov D and Spasov K 2018 Geometric tolerances in execution of concrete structures *X Jubilee International Scientific Conference Civil Engineering Design and Construction (Science and Practice) Sept. 20-22 2018 Varna (Bulgaria)* pp 792-801

IOP Conf. Series: Materials Science and Engineering 1276 (2023) 012019 doi:10.1088/1757-899X/1276/1/012019

Stability behaviour of concrete arch bridge structures in relation to the load direction out of the arch's plane

V T Nikolov

Faculty of Structural Engineering, Reinforced Concrete Structure Department, University of Architecture, Civil Engineering and Geodesy, 1 Hr. Smirnenski Blvd, Sofia 1046, Republic of Bulgaria
vtnikolov_fce@uacg.bg

Abstract. Arches are one of the predominant structural elements for bridges with big spans. With the current technological advancement, they are applicable in pure form, without any transverse stabilizing systems. The slender form of the arch elements leads to reduction in their capacity due to stability failure. Their behaviour is directly related to the presence of first and second order effects – imperfections and deformed state. The typical bridge loads and combinations, often invoke effects outside of the elements' plane of reference. Such conditions are reducing the capacity of the structure in relation to magnitude of the effect. Analysis of such effects is complex task including interaction between complex stress state, 3-dimensional geometry, and non-linear material behaviour. In the following work are presented series of geometrical configuration in condition of material and geometrical non-linearity for the determination of critical load leading to stability failure. The results display trends in the stability behaviour of reinforced concrete arch bridge elements.

1. Introduction

The stressed state of arch structure is predominantly axial with significant sensitivity for bending effects inside the plane of reference. With proper design the bending moments occurring in the structure can be fully controlled according to the structure's capabilities. Considering bridge structures, we often deal with significant horizontal load and asymmetrical loadings. Those effects lead to disrupting forces outside of the plane of reference for the structural elements. The resulting sectional forces can be taken care of with the structure capacities itself and do not present major issues for the structure. Due to the massive sizes of the bridge structures, the expected issues deriving from those horizontal loads comes from stability standpoint.

Analysis of stability is advanced task and often requires great amount of idealization. Structure stability procedure for the determination of out of plane stability of arch elements are rare. The full analysis of arch involves calculate the structure`s statics in deformed conditions. This leads to non-linier relation between the in-plane and out-plane loads. In the current work full analysis with geometrical and material non-linearity are solved for set of geometrical conditions to find relations and trends in the structure`s behaviour. Finite element method is used to achieve reliable results for expected critical load leading to stability failure.

The paper presents the results for the above-described conditions in graphical variant for series of geometrical conditions in order to determine a "behaviour" curve of the critical loads in relation of the angle of the resultant force. Each graph contains the solutions for wide range of possible force directions, representing a "behaviour" curve described by the forces – displacement relation.

Initial analysis are held with concrete arch with 60m span and 30m rise. The form of the arch is circular with fully restrained support conditions (Fixed support). The analysed cross section is box section with 3.2m height, 2.2m width and 0.4m thick walls. Considered loads on the system are taken as uniformed distributed load of 100kN/m.

2. Analytical model for determination diapason of angle deviations

Model is build based on the forementioned geometry with the condition of material and geometrical non-linearity. Uniformed distributed load is presented in the structure with different angles in relation to the plane of the arch. A model analyse is held to define the expected deformation state whit dominant out of plane deformation. Further analyses are done with this initial data with different initial angle of the loading.

Angle of the load relevant to the ache`s plane is definitive factor for the current analysis. All combinations of horizontal and vertical external load can be represented with resulting force defined with specific angle. The difference in angle loading will lead to non-linear function of the structures critical load.

For determination of a limit for the angle an iterative process is used in the model. In this diapason, the stability failure is leading factor, defining the capacity of the structural element. In order to determine the actual diapason, the forementioned analytical model is solved for every possible angle between 0°÷45° and increment of 1°.

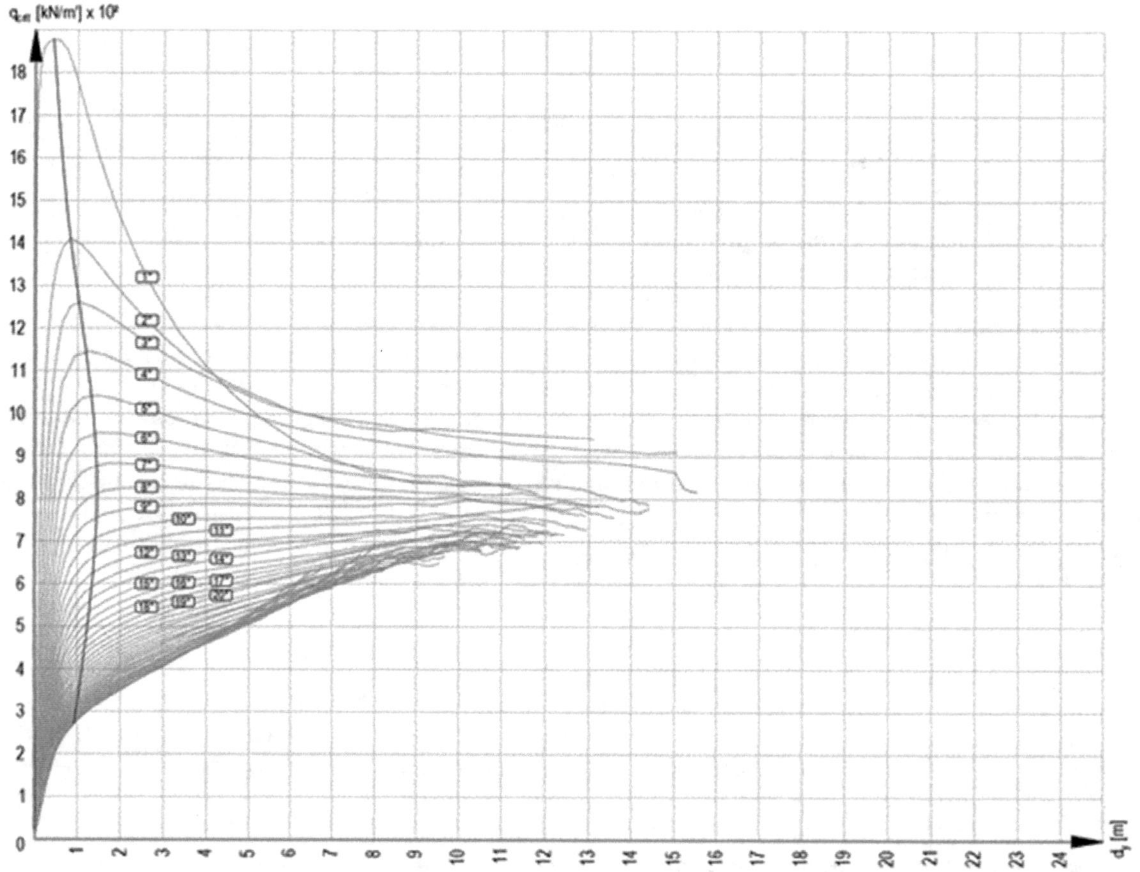

Figure 1. Graphical interpretation of the load-displacement relations of the arch.

Based on the results, diapason between 0° and 11° can be assumed as relevant. Deviation angle above 11° results in bifurcation point stiffness deterioration of the system. Results for this diapason experience progressive increase of error in the determination of the actual position of the critical force. For loading angles between 11°÷18°, critical load can be determined, despite of the considered inaccuracy of the results. With the analysis of the elements with load angles above 18° is observed overlapping of results. This phenomenon is direct indicator for key change in the behaviour of the structure. The analytical model changes its mechanism of resistance with one that is not sensitive to structural stability failure. The non-linear behaviour of the deformation in this diapason is result of the material nonlinearity considered in the model.

According to the presented results, the behaviour of the arch elements in relation to their stability can be divided in 3 groups:

- Direct stability failure of the system – angles between 0°÷11°;
- Conditional stability failure of the system – angles between 11°÷18°;
- Stability independent failure of the system – angles above 18°;

In relation of the cross-section geometry and capacity the precise ranges of the 3 groups can vary up to ±2° fluctuation. For global geometrical changes, such as span length, rise height, arch form, or drastic changes in the cross section the 3 group`s ranges can change drastically. In some geometrical configuration some of the groups can disappear completely.

As limitation of the theories considered in the presented analysis, cracking of the concrete and other stiffness reducing effects cannot be directly evaluated. In more realistic model, where such effect can be taken into consideration, the deterioration of stiffness has more pronounced effects, after the critical load is reached. The behaviour of the structure can only be analysed with more complex models and require further research on the topic.

The current results focus only on the behaviour before and at the moment of stability failure. In other to find the scope of angle deviation additional models are analysed. Results for the models, with variation in geometry are presented to determine trends in the behaviour of the structures.

3. Behaviour curves for geometrical variant

In other to validate the results for the displayed trends, additional models with different geometrical conditions are used:

3.1. Thickness of the box walls

The thickness of the walls is a key factor for the stiffness of the structures. In order to determine the relevance of the cross-section stiffness to the results two version of the cross section is analysed. The standard presented section with wall thickness of 0.4m and 0.8m. For better consistency of the comparison, the original presented section of 2.2x3.2 is changed to 2.0x2.0m. This is done to present a box section providing equal contribution of the change in wall thickness for both horizontal and vertical stiffness.

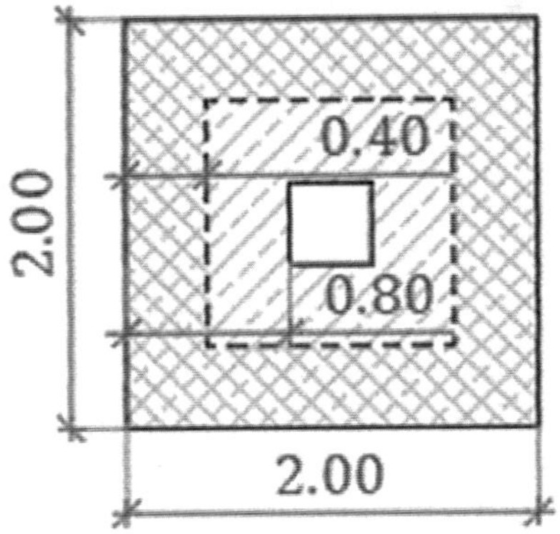

Figure 2. Variants of cross-section geometry.

3.2. Rise of the arch

Two variants of the arche`s rises are considered, in other to present valid range from the lowest to the highest. In these conditions a relevant proportion to the arch span is considered as raises ½ and ¼.

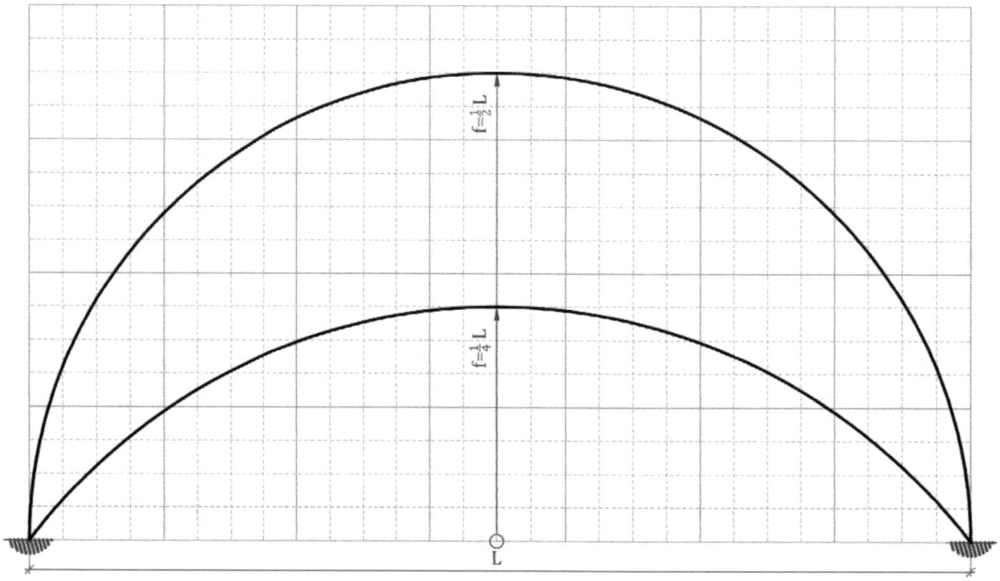

Figure 3. Variants of arch elements rises.

3.3. Form of the arch

Dependent on the form of the arches there can be significant difference to the stability behaviour of them. To validate this difference two types of acres are uses in the following paper: parabolic form and circular form. The two variants are picked based on the fact that those are one of the most narrow and most wide crown geometry for arches used in the current and historical designs.

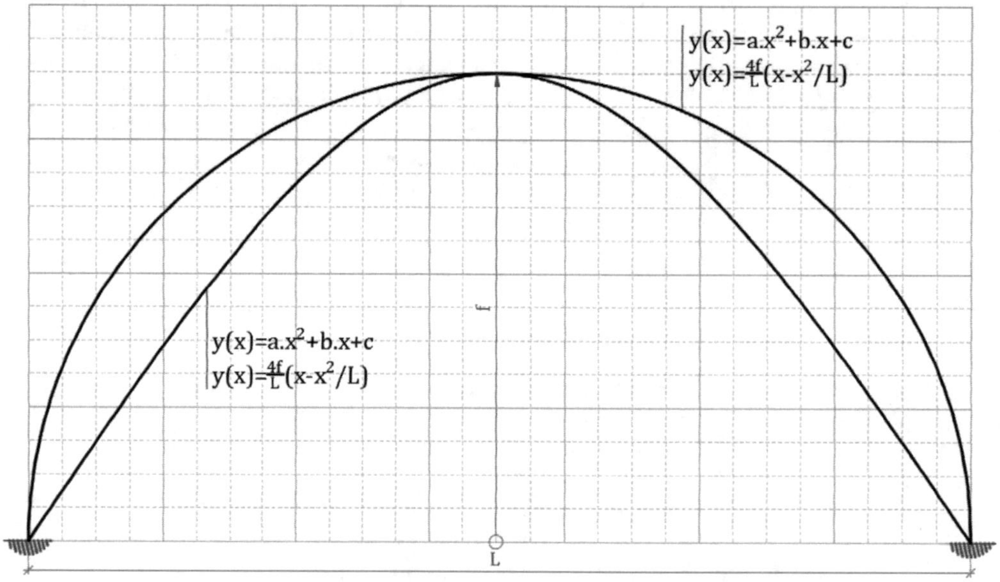

Figure 4. Variants of arch elements geometry definition.

3.4. Comparison of results

The results are presented in graphical form as set of curves presenting the force-displacement relation for each angle. From the series a single curve can be established to present the behaviour of the stability as form of critical force-angle deviation relation.

As indication the following codes are used:

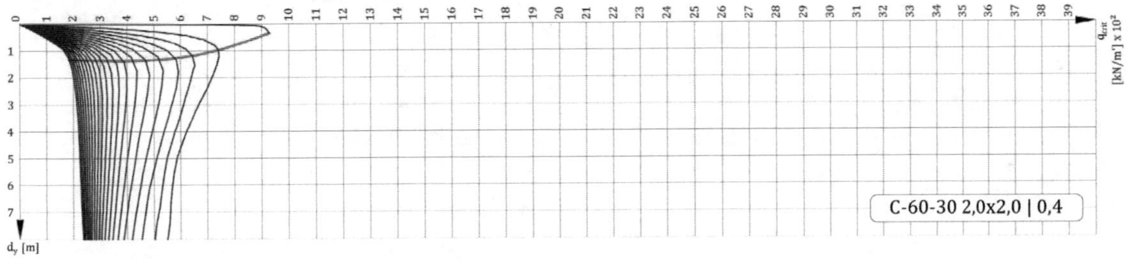

Circular arch with 60m span and 30m rise, cross section 2.0x2.0m and wall thickness of 0.4m

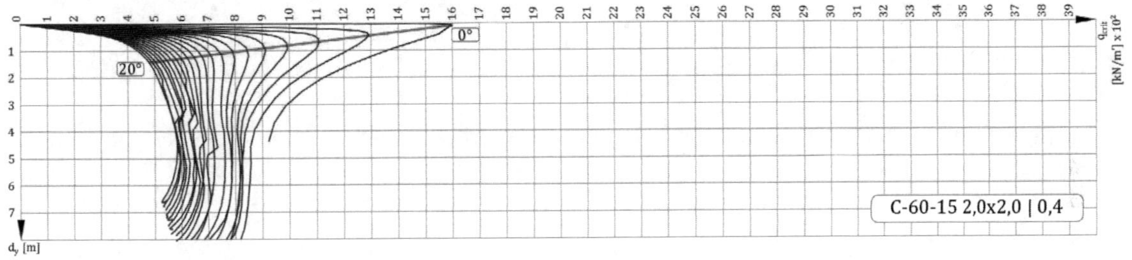

Circular arch with 60m span and 15m rise, cross section 2.0x2.0m and wall thickness of 0.4m

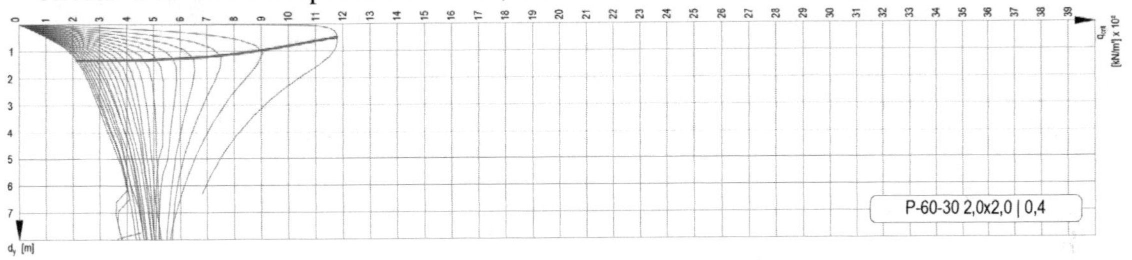

Parabolic arch with 60m span and 30m rise, cross section 2.0x2.0m and wall thickness of 0.4m

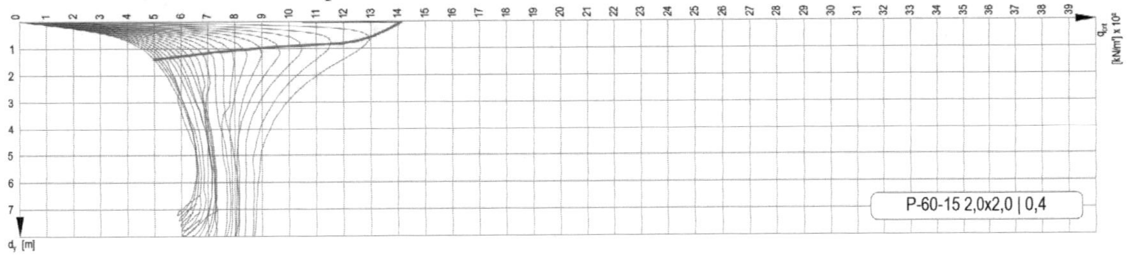

Parabolic arch with 60m span and 15m rise, cross section 2.0x2.0m and wall thickness of 0.4m

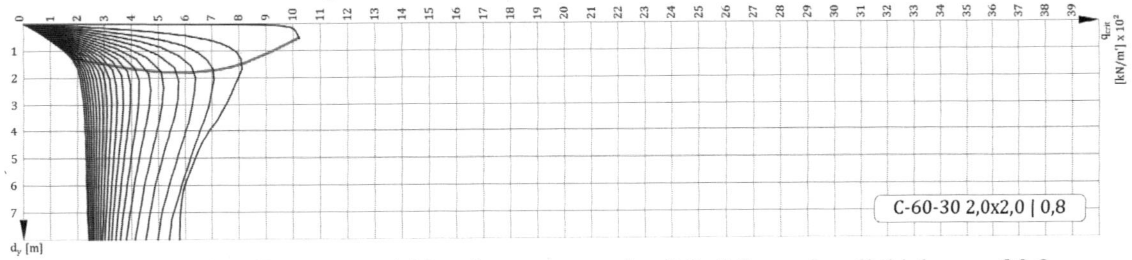

Circular arch with 60m span and 30m rise, cross section 2.0x2.0m and wall thickness of 0.8m

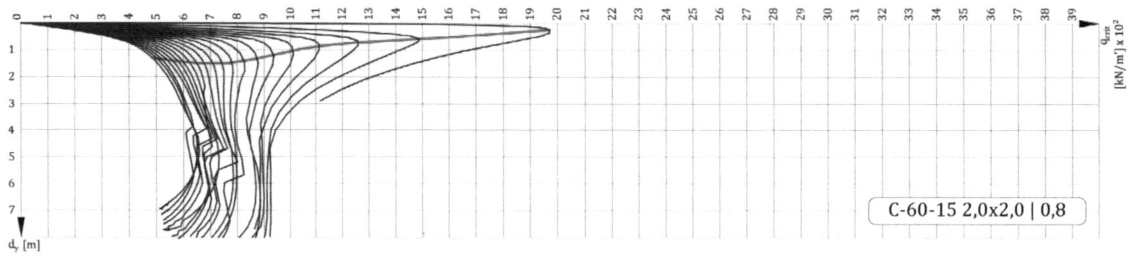

Circular arch with 60m span and 15m rise, cross section 2.0x2.0m and wall thickness of 0.8m

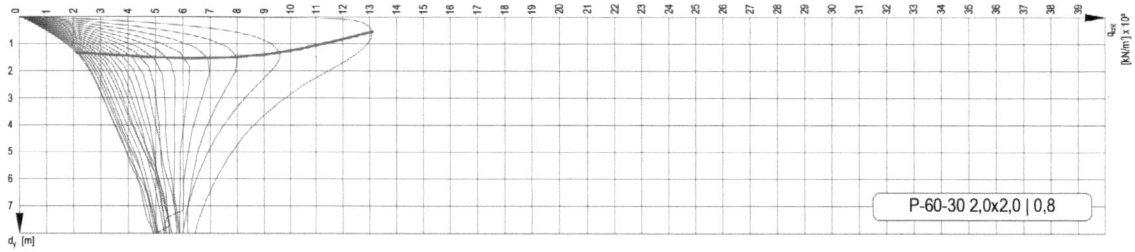

Parabolic arch with 60m span and 30m rise, cross section 2.0x2.0m and wall thickness of 0.8m

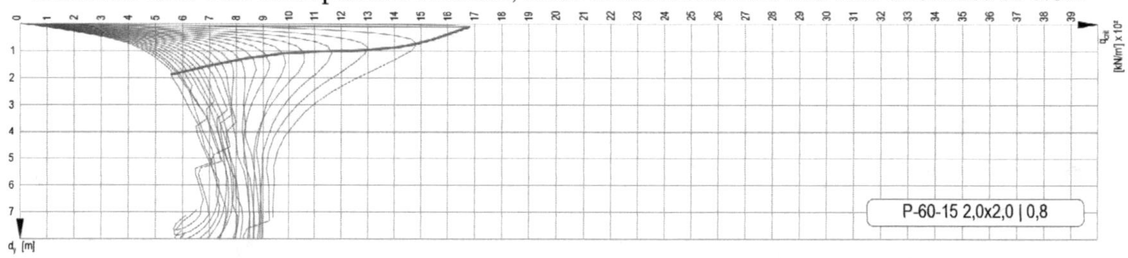

Parabolic arch with 60m span and 15m rise, cross section 2.0x2.0m and wall thickness of 0.8m

Figure 5. Results of the analysis.

In the range specified as "direct stability failure of the system" we can observe the following trends.

Comparing the solution for the different thickness it is observed that the bigger section leads to higher capacity against structural stability failure. Otherwise, most of the other behaviour do not experience any drastic changes, such as stiffness deterioration after the point of bifurcation and deformations related to the bifurcation point.

Comparing high-rise and low-rise arches it is observed that low-rise arches possess greater capacities however the residual capacity due to stiffness deterioration is greater percentagewise compared to the critical load itself. Furthermore, the low-rise arches failure occurs at significantly less deformed state than the high-rise specimens. This leads to increased fragility of the low-rise system, with significantly higher capacities.

Comparing circular to parabolic geometry we observe slightly increased critical force capacity and faster convergence to unified capacity value at the end of the stiffness deterioration segment of the load-displacement curve in favour of the parabolic geometry of the arch.

IOP Conf. Series: Materials Science and Engineering 1276 (2023) 012019 doi:10.1088/1757-899X/1276/1/012019

4. Conclusion

Evaluating the effects of the angle deviation in relation of the plane of reference it is concluded that angled below 8° to 12° the stability is predominant criteria for structure failure. Angled below 15° to 20° have to be evaluated for stability failure. For angles above 20° the structures are highly sensitive, and mechanisms or optimization are needed to create structure that is reliable and cost effective.

As parameters affecting the stability of the arch structure is observed that the capacity and stiffness of the cross section leads the significant benefits to the structural critical force capacity. Lower-rise arches are less susceptible to stability failure and provide significant stiffness increase but reaches not stability failures faster. The arch geometry is relevant for the structure stability with the observed results that with sharper geometry forms such as parabolas, are less affected by other variables such as cross section and rise of the arch. With the parabolas those variable still provide influence over the critical loads, but significantly less than the more sensitive circular form.

References

[1] The European Standard EN 2007 Eurocode 1992-1 - General rules and rules for buildings

[2] The European Standard EN 2007 Eurocode 1992-2 - General rules and rules for bridges

[3] Karnovsky I., 2009 Advanced Methods of Structural Analysis, ISBN: 978-1489983350, Springer

[4] Karnovsky I., 2011 Theory of Arched Structures: Strength, Stability, Vibration, ISBN: 978-1461404682, Springer

[5] Hendy C R and Smith D A 2007 Designers` guide to EN 1992-2. Eurocode 2: Design of concrete structures. Part 2: Concrete bridges, ISBN: 978-0-7277-3159-3 Thomas Telford

[6] Chen W., Duan L., 2014 Handbook of International Bridge Engineering, ISBN: 978-1-4398-1030-9, CRC Press

[7] Lacarbonara W., 2013 Nonlinear Structural Mechanics: Theory, Dynamical Phenomena and Modelling, ISBN: 978-1441912756, Springer

[8] Farshad. M, 1994 Stability of structures, ISBN: 978-0444816986, Elsevier Science Ltd

[9] ©LUSAS authors collective, 2021 Theory Manual Volume 1 Version 19.1 - Issue 1, Distributed worldwide by LUSAS - Finite Element Analysis Ltd

[10] ©LUSAS authors collective, 2021 Theory Manual Volume 2 Version 19.1 - Issue 1, Distributed worldwide by LUSAS - Finite Element Analysis Ltd

[11] ©LUSAS authors collective, 2021 Element Reference Manual Version 19.1 - Issue 1, Distributed worldwide by LUSAS - Finite Element Analysis Ltd

Service condition of several pedestrian cable-supported bridges in the Rhodopi Mountain of Bulgaria

S D Ivanov[1] **and L D Georgiev**[2]

[1] Department of Steel, Timber and Plastic Structures, Faculty of Structural Engineering, University of Architecture, Civil Engineering and Geodesy, Sofia, Bulgaria
2 Department of Road construction and Transport Structural Facilities, Faculty of Transport Engineering, University of Architecture, Civil Engineering and Geodesy, Sofia, Bulgaria

sdenkov_fce@uacg.bg, lazar_fte@uacg.bg

Abstract. This paper describes the service condition of few long span pedestrian suspension bridges in Rhodopi Mountain of Bulgaria. They were built more than 50 years ago and were reconstructed after 35 years of exploitation. Visual inspection was made in the summer of 2021 by the authors. Typical design shortcomings are presented. In conclusion some ideas for the improvement of their condition with respect to their future save exploitation are suggested.

1. Introduction
Rhodopi Mountain is situated in the south-eastern part of Bulgaria. The terrain is crossed by relatively wide riverbeds and ravines. Number of dams have been built in the past in that area as well. All these makes it hard and expensive to develop regional infrastructure.

Even though most of the population in that region is settled in the several cities centers still some outlying villages with poor access do exist. Old pedestrian routs and shortcuts are still utilized by the locals in order to reach their homes.

More than 50 years ago, along that paths, over the natural obstacles, long span cable supported bridges were built. Despite their relatively large spans, up to 260m, these bridges usually lack project documentation and were craft made utilizing steel cables from the regional mining industry. Adding poor or almost no maintenance makes them even dangerous to cross.

Three bridges are presented in this paper, all of them crossing the river Varbitza at different locations. Their service condition is described based on visual inspection, performed by the authors in the summer of 2021.

2. Description of the bridges
The three bridges that are in the focus of this paper are all suspension pedestrian bridges. Their main parameters are summarized in table 1. According to the authors' assumption and the data found on one of the bridges, they were originally built in the period 1960-1970. In the period 2000-2010, after nearly 40 years of exploitation, they were reconstructed. During the reconstruction the whole superstructure, as well as the cables, were changed. Only the substructure, pilons and anchor blocks, from the original design, were utilized.

Table 1. Main parameters of the bridges.

Bridge	Main span (m)	Side spans (m)	Type and Diameter of Main cable (mm)	Distance b/n two main cables (mm)	Distance b/n the hangers (m)
Zagorsko	210	1 X 71	12 open spiral strands Φ14	1420	4,00
Veligdenche	122	no	6 open spiral strands Φ11	1420	4,05
Podkova	260	no	16 open spiral strands Φ14	1420	4,00

2.1. The Bridge at Zagorsko village

The first bridge is situated close to the village of Zagorsko. It is a suspension bridge with main span of 210m. From one side of the main span there is no side span envisaged, from the other side there is a side span 71m long. General view of the bridge is presented in figure 1.

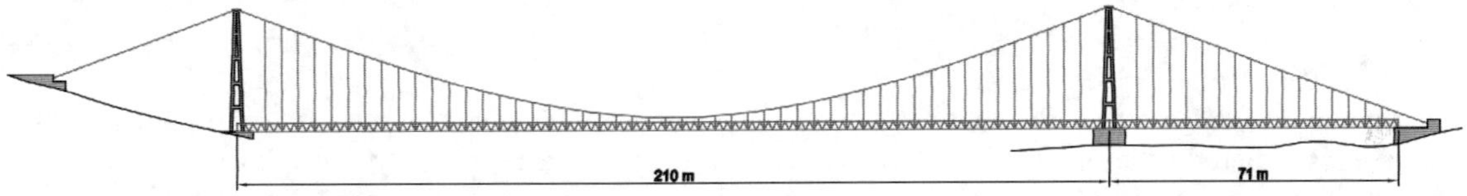

Figure 1. General view of the bridge at Zagorsko village.

The distance between the axis of the two main cables is 1.42m. Each main cable consists of 12 separate hot dip galvanized open spiral wire strands with a diameter of 14mm. The hangers are two different types - open spiral wire strands and solid steel rods with diameter 14mm. They are spaced 4.0m apart.

The main girders are Warren type steel trusses with structural height of 1.0m (slenderness h/L=1/210). Their chords are made from thin-walled cold form U steel profiles. The main girders are used as parapets for the deck.

The deck is 1.50m wide, measured from the outer edges of the bottom chords of the main steel trusses. It consists of cross beams, over which profiled steel sheeting is place. It is used for permanent formwork for the concrete slab cast on top.

The pilons are made from concrete. They have frame like structure with four inclined legs that connect two by two at the top, figure 2. Five cross beams connect the legs in transversal direction. Information for the saddles on top of them is missing and during the visual inspection they were not accessible. The anchors blocks are massive concrete blocks, figure 3.

IOP Conf. Series: Materials Science and Engineering 1276 (2023) 012020 doi:10.1088/1757-899X/1276/1/012020

Figure 2. Concrete pilon.

Figure 3. Concrete anchor block.

2.2. The Bridge at Veligdenche village

The bridge that is close to Veligdenche village is a suspension bridge with main span of approximately 122m. It is situated in a beautiful area over the meanders of Varbitza river and crosses it close to the railway stop. From both sides of the main span no side spans are envisaged (see Figure 4).

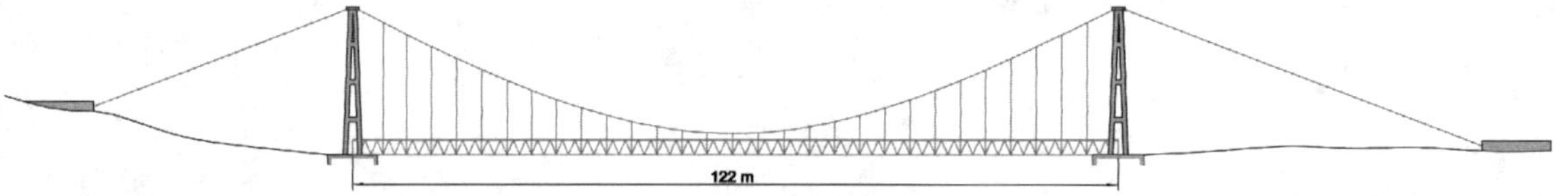

Figure 4. General view of the bridge at Veligdenche village.

The distance between the axis of the two main cables is again 1.42m. Each main cable consists of 6 separate hot dip galvanized open spiral wire strands with a diameter of 11mm. The hangers are two types - open spiral wire strands and steel rods with diameter varying from 11mm to 18mm. They are spaced 4.05m apart.

The main girders are Warren type steel trusses with structural height of 1.05m (slenderness of h/L=1/116). Their chords are made from thin-walled cold form U profiles. The main girders are used as parapets for the deck.

The deck is 1.50m wide, measured from the outer edges of the bottom chords of the main steel trusses. It consists of cross beams, over which profiled steel sheeting is place. It is used for permanent formwork for the concrete slab cast on top. Wind bracing of "X" type is developed under the steel sheeting, in which the diagonals are made from reinforcing steel bars.

The pylons and the anchor blocks are made from concrete. The pylons have frame like structure with four inclined legs that connect two by two at the top. Four cross beams connect the legs in transversal direction. Over the last one (on the top) the saddles for the cables are situated.

2.3. The Bridge at Podkova railway station

The last of the three bridges is crossing the river Varbitza close to the end of the railway line No 4 at the final railway station of Podkova. It is a suspension bridge with main span of approximately 260m without side spans. General view of the bridge is presented in figure 5.

Figure 5. General view of the bridge at Podkova railway station.

The distance between the axis of the two main cables is the same as with the other two bridges 1.42m. Each main cable consists of 16 separate hot dip galvanized open spiral wire strands with a diameter of 14mm. The hangers are two types - open spiral wire strands and steel rods with diameter varying from 14mm to 20mm. They are spaced 4.0m apart.

The main girders are Warren type steel trusses with structural height of appr. 1.15m (slenderness h/L=1/226). Their chords are made from thin-walled cold form U profiles. The main girders are used as parapets for the deck.

The deck is 1.50m wide, measured from the outer edges of the bottom chords of the main steel trusses. It consists of cross beams, over which profiled steel sheeting is place, which is used for permanent formwork for the concrete slab cast on top. Wind bracing of "X" type is developed under the steel sheeting, in which the diagonals are made from reinforcing steel bars. The pylons and the anchor blocks are made from concrete. No data was available from their original design.

3. Visual inspection findings

All three bridges have been reconstructed 10-15 years ago, when the deck, stiffening girders and cables were changed with new ones. The substructure, pylons, and anchor blocks of each bridge were kept from their original construction dated back to the 60-ties of the 20th century. The visual inspection, made in the summer of 2021, focuses mainly on the superstructures. More information on shortcomings in the elements of the substructure may be found in [1], where results from inspection made in 2004 of similar bridges in that region are presented.

Since the reconstruction of the three bridges is made from the same contractor and designer their superstructures are quite similar. In this respect their deficiencies are going to be discussed in general (not individually bridge by bridge).

3.1. Cable system

During the reconstruction old full-locked coil main cables were cut and changed with open spiral strands for all three bridges. Thus, the new main cables consist of several separate wire strands. Wires are hot dip galvanized, which correspond to the first corrosion protection level/stage according to [2]. Nevertheless, the measures to ensure the durability of the main cables seems insufficient since other protection levels are missing. This is especially valid near the anchorages as well as at the connecting details with the hangers.

The termination of each strand from the main cable is carried out with U-bolt wire rope grips with thimbles, figure 6. Passing of the strands around the thimble has let to their unweaving, which is either due to poor execution or to noncompliance with the dimensions (diameter) given in [3]. Even though the number of grips seems sufficient, in most cases the terminations do not comply with the fitting instructions of [4]: position of the first grip immediately against the thimble, distance between the grips, position of the bridge element on the load bearing part of the rope etc. Very important issue is also the control of the tightening torque for the grips.

Figure 6. strands of the main cable termination.

Figure 7. Connection between hanger and main cable.

The connection detail between the hangers and the main cables is also made with several U-bolts. Each U-bolt envelops 2 to 4 strands, and the bridging element is substituted by a steel flat plate. In the case of Zagorsko bridge normal bolts were used with steel flat plates from both sides, figure 7. This makes it hard to control the transverse pressure on the strands on the one hand and ensure slipping on the other. The detail is also prone to corrosion, capturing moisture and dirt, which greatly reduces its durability.

3.2. *Stiffening girder*

The stiffening girders are Warren type steel trusses made from cold-formed thin-walled U-profiles without gusset plates. Cold-formed thin-walled profiles reduce the quantity of steel and the self-weight of the superstructure, but they are not a common solution in bridge construction.

The stiffness provided by the two steel trusses is considered as insufficient, especially in the case of Podkova bridge. This assumption is since crossing this bridge from a single pedestrian leads to considerable vertical and torsional vibrations. The comfort criteria, according to [5] and [6], are not met.

The stiffening girder is divided in appr. 4.0m long assembly elements, which corresponds to the distance between the hangers. The on-site assembly connections of the truss chords are made with the help of one-sided splice, connecting only the web of the U-profile. The splice plate is welded to one of the assembly elements and bolted to the next one with 4 bolts class 5.8, figure 8.

The connection between the hangers and the steel truss girders in all three bridges is realized by simply passing the hanger end through a hole into the web of the horizontally laid U-profile of the top chord, figure 9. From the other side a nut is holding the hanger. As can be seen in figure 9, such detail is quite unreliable for pull out failure. In addition, the normal force in the hanger is transferred through bending of the web of the thin-walled cold formed U-profile out of its plane.

Figure 8. Connection of the top chord.

Figure 9. Pull out failure of a hanger.

Considering the corrosion protection of the steel trusses in all three bridges a significant level of corrosion has been recognized. The corrosion is already affecting all truss elements, figure 10. This has led to the conclusion that the applied initial surface protection does not correspond to the environment category. Maintenance measures are also missing.

3.3. *Deck and wind bracing*

The deck of all bridges consists of concrete plate laid over steel sheeting supported on transversal beams. Composite action has not been realized. The concrete plate is with approximately 4-5cm thickness and with no or very coarse reinforcement. This has led to considerable crack formation and at some places its complete wear out, figure 11. Moreover, there is no waterproofing envisaged on the deck for which the cross beams and the steel sheeting below the concrete plate were having considerable level of corrosion as well.

Wind bracing was envisaged below the concrete plate (at deck level). It has been realized with pair of reinforcing bars crossing each other between the cross beams and welded at their ends. The wind bracing is supported in horizontal direction only at the pylons, which leads to height to span ratio between 1/85 (for Veligdenche bridge) and 1/180 (for Podkova bridge). Horizontal ropes for additional intermediate supports were not envisaged.

Considering the condition of the concrete plate and the high slenderness of the "X"-bracing one could make the conclusion that the stiffness and also load bearing capacity of the superstructures for horizontal transversal loads is insufficient.

Figure 10. Corrosion level of the stiffening girder.

Figure 11. Wear out of the concrete plate.

4. Summary and Conclusions

During the visual inspection of three pedestrian suspension bridges, provided in the summer of 2021, some structural deficiencies of their superstructures have been found. They can be summarized as follows:

- Insufficient stiffness and load bearing capacity in transversal horizontal direction;
- Poor craft made detailing, with load bearing capacity that is hard to evaluate;
- Inappropriate concept design of the decking structure and surfacing;
- Insufficient corrosion protection measures both for the steelwork and for the cables;
- Lack of maintenance;

Based on the above deficiencies the following conclusions are drawn:

Even though the superstructures of the considered three pedestrian bridges have been totally reconstructed around 15 years ago their condition can be classified as unsatisfactory. Their design is far away from the good engineering bridge design practice of our time. Execution and maintenance are also classified as poor.

Nevertheless, these bridges are situated outside urbanized area, in mountains region, where traffic intensity is low, a compromise with their reliability is inadmissible. Qualified personal including bridge engineers with enough experienced is a must when designing cable supported bridges of such spans. Quality control of the execution as well as regular maintenance measures are another important aspect for providing the required reliability. Unfortunately, all these points were missing in the described bridges, and they are dangerous for exploitation and probably with significantly reduced service life in comparison with the requirements of EN 1990.

References

[1] Dulevski E and Ivanov S 2004 Assessment of the condition of pedestrian suspension bridges in the Rhodopi Mountain *Proc. 3-rd Nat. Conf. on Design and Construction of Buildings and Facilities*, Plovdiv, Bulgaria, pp.178-187, in bulgarian

[2] EN 1993-1-11:2006, Eurocode 3 - Design of steel structures - Part 1-11: Design of structures with tension components, 2009

[3] EN 13411-1:2002+A1:2008, Terminations for steel wire ropes – Safety - Part 1: Thimbles for steel wire rope slings references

[4] EN 13411-5:2003+A1:2008, Terminations for steel wire ropes – Safety - Part 5: U-bolt wire rope grips

[5] EN 1990:2003/A1:2006, Eurocode – Basis of structural design

[6] Kasapova K and Dinev D 2020 Human-induced vibrations on footbridges. Current codes of practice – overview *Journal of Theoretical and Applied Mechanics* (Bulgarian Academy of Sciences) vol. 50 Issue 2 pp. 142-157

IOP Conf. Series: Materials Science and Engineering 1276 (2023) 012021 doi:10.1088/1757-899X/1276/1/012021

Composite Girders of Classes 1 and 2 With Partial Plasticization

Encho Dulevski

University of Architecture, Civil Engineering & Geodesy, Faculty of Transport Engineering, 1 Chr. Smirnenski Boulevard, 1046 Sofia, Bulgaria

emd_fce@uacg.bg

Abstract. A case of a bridge composite girder is analysed. According to EC4 rules, its cross section is coinciding into class 1 and should be able to resist the design moment in ULS. But after tracking the loading history with gradually increasing the live load up to the design level, it is established, that an early web instability exists before the loading ends. This effect contradicts to some code settings and the case should be considered as a warning to the designers working in similar circumstances.

1. Introduction

The cross section classification in [5] is based on the capability of the girder parts in compression to have local stability at corresponding ultimate stress level. This is established by Table 5.2. The other codes ([6] - [9]) direct the designer to the same table with some additional explanations and comments. Generally, for 1 and 2 cross section classes the design is focused on proving by the formula $M_{plRd} \geq M_{Ed}$ that the girder is able to resist the design loads set by the codes [1] - [3]. In most cases it is enough, but not always. During the construction and in service life they fall into multiple intermediate conditions (loading history). Some of them could be critical for the web stability before the plastic bending moment is reached. During some of those conditions the neutral axis is found sufficiently distant from the upper flange, so that the web becomes locally unstable. There are doubts about the absolute applicability of clause 5.4.2.4(2) from [9]. This statement will be justified bellow by an example of a road bridge composite girder.

2. Description of the structure

The cross section of the superstructure, the object of the analysis, is shown in figure 1. The RC deck plate is supported by cross girders at distance of 2.5 m. Main girders are simply supported (40m span). During pouring the concrete the girders should resist dead weight and technological loads without propping in the span. Materials: structural steel grade S355; concrete class 30/37. Full interaction between the steel girder and the deck plate is accepted.

We are cautious of discussing the quality of the structural concept. But it is practically possible and is enough to support the author's intention.

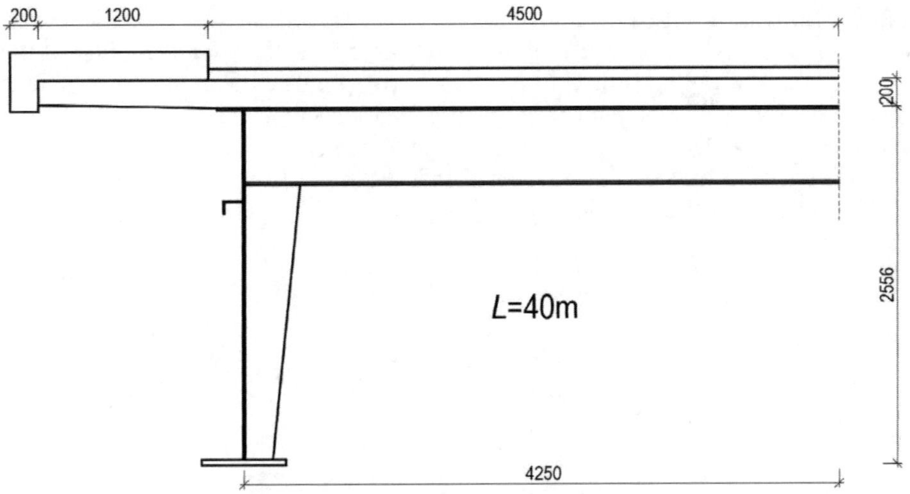

Figure 1. Superstructure cross section.

The structure is analyzed for dead weight, live load - (LM 1) and sidewalk load, shrinkage, creep, temperature difference - ([2], Figure 6.2b – simplified procedure). Construction stages, SLS and ULS are investigated. As a first step the cross girders are calculated and the profile HE550AA is proven enough for their function. Further the attention is focused on the main girders. Its cross section is drawn in figure 2a.

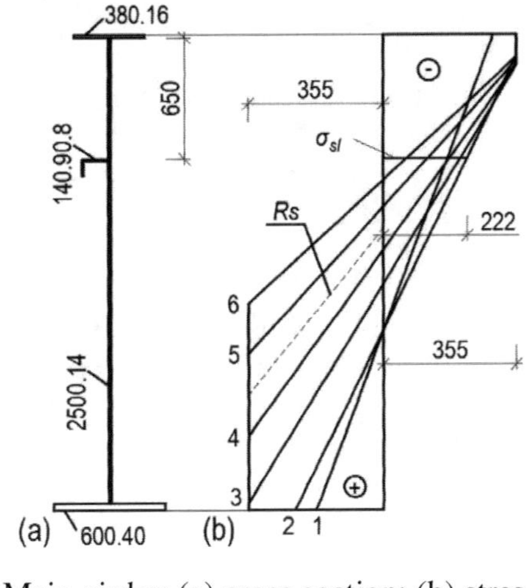

Figure 2. Main girder: (a) cross section; (b) stress diagrams.

3. Construction stage

The formwork is supported by the steel structure using light scaffolding. Wet concrete, $1\ kN/m^3$ heavier than the hardened one [1], self-weight, shuttering and scaffolding are the loads considered. The girder mid-cross section direct stresses, obtained for both edges are $-292\ MPa$ and $177\ MPa$. The overall stability of the upper girder flange and of the longitudinal stiffener are checked according to paragraph 6.3 from [5]. Annexes A1 and A2 from [6] are applied. According to A.2.1(1) the web stiffener stability check replaces the local stability check of the web. Aforementioned checks presented in the form of mathematical inequalities $\sigma_{cr} \geq \sigma_{Ed}$ are respectively: $292 > 280\ MPa$ and $209 > 170\ MPa$.

4. Serviceability limit state (SLS)

In this state the girder cross section is composite. For each girder, the effective width of the deck plate comprises the half deck width according to part 5.4.1.2 of [8]. The direct stresses at distinctive cross section levels are determined before and after the effects of the creep appear.

For the loads mentioned above, corresponding calculations are made and results for the mid-cross section stresses are shown in Table 1. It is noted that the loads from shuttering, scaffolding and the concrete additional humidity are downloaded (g_1').

Table 1. Stresses at SLS (MPa).

Load\Stress	$\sigma_{c,sup}$	$\sigma_{a,sup}$	$\sigma_{a,inf}$
g_1		-215.0	130.0
g_1'	0.38	1.7	-9.8
$g_2(0)$	-1.15	-4.9	29.0
$g_2(\infty)$	-0.92	-13.2	30.7
P	-4.30	-18.5	108.5
Shrinkage	-1.83	-55.3	12.8
Δt	-0.36	-11.0	2.2
Total	-7.26	-311.3	274.4

g_1 - dead weight at constr. stage; g_1' - formwork, humidity; g_2 - pavement, sidewalks, others; (0), (∞) - before, after creep; P- live loads (LM1 + sidewalk live load).

The total edge stresses are below the design stresses. After obtaining the stresses, a check for the web local stability is made. The procedure applied for the construction stage is applied again. The inequality $\sigma_{sl,cr} = 214 > 174 = \sigma_{sl,Ed}$ shows the web is locally stable. The value of $174\ MPa$ excludes the influence of the live load which turns out favorable (the neutral axis moves upwards).

5. Ultimate limit state (ULS)

At the beginning, the composite section class is checked. The ultimate condition of full cross section plasticization according to EC4 is shown in figure 3b. The stress diagram is too uniform and that is why only its upper part is drawn. The diagrams of the different materials are plotted in different scales for clarity. Compressed girder part includes the upper flange and only 4mm of the web. Figure 3b gives the next web parameters according to [5], Table 5.2: $\alpha = 4/2500 = 0.0016$, $c/t = 2500/14 = 178.6 < 36 \cdot 0.81/0.0016 = 182250$. Therefore, the web section falls into class 1 without even taking into account the presence of the longitudinal stiffener. The upper steel flange before casting of the concrete falls into class 3, but after concrete hardening according to clause 5.5.2(1) from [8], is transferred into class 1. Thus, the entire cross section remains into class 1. The plastic bending resistance $M_{pl,Rd} = 40906\ kNm$ is calculated. The obtained total design moment from loads g_1, g_2, P is $M_{Ed} = 32570\ kNm$ (the influences of shrinkage, creep and temperature are excluded according to clauses 5.4.2.2(7) and 5.4.2.5(2) from [9]). As the difference between M_{Ed} and M_{plRd} is significant for the loads considered, the girder cross section should be plasticized partially.

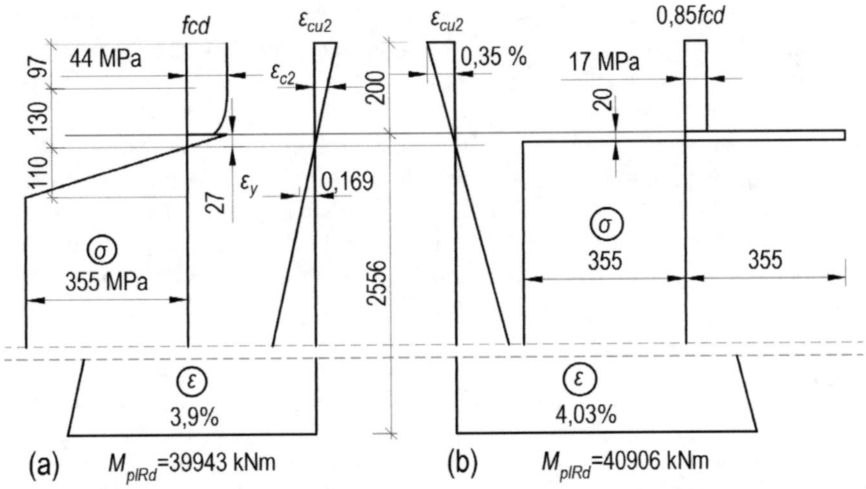

Figure 3. Strains and stresses at "full" plasticization, (a) with elastic core; (b) according to EC4.

It is interesting to know the girder behavior during a gradually live load increase up to the design level and further up to the moment when $M_{pl,Rd}$ is reached. The stress diagram is achieved in several steps. The first step includes the design value of load g_1 (figure 2b). Step 2 contains stresses from entire dead load, full influence of shrinkage, creep and temperature. The stiffener's stress is $222\ MPa$ – beyond the web's possibility to remain stable. It is calculated that the web would be stable (not plotted) under application of the entire dead load, creep and 81.6% of shrinkage. This strange percent value determines the condition at which the design stress appears at the upper web edge. Then the web stability is available: $\sigma_{sl} = 205 < 212 = \sigma_{sl,cr}$. Step 3 adds stresses from live load part in which the bottom flange plasticizes. At each subsequent step, the plasticized part of the previous step is excluded from the cross section. In code [6] instructions for determining the web stability for the case of partially plasticized web are missing, but is evident that the step 2 condition should not be allowed.

A question arises: why creep, shrinkage and temperature are involved into the calculations since for ULS EC4 does not require that? When the full cross section is plasticized the effect of those influences disappears. However, for partially plasticized cross section it exists, applied on the cross section part remaining in elastic condition.

Subsequent loading steps (figure 2b) are characterized by further plasticization inward of the web edges; for the upper edge it continues up to step 3 then goes back. The stress distribution corresponding to the design loading falls between the diagrams of steps 4 and 5 (drawn by dotted line and marked R_s).

Table 2. Loading steps, moments, stiffener stresses.

Steps	1	2	3	4	5	6
$M_{ed}(kNm)$	11000	14600	26272	30000	34000	36000
$\sigma_{sl}(MPa)$	170	222	205	174	117	62

In figure 2 the stress diagrams are extended only over the steel section because the concrete part remains in elastic condition and the accepted scale is inappropriate for clear perception of concrete stresses. More details from this series of calculations are presented in Table 2.

The plasticization process should be controlled, so that the edge strains to remain below the ultimate values determined in the codes; ε_{cu2} from [4] and ε_u for the structural steel (see C.8 from [6]). Figure 3 presents the condition of "full" plasticity. Inside the strain diagram (3a), the points ε_{c2} and ε_y

corresponding to the design stresses are marked. With their help, the stress diagram is drawn. Around the neutral axis it contains an "elastic" stress core, which is unavoidable due to the low strain level there. The term "full plasticization" is attached conditional. EC4 is accounting the core effects only upon the concrete part, decreasing the design stresses by 15% but neglects them for the steel section. Because of that the neutral axis is moved up and M_{plRd} becomes larger. The difference is small for the case considered but when the neutral axis is located deeper inside the steel section the opposite trend could be observed. This is due to the influence of the elastic core which is decreasing in the concrete part as well as in the steel part. EC4 does not give other solution for the distribution of the concrete part stresses. But the effect can be demonstrated even in the current example. If the concrete diagram from Figure 3a is averaged, the stress will be $0.92f_{cd}$ instead of $0.85f_{cd}$ as seen for the EC4 diagram. Furthermore, the first one may increase close to $1.0f_{cd}$.

In the contrary to the initial section classification into class 1, it turns out that the cross section may be associated with different classes depending on the loading level. The situation for class 1 cannot be reached because the next classes (3, 4) become critical at load levels below the level corresponding to plastic bending resistance. Therefore, the girder cross section should be adapted for each possible load level, below the design level in such a way, that the web stability is always insured. The plastic bending resistance remains imaginable albeit it is allowed according to Table 5.2 from [4]. The designer should be aware!

The girder (same cross section) was checked for another construction method – propped structure during concreting and hardening. In this case the girder remains in elastic condition after applying the design loads for ULS. The maximum compression direct stress in the longitudinal stiffener of 32.8 MPa is obtained, sufficient for the web's local stability.

6. Conclusion

Here is made an in-depth study of a simple engineering task. Guided only by the code instructions, sometimes the designer can easily make a mistake with unforeseen consequences. The ULS design situation may occur only once, after that the structure is not able to perform its functions. The situation played out above shows that critical load may not be the design load. In other words the application of clause 5.4.2.4(2) from [9] to each designed composite girder, classified in 1 or 2 classes, may not provide always sufficient reliability. The provisions of the code precedent clause should be valid for all cross section classes is the author's opinion. Furthermore, in both clauses are used terms "staged construction" and "sequence of construction". Its meaning should be explained separately if they are charged with some difference. A structure consisting of class 1 or 2 composite girders should be checked for possible loading situations (including part of the design live load) in order to prove its local stability.

Different ways exist, well known to the qualified designer, to solve the case considered with the available tools now (accepting suitable assumptions). One of them is described above. The point was to demonstrate a special design situation. Knowing about its effect, the designer should put more thought during his responsible job to create a reliable structure. Similar situations may exist also at composite girders under negative moments. Such case, normally including considerable shear forces, seems to be even more complicated.

IOP Conf. Series: Materials Science and Engineering 1276 (2023) 012021 doi:10.1088/1757-899X/1276/1/012021

References

[1] EN 1991-1-1. Eurocode 1: Actions on structures – Part 1-1 : General actions – Densities, self-weight, imposed loads for buildings. 2002.
[2] EN 1991-1-5. Eurocode 1:Actions on structures – Part 1-5 General actions – Thermal actions. 2003.
[3] EN1991-2. Eurocode 1: Actions on structures - Part 2: Traffic loads on bridges.2003.
[4] EN1992-1-1. Eurocode 2: Design of concrete structures – Part 1-1: General rules and rules for buildings. 2005.
[5] EN1993-1-1. Eurocode 3: Design of steel structures – Part 1-1: General rules and rules for buildings. 2005.
[6] EN1993-1-5. Eurocode 3: Design of steel structures – Part 2-5: Plated structural elements. 2006.
[7] EN1993-2. Eurocode 3: Design of steel structures – Part 2: Steel bridges. 2006.
[8] EN1994-1-1. Eurocode 4: Design of composite steel and concrete structures – Part 1-1: General rules and rules for buildings. 2004.
[9] EN1994-2. Eurocode 4: Design of composite steel and concrete structures – Part 2: General rules and rules for bridges. 2005.

On the need to include prestressed steel structures in the standard EN 1993. Load-bearing capacity of steel tensioners according to the rules of EN 1993

T Georgiev*

University of Structural Engineering and Architecture "Lyuben Karavelov" Bulgaria, Faculty of Construction, Department "Building Structures", 1373 Sofia, 175 Suhodolska str.

*E-mail: georgiev.todor.d@gmail.com

Abstract. This report draws attention to the absence of prestressed steel structures in the EN 1993 standard. He pleads for the correction of this omission. As an example of the difficulties in the transition of research to the statements of the European Guides, the problem of determining the bearing capacity of tensioners on prestressed structures has been developed.

1. Introduction

Prestressed steel structures experienced the most turbulent period in their development from the mid-1950s to the mid-1980s. The design and construction of the prestressed main farm of an aircraft hangar near Brussels, designed by Prof. Manel in 1953, is considered to be the starting point [1].

Within 12 years, 5 major international scientific conferences dedicated entirely to prestressed steel structures were held: in 1963 in Dresden; in the town of Tale in Czechoslovakia in 1966; in Leningrad (St Petersburg now) in 1971; in 1972 in Amsterdam (IVBH Congress), in Warsaw in 1974 and in Bratislava in 1975. [2]

In the 1960s, the first instructions and standards for the design of prestressed steel structures were issued in Europe: in 1963 in the former USSR and in 1969 in the former Czechoslovakia [1,3].

Many prestressed steel structures have been realized, but they remain unique and are rarely repeated in other buildings or facilities. Gradually, interest in this type of construction is waning, although the geography of researchers working on this issue is expanding and practically covers the whole world. However, the number of researchers is gradually declining.

At the same time, in precast concrete structures, prestressing finds applications that allow serial production of structural elements: prestressed roof and floor panels, prestressed floor and crane beams; prestressed roof trusses and Virindel beams [4 , 5], etc .. In recent years, prestressing has begun to be applied to reinforced pavements and reinforcement of various structural elements [6, 7]. In addition to the hidden tensile elements typical of reinforced concrete structures, in recent years the use of external tensile tensioners has begun, mainly for the purpose of reinforcement [8, 9].

Technologies for the production of the reinforced concrete elements described above are being created and improved, which in turn allows to reduce production costs. In addition to the typical reinforced concrete performance advantages, prestressed reinforced concrete structures are widely used. This in turn leads to the study of this type of construction in the basic courses at universities, while prestressed steel structures are studied only in special courses.

2. Eurocodes and presstressing

When creating and implementing the system of structural Eurocodes, which lasted more than 35 years, prestressed steel structures do not find a place in the standard EN 1993.

The same applies to the standards EN 1995 (wooden structures) and EN 1999 (aluminum structures) in which there is no prestressing too. In these constructions, prestressing is either rarely applied [10].

At the same time, prestressing is included in the standards for the design of reinforced concrete, combined reinforced concrete structures and even for masonry structures [11].

This non-inclusion of prestressed steel structures may also be due to the fact that EN 1993 is the highest of the standards in the Eurocode system.

The absence of prestressed steel structures in the EN 1993 standard is an additional, serious obstacle to their study and application in construction practice. Since pre-stressed steel structures have existed (with some exceptions), they have been developed in the absence of a regulatory framework. When it comes to designing such structures, designers work only on literature data that may contain inaccuracies and this poses a risk to designers.

Since the entry into force of the standard Eurocode codes, there has been a need to "translate" all knowledge in the field of prestressed steel structures into the "language" of Eurocodes and to bring them into line with their provisions and accompanying standards. This is a complex task and not everyone will decide on such a step. We must not forget the fact that each Eurocode explicitly describes when the statements in it are valid. Prestressed steel structures use materials, some of which do not appear in the reference standards. The same applies to some technological factors, coefficients, etc ..

Arguing for the inclusion of prestressed steel structures in the EN 1993 standard, the author realizes that this will be neither easy, nor cheap, nor can it be done quickly. The reason is that the variety in this type of steel structures is greater than in ordinary and thin-walled structures combined. But as the saying goes, "the longest way is taken in small steps".

As an illustration of the difficulties of the above-mentioned "translation", it is suggested below how the design capacities of the pre-tensioned steel tensioners of prestressed steel structures can be determined in accordance with the provisions of the Eurocodes.

3. Types of tensioners according to their design

The material from which a tensioner is made is usually high-strength steel, but it can also be high-strength steel ([1, 12]). or high-strength fiberglass (eg polyaramide [13]). The case of steel tensioners will be considered here.

Tensioners of various designs are used [1, 12]. The tensioner made of high-strength steel can be made of various steel products - wires, strip steel, rod reinforcement. From the wires (strings) are made (see Fig. 1): dense bundles of high-strength wires, annular bundles of wires, ropes of various designs, tensioners of coiled wire.

In dense and annular bundles, the wires are parallel and not twisted, and the ropes are always twisted. The rod reinforcement is applied independently.

Several types of high-strength steel wires with different cross-sections are used for the production of different ropes (see Fig. 2).

Steel tensioners made of twisted 7 string ropes are made of high-strength steel wires (strings) with small diameters - up to 8 mm.

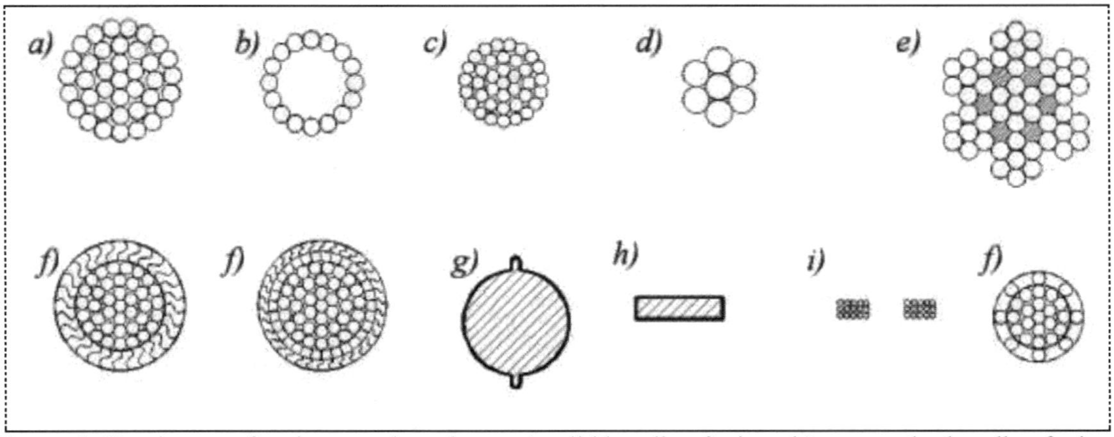

Figure 1. Tensioners of various steel products: a) solid bundle of wires; b) an annular bundle of wires; c) rope of spirally twisted wires (single twisted three-layer rope); d) rope of 7 twisted bodies; e) double twisted rope of twisted seven-stringed ropes; f) rope with closed section; g) high-strength bar reinforcement; h) strip steel; i) wound wire [1,12,14]

What is special about the tensioners is that they have a different construction and therefore have different load behavior.

Figure 2. Sections of high-strength wires: a) round (circular); c) segment of a circular wreath (wedge-shaped section); c) S - figuratively. [1, 12]

In [15, 16 и 17], the structures with stretched elements (hanging structures of all kinds) are considered. Their tensile elements (including when prestressed) are classified into 3 groups.

Group **A** includes rods with a dense cross section and thread at the ends.

Group **B** includes twisted ropes and combinations thereof.

Bundles of parallel wires are included in group **C**. The elements of groups **B** and C must be made of high-strength wires.

4. Tensile strength characteristics of tensioners

The strength properties of tensioners depend on many factors: the chemical composition of the steel, the heat treatment, the diameter of the wires, the cross-sectional shape of the wires. As prestressed steel structures are not included in the Eurocodes design system, information on the strength properties of tensioners must be obtained from other normative documents, from the specialized literature or from the technical specifications of the manufacturers. The closest to the considered problem are the normative documents [15, 16, 17, 18], the standards [19, 20 and 21] and the literary sources [1, 12, 4 etc.].

According to data from [22], citing the IS 1785 standard, the strength properties of high-strength wires decrease with increasing wire diameter.

The situation is similar with the wires included in modern European standards [20] - the tensile strength of wires ranges from 1860 MPa for wires with diameters of 4 and 5 mm, to 1680 MPa for wires with a diameter of 8 mm.

The elements of group A have the strength properties of the steels from which they are made. The practice is to use reinforcing bars with increased strength or high strength. The characteristic values of their tensile strengths and tensile strength are accepted by the EN 1993-1-1 [15], or by the EN 1992-1-1 [18].

High-strength wires must comply with standard EN 10264 [20]. The ropes must comply with standard EN 12 385 [21].

In [42 and 43] the maximum values of the characteristic tensile strength are given as follows:

- for round steel wires $\boldsymbol{max\ f_u}$ = 1770 MPa
- for Z-shaped steel wires $\boldsymbol{max\ f_u}$ = 1570 MPa
- for stainless steel wires $\boldsymbol{max\ f_u}$ = 1470 Mpa

5. Load-bearing capacity of steel tensioners according to the rules of EN 1993

For tensioners of group "A" (bars with solid cross-section) according to [15] the sizing should be carried out in the usual way, given in BDS EN 1993 [23] and [24] (stainless steels) depending on the type of steel used:

$$N_{Ed} \leq N_{t,Rd} \tag{1}$$

Unfortunately, this guide is too general and needs to be clarified. Two possible approaches to this problem are outlined below.

Reduced to the notation of [15], formula (1) takes the following form:

$$N_{Ed} \leq F_{Rd} \tag{2}$$

or:

$$N_{Ed} = max\, N_{o,Ed} \leq F_{Rd} = N_{o,Rd} \tag{3}$$

where:

$N_{Ed} = max\, N_{o,Rd}$ is the maximum tensile force that can occur in the tensioner;

$F_{Rd} = N_{o,Rd}$ is the design value of the tension resistance of the tensioner.

In the first approach, according to the author, this bearing capacity should be obtained analogously to the bearing capacity of tensile elements with holes for couplings (ie with weakening) from [23], or analogously to the bearing capacity of tensile strength of unstressed bolt [24]:

$$F_{t,Rd} = 0{,}9.A_s \frac{f_u}{\gamma_{M2}} = 0{,}72.A_s.f_u \tag{4}$$

where:

$\gamma_{M2} = 1{,}25$ is the partial factor for resistance of cross-sections in tension.

A_s is net area of a cross section of the rod through the thread at both ends;

f_u characteristic value of tensile (ultimate) strength of the steel from which the tensioner rod is made.

In [25] the formula (4) has the same form, with the difference that the coefficient ***0,9*** is calculated by taking into account more precisely the type of weakening.

In the second approach, the bearing capacity can be determined by analogy with the magnitude of the tensile force of pre-tensioned high-strength bolts, again from [24], when the thread at the ends of the tensioner meets the requirements of BDS EN 1990:

$$F_{t,Rd} = 0{,}7.A_s.f_u \tag{5}$$

The problem with the calculated tensile strength of the net cross section for steel grades above S 460 to S 700 (these are steels with increased strength) is also addressed in [26] in the form of an additional rule to EN 3-1-1, item 6.2. 3 (2):

$$N_{t,Rd} = 0{,}9.A_{net} \frac{f_u}{\gamma_{M12}} \tag{6}$$

where:

A_{net} is net area of a cross section

$\gamma_{M12} = \gamma_{M2} = 1{,}25$

After unification of the designations and abbreviations, the following is obtained again:

$$F_{Rd} = N_{o,Rd} = 0{,}72.A_{net}.f_u = 0{,}72.A_s.f_u \quad (7)$$

This result completely coincides with formula (4).

Finally, for tensioners of group "A", made of steels with increased strength up to class S 700 inclusive, it is possible to offer the bearing capacity to be calculated as follows:

$$F_{Rd} = N_{o,Rd} = 0{,}7.A_s.f_u \quad (8)$$

The question of how to dimension high-strength steel tensioners with a dense cross-section and threads at the ends (made of high-strength wires with a diameter in the uppermost available range) remains unregulated (unclear).

Note: according to [23], tensile elements have, etc. "Plastic bearing capacity", but in prestressed steel structures, tensioners are required to always work elastically, which means that the maximum stresses in them do not exceed the stress at which a residual deformation of a maximum of 0.1% - (see also [15 and 17]).

For tensioners of groups "**B**" and "**C**", when they are prestressed, in [15] there are more detailed instructions, in comparison with those for tensioners of group "A", which, however, need some supplementation to be usable.

The ultimate limit capacity test we are carried out according to the formula:

$$F_{Ed} \leq F_{Rd} \quad (9)$$

Reduced to the notations accepted here, it takes the form:

$$F_{Ed} = max\, N_{o,Ed} \leq F_{Rd} = N_{o,Rd} \quad (10)$$

The lower of the following two values is considered to be the tensile strength of the tensioner:

$$F_{Rd} = N_{o,Rd} = min\left(\frac{F_{uk}}{1{,}5\gamma_R}\, or\, \frac{F_k}{\gamma_R}\right) \quad (11)$$

where:

F_{uk} is the characteristic value of the breaking force.

F_k is the characteristic value of the effort at a certain percentage of the elongation; for tensioners of group "**B**": $F_k = F_{02,k}$, and for tensioners of group "**C**": $F_k = F_{01,k}$.

γ_R is a private factor which depends on whether constructive measures have been taken to reduce the bending stresses of the tensioner in the anchor devices as a result of its rotation; if such measures have been taken $\gamma_R = 0{,}9$; if no measures have been taken against the said bending, then $\gamma_R = 1$. Because the structures deform when the tensioners are tensioned, the angle at which the tensioner is anchored and the tensioner changes, so it is better to accept $\gamma_R = 1$.

When F_k is lesser than the two values in condition (11), this is an indication that the tensioner will work elastically when the impacts on the beam reach their calculated values, thus fulfilling the basic requirement for tensioners for elastic work throughout the period of operation .

For tensioners of group "**C**", the characteristic value of F_{uk} is calculated as follows:

$$F_{uk} = A_m.f_{uk} \quad (12)$$

where:

A_m is the cross section area of the metal;

f_{uk} is the characteristic value of the tensile strength of the wires from which the bundle of parallel wires is made.

The situation is more complicated with the tensioners from group "**B**", which is due to their complex construction - they are twisted ropes. They consist of wires that can be grouped by twisting into bundles, and the bundles can be twisted around a core, which can also be twisted into a bundle.

Not all wires in the rope are necessarily the same diameter.

In group "**B**":

$$F_{uk} = F_{min}.k_e \tag{13}$$

where:

k_e is the coefficient of loss, which depends on the type of anchoring at the ends of the twisted rope, as follows: for anchoring in an anchor cup filled with resin or metal $k_e = 1$; for a loop provided with a ferula $k_e = 0,9$; for anchoring with a stamped anchor cup $k_e = 0,9$; for anchoring with "U" - bolt $k_e = 0,8$.

$$F_{min} = \frac{K.d^2.R_r}{1000} \quad [kN] \tag{14}$$

where:

K is the experimental coefficient for the minimum breaking force taking into account the loss of rotation;

d is the nominal diameter of the rope in $[mm]$;

R_r is the class of rope in $[kN/mm^2]$;

The three quantities K , d , and R_r are clarified in [21].

$$K = \frac{\pi.f.k}{4} \tag{15}$$

where:

k is the experimental coefficient of loss when spinning the rope; depends on the class and construction of the ropes; k is equal to the ratio of the experimental breaking force to the calculated breaking force.

f is the coefficient of filling:

$$f = \frac{A}{A_u} \tag{16}$$

where:

A is the sum of the cross section area of the wires in the rope (metal section);(above the designation A_m was for tensioners of group "C").

A_u is the cross section area of the circle in which the rope fits snugly.

The value of R_r have to provided by the rope manufacturer; the strength class of the wires does not have to be the same as the rope class.

The conclusion for tensioners from group "**B**" is that the manufacturer must give several important values needed for the calculations. It is obvious that regulation of standards and obligations must be introduced.

6. Conclusions

The absence of prestressed steel structures in the EN 1993 standard is an additional, serious obstacle to their study and application in construction practice. Since prestressed steel structures have existed (with some exceptions), they have been developed in the absence of a regulatory framework.

To calculate the load-bearing capacity of tensioners on prestressed steel structures, the designer has to look for data in several standards. These data are not detailed enough. In addition, the designer must have important data from the manufacturer. This applies especially to tensioners from group "B", which are the most convenient to use.

References

[1] Ferenchik P and Tohachek M 1979 Prestressed steel structures M Stroyizdat
[2] Georgiev T 2018 The Bulgarian contribution in prestressed steel structures. Suspended structures - part 1 - realizations. UACEG Yearbook, Volume 51, Issue 4; pp. 139 - 149. Sofia, ISSN 1310-814X
[3] Belenya E and Gaidarov 1963 Guidelines for the design of prestressed steel structures. M, Goststroyizdat
[4] Dimitrov D 2013 Design of prestressed reinforced concrete structures according to Eurocode 2, ed. "Continuous improvement" Sofia
[5] Tsvetkov S 2019 Constructors for civil engineers - prefabricated one-storey buildings of reinforced concrete elements. VSU "L. Karavelov", Sofia
[6] Hrischev L and Naydenov V 2010 Materials for application of post-tensioned concrete slabs-on-ground. Technological card for construction. Part I – Materials, Conference Proceedings *International Conference on Civil Engineering Design and Construction 2010, Varna, Bulgaria,* pp. 399-404, ISSN 1314-6955
[7] Hrischev L, Naydenov V and Popivanov H 2010 Materials for application of post-tensioned concrete slabs-on-ground. Technological card for construction. Part II – Construction Technology, Conference Proceedings *International Conference on Civil Engineering Design and Construction 2010, Varna, Bulgaria,* pp. 405-410, ISSN 1314-6955
[8] Brozig C Experimental study of reinforced concrete beams elements reinforced with external prestressed reinforcement , 2017 Yearbook of UACEG - SOFIA, Volume 50, Issue 1
[9] Ivanov G 2019 Experimental study of the behavior of reinforced concrete slabs reinforced with externally stressed reinforcement. Yearbook of the University of Architecture, Civil Engineering and Geodesy, Volume 52, Issue 3, Sofia, Bulgaria, ISSN 1310-814X (print edition); ISSN 2534-9759 (online edition), pp. 901-912.
[10] Prestessed laminated wood beams, 1964 US forest service, research paper, FPL 8, January
[11] Barakova N. and Barakov T, 2014 Masonry structures, Eurocode 6, Sofia, "Continuous Improvement" Ltd
[12] Belenya E 1975 Prestressed load-bearing metal structures. M., Stroyizdat
[13] Burgoyne C J 1992 Tests on beams, prestressed with polyaramid tendons, 1-st international conference on advanced composites materials in bridges and structures, Sherbrooke Qebeck
[14] Mladzhov RV 1987 Hanging structures. Sofia. Technique 156 p.
[15] BDS EN 1993-1-11 2008 Design of structures with stretched elements, Sofia, BIS
[16] BDS EN 1993-1-11 / NA 2011 Design of structures with stretched elements, Sofia, BIS, March
[17] BDS EN 1993-1-11 / AC 2014 Design of structures with stretched elements, (Corrigendum), Sofia, BIS, September
[18] BDS EN 1992-1-1, 2007 Design of concrete and reinforced concrete structures, General rules and rules for buildings, Sofia, BIS
[19] pr BDS EN 10 138 2022 Prestressed steels. Sofia, BIS
[20] EN 10 264, 2012 Steel wire and wire products Steel wire for ropes, Sofia, BIS
[21] EN 12 385-2 2008 Steel wire ropes. Safety, Sofia, BIS.
[22] Sengupta A and Menon D 2012 Prestressed Concrete Stuctures, Indian institute of technology, Madras E Learning cours
[23] BDS EN 1993-1-1, 2011 General rules and rules for buildings, Sofia, BIS
[24] BDS EN 1993 1-8: 2005; Eurocode 3: Design of steel structures. Part 1-8: Design of nodes
[25] BDS EN 1993 1-4: 2007; Eurocode 3: Design of steel structures. Part 1-4: General rules. Additional rules for stainless steels
[26] BDS EN 1993 1-12: 2007; Eurocode 3: Design of steel structures. Part 1-12: Additional rules for extending the scope of EN 1993 to steel grade S 700.

Numerical modeling of mechanical behavior of a reinf orced concrete column-short corbel assembly strengthened by bonding a CFRF

V Stankov[1,2], I Ivanova[2], J Assih[1] and D Dontchev[2]

[1] University of Reims Champagne-Ardenne, Department of Civil Engineering, France,
[2] University of Chemical Technology and Metallurgy, Department of Applied Mechanic, Sofia, Bulgaria

veselin.stankov@etudiant.univ-reims.fr

Abstract. In this study a numerical simulation is developed using finite element (FE) software Abaqus, to assess and simulate the structural behavior of reinforced concrete column - short corbel assembly strengthened by bonding composite materials The objective of the presented simulation is to reproduce the behavior of the considered structure as the displacements, strain, the mode of damages, the cracks propagation, the ultimate force, and the behavior of the main reinforcements, obtained from the experimental study. "Concrete Damage Plasticity" (CDP) has been used for modeling concrete. It combines the damage and the plasticity models and it reproduces failure modes that are based on tensile cracking and compression crushing. The perfectly plastic elastic behavior of steel was introduced in this study. Carbon Fibre Reinforced Fabrics (CFRF) was assumed as a linear elastic orthotropic material. The mechanical behavior of the composite material, in the elastic stage, was defined as "Lamina" and the properties were obtained by the "Rule of mixture". The "Hashin Damage" failure criteria were also used for composite materials. The surface contact between the concrete and the composite material was developed through the cohesive zone model. The loading of the structure was done by displacement control for all developed models. This allows us to go further and study the influence of various parameters on the behavior such as the number of layers of bonding reinforcement, and the type of composite material. The optimal reinforcement is obtained by the wrapping of three layers of unidirectional carbon fiber fabrics.

1. Introduction

The reinforced concrete (RC) column-short corbel assembly is often used in civil engineering structures for manufactory, commercial buildings, bridges, etc. During his life, these structures are subjected to different conditions as aggressive environment, excessive load, fatigue, etc. which reduce their bearing capacity. One of the solutions is to strengthen the RC structure [1], [2], [3]. The carbon fiber materials have lightweight, good corrosion resistance, and it is easy to implement. Many researchers have been working on RC short corbel and column separately [1], [2], [4] [5], [6], etc. These structural elements have very complex mechanical behavior – the short corbel where shear span-to-depth ratios are less than unity is subjected to traction, compression, bending, and shear forces while the columns are subjected to compression and buckling. That`s why we are interested to investigate the assembly of these two structural elements. In the literature review, little information about the mechanical behavior of considered structural assembly was found. First, the mechanical behavior of RC column-short corbel assembly without strengthening is studied. After that, the strengthened assembly by wrapping by a different number of Carbon Fibre Reinforced Fabrics (CFRF) layers is investigated. The experimental study is a resource and time-consuming process. Therefore, it is economically recommended to use a combination of experimental investigation and numerical simulation (non-linear FE analysis). One of the important parts in the development of the FE simulation is the choice of an appropriate type of analysis and model of the behavior of the constitutive materials [7], [8]. For the nonlinear behavior of concrete, the “Concrete Damaged Plasticity” (CDP) model is implemented [7], [9], [10], also known as the "Barcelona model". The CDP model was developed by Lubliner et al [9] for monotonic load and then it was extended by Lee & Fenves [10] for cyclic loading. The damage mechanisms included in the model are tensile cracking and compressive crushing. This model considers the irreversible damages that can occur in the fracture process it can control the stiffness recovery effects. The reinforcement with elastic-perfectly plastic behavior was embedded as truss elements in concrete. Many researchers used the "Rule of Mixture" while defining the properties of CFRF [11], [12], [13]. We also used that rule. The behavior of CFRF is represented by Hashin's damage initiation criterion [7], [14]. A model of RC short corbel strengthened by bonding of CFRF was developed and validated using the experimental data from [1], [2], [4]. After the validation with these properties, the column was extended to obtain our model of reinforced concrete (RC) column-short corbel assembly. All FE model was developed, controlling the displacements. 30 mm ramped with smooth step displacement was introduced for all models. The value of the displacement was chosen in such a way, to cause a full collapse of the structure. The FE simulations showed with good agreement that the global mechanical behavior, crack pattern and bearing capacity of the structures were successfully reproduced.

2. Experimental program

2.1. Dimension of structure

The reference specimen of RC column short corbel assembly without strengthening is fabricated in real dimension. The assembly is composed of two short trapezoidal corbels cantilevering encastred symmetrically to the lathery sides of the column. The cross section of the column is 300mm in width and 150mm in height. The length of the column is 2660mm. The RC corbels have a cantilever projection length of 250mm, 150mm width, a total depth of 360mm at the column face, and 200mm at the end. The column was reinforced by four steel bars having a diameter of 14mm and stirrups with 6 mm diameter placed at 90mm distance. The main reinforcement of the corbel has a 10 mm diameter, placed at the top of the corbels having an effective concrete cover of 25 mm. The two main reinforcement bars are welded with a 10mm cross bar, close to the end of each corbel to provide additional anchorage. A horizontal closed stirrup having a diameter of 6mm is used as secondary reinforcement of corbels placed at 110mm from the main tension reinforcement. The dimension and the steel reinforcement of the structure are presented in figure 1.

2.2. Materials

The materials used in this study are experimentally characterized. Concrete cylindrical specimens with 160mm diameter and 320mm height were tested in simple compression. Steel reinforcement bars HA14, HA10, and Ø6 were tested in traction. The adhesive and the CFRF were tested also in traction. The concrete, steel reinforcements bars, and unidirectional carbon fiber reinforced polymers (CFRF). All results from the material characterization are presented in Table 1.

Table 1. Properties of materials

Materials	Young's Modulus (GPa)	Ultimate Strength (MPa)	Poisson's ratio
Concrete	30±2	33.2±2	0.25
Steel bars	200±1	610±10	0.3
Adhesive	4.1±1	36±1	0.41
Fabrics	86±1	775±65	0.45

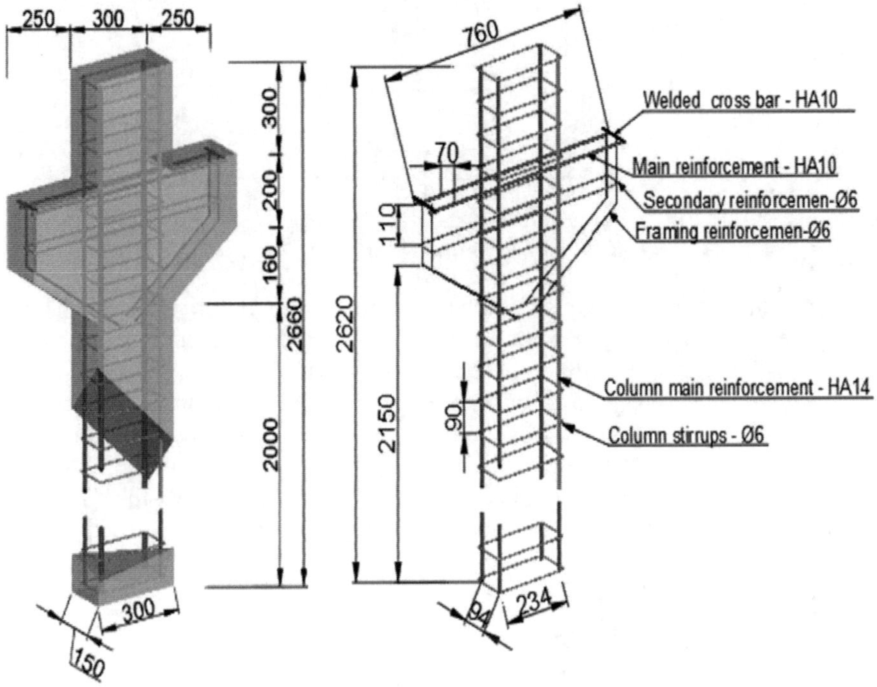

Figure 1. Geometry and steel reinforcement of RC column short corbel assembly.

2.3. Configurations of experimental study

The RC column-short corbel assembly was tested on a test slab in the civil engineering laboratory of the University of Reims Champagne-Ardenne, France - figure 2. The test machine has a hydraulic jack with a nominal capacity of 1000 kN. The structure was subjected to static load till its total collapse in reverse position compared to reality with a loading speed of 0.02kN/s. The shear span to depth ratio of the short corbel is 0.45. The load is applied to the base of the column and it is redistributed to the short corbel as its reaction. A hardwood plate is placed between the support and the concrete section to avoid the local crushing of concrete. The structure was instrumented with a strain gauge glued to the steel reinforcement and the strain data was captured every 1.0s. To protect the strain gauge from the humidity of concrete, special silicon was used figure 3a.

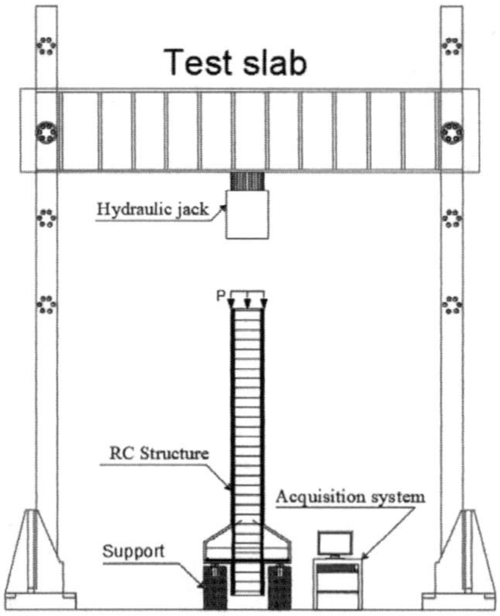

Figure 2. Test slab in the civil engineering laboratory of the University of Reims Champagne-Ardenne, France.

Figure 3a present the 5mm strain gauge glued to the steel bar of 10 mm diameter. The contact surface is carefully prepared in a way to have a good contact between the steel and strain gauge surfaces. The complete structure is instrumented with eight strain gauges symmetrically paced from the vertical plane of symmetry. Thus, it allowed us to confirm the right position of structure at the test slab-because it is important to apply the load as close as possible to its vertical ax. The position and names of the strain gauge are shown in figure 3b. The developed FE model presents only one-quarter of the structure, which means we will use-half of them because there is no big difference between the data obtained from the strain gauges placed at the right and left part of the structure relative to the vertical axis of symmetry.

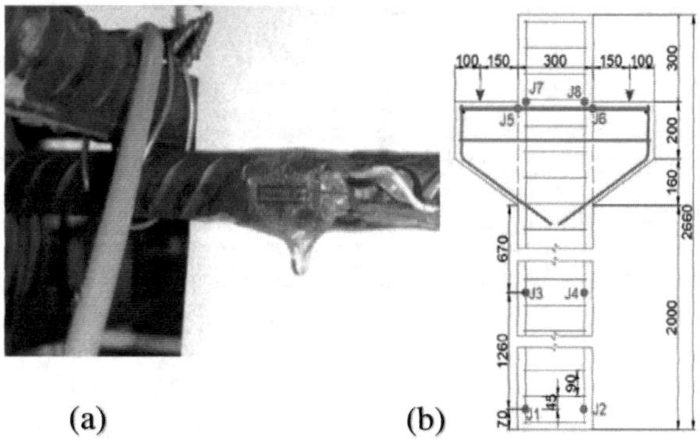

(a) (b)

Figure 3. a) Strain gauge glued at the steel reinforcement with a silicone protection and position of strain gauge at the structure steel reinforcement, b)Position and names of the strain gauge.

3. Finite element modeling

3.1. Material properties

3.1.1. Concrete

For our research, the most suitable model for concrete is the CDP model (Concrete Damaged Plasticity) proposed by LUBLINER, et al., 1989 [9] for a monotonic loading and later modified by LEE, et al.,

1998 [10] for cyclic and dynamic loads. In this model, the conception of isotropic damaged elasticity in combination with isotropic tension and compressive plasticity representing inelastic behavior is used. This model allows the reproduction of irreversible deformations. It assumes that the main failure mechanisms are tensile cracking and compressive crushing of the material. The constitutive model is represented by equation (1) as follows:

$$\sigma = (1-d)\mathrm{E}_0(\varepsilon - \widetilde{\varepsilon^{pl}}) \quad (1)$$

Where ε —strain, $\widetilde{\varepsilon^{pl}}$ – Plastic equivalent strain, d – Scalar damage variable and, E_0 – initial Young`s modulus of undamaged concrete.

The CDP model contains five important parameters to be introduced. The first parameter is the dilation (dilatancy angle) angle ψ measured in the p–q plane at high confining pressure. To determine the proper value of this parameter sensitivity analysis is to be done. For this study, the taken value is 36 [15]. The second parameter is ϵ - flow potential eccentricity. This parameter defines the rate at which the hyperbolic flow potential approaches its asymptote [7]. The default of ϵ=0.1 is used. The next parameter is the ratio between the biaxial compressive yield strength f_{b0} and the uniaxial compressive yield strength f_{c0}. The default value $f_{b0}/f_{c0} = 1.16$ is taken for this study [7]. The fourth parameter is the ratio of the second invariant of the stress on the tensile meridian, defining the form of the yield surface. It must satisfy the following condition $0.5 < Kc < 1$. The value suitable for this study is $Kc = 2/3$ [7]. The last parameter to define is the viscosity parameter μ. This parameter is very useful for Implicite analysis, for visco-plastic regularization, but it is ignored in Explicite analysis [7]. Many researchers have been working on defining the compression, tension behavior, and damage parameters for concrete that needed to be introduced in the FE analysis [16] [17] [18] [19]. Each one represents with a good agreement the ultimate load, crack pattern, and mechanical behavior. In our case, the most suitable model was proposed by [17].

The model proposed by F. ALFARAH [17] is based on Z. BAZANT and B. OH [20] and D. HORDIJK [21]. In general, the prescription of CEB-FIP 2010 [22] was taken to determine the needed parameters. Using this prescription, the obtained global behavior of the structure was closer to the real one.

3.1.2. Steel reinforcements

The steel reinforcement of the structure is with a diameter from 6 to 14mm and with yielding stress 557 MPa. The behavior of the material is assumed elastic perfectly plastic. The stress-strain curve is presented in figure 4.

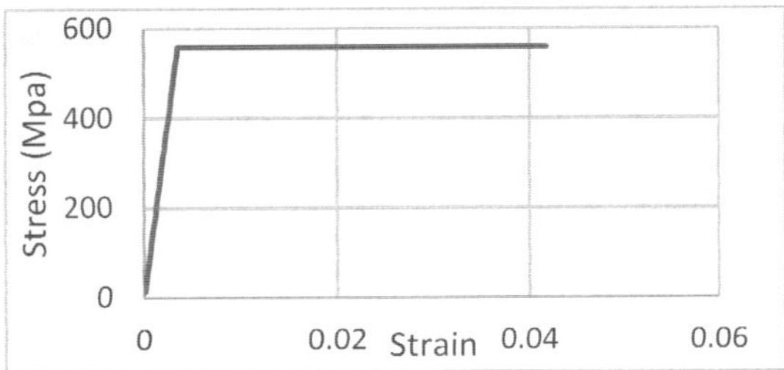

Figure 4. Stress-strain curve of steel reinforcements.

3.1.3. Carbon Fibre Reinforced Fabrics (CFRF) properties

A multiply unidirectional CFRF sheet is used for strengthening the reinforced concrete column-short corbel assembly. The thickness of CFRF and adhesive is assumed constant and perfect contact between them is also assumed. The Young's modulus along the fibers direction was determined by the mixing rule, where it is assumed, that when applying an axial load in the longitudinal direction, the load will be shared between the fiber and the matrix [23].

This condition can be represented by equation (2) and (3) as follow:

$$\varepsilon_c = \varepsilon_f = \varepsilon_m, \tag{2}$$
$$P_c = P_f = P_m \tag{3}$$

Where $\boldsymbol{\varepsilon_c}$ is the strain of CFRF, $\boldsymbol{\varepsilon_f}$ is the strain of fiber, $\boldsymbol{\varepsilon_m}$ is the strain of matrix, $\boldsymbol{P_c}$ is the stress of CFRF, $\boldsymbol{P_f}$ is the stress of fiber, $\boldsymbol{P_m}$ is the stress of matrix. According to [24], [25], and [26], CFRF bonded with adhesive provides an equivalent layer, but the adhesive layer between the concrete surface and the composite material is excluded from the equivalent layer. Therefore the mechanical properties can be estimated as follow:

The Young's modulus along the direction of the fibers was determined by:

$$E_1 = E_f V_f + E_m V_m . \tag{4}$$

Where E_f is the fiber modulus, V_f is the fiber volume fraction, E_m is the matrix modulus and V_m can be determined by the equation (5).

$$V_m = 1\text{-}V_f \tag{5}$$

The Young's modulus transverse the direction of the fibers was determined by:

$$E_2 = \frac{E_f E_m}{V_m E_f + V_f E_m}, \tag{6}$$

where $\frac{1}{E_2} = \frac{V_m}{E_m} + \frac{V_f}{E_f}$. (7)

3.1.4. Adhesive

The main data used for material characterization was given by the manufacturer. The used adhesive is a 2-part epoxy impregnation resin Sikadur®-330. The main material characteristic is tensile Strength - 30 N/mm^2, flexural E-Modulus -3800 N/mm^2, tensile E-Modulus - 4500 N/mm^2, elongation at the break - 0.9%. To simulate the contact between the concrete surface and the composite material we use the "Cohesive zone model" (CZM), by traction-separation law. To satisfy the main condition of this model, we define an additional special interface FE named cohesive element. The CZM consists of three separation components - one normal t_n and two parallel to the interface t_s and t_t. The cohesive elements have only translation degrees of freedom. The used criteria for damage initiations is quadratic traction damage (QUADS) for both modes I and mode II loading. This criterion can be represented by equation (20):

$$\left\{\frac{\langle\varepsilon_n\rangle}{\varepsilon_n^0}\right\}^2 + \left\{\frac{\varepsilon_s}{\varepsilon_s^0}\right\} + \left\{\frac{\varepsilon_t}{\varepsilon_t^0}\right\}^2 = 1 \tag{8}$$

It is useful to introduce an effective displacement to describe the evolution of damage under a combination of normal strain and shear across the interface [27].

$$\delta_m = \sqrt{\langle\delta_n\rangle^2 + \delta_s^2 + \delta_t^2}. \tag{9}$$

To deal with the general loading cases such as modes I and mode II, Mixed mode CZM is to be defined. The work done by the tensile force and their related relative displacements in the normal, first, and second shear directions, are G_n, G_s , and G_t respectively. The behavior of the cohesive element is assumed elastic linear until the damage initiation and can be presented as follow:

$$\begin{Bmatrix} t_n \\ t_s \\ t_t \end{Bmatrix} = \begin{bmatrix} K_{nn} & 0 & 0 \\ 0 & K_{ss} & 0 \\ 0 & 0 & K_{tt} \end{bmatrix} \begin{Bmatrix} \delta_n \\ \delta_s \\ \delta_t \end{Bmatrix}. \tag{10}$$

IOP Conf. Series: Materials Science and Engineering 1276 (2023) 012023 doi:10.1088/1757-899X/1276/1/012023

Where *Knn, Kss,* and *Ktt* are the elastic stiffness of the adhesive in the normal and the two shear directions respectively – figure 5

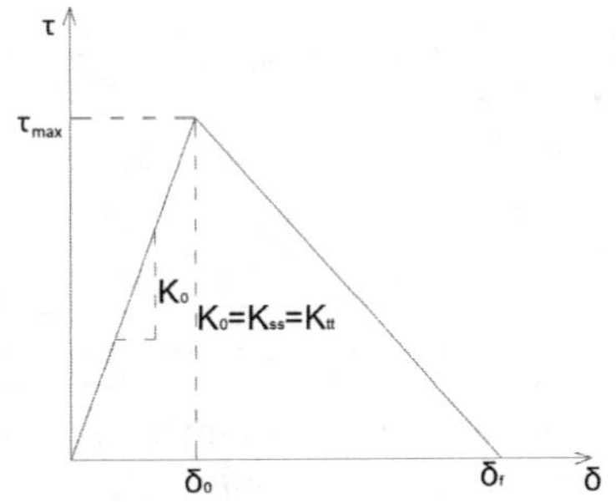

Figure 5. Traction-separation response.

Then mode I loading and can be expressed by equation (11).

$$K_{nn} = \frac{E_a}{T_0}. \tag{11}$$

And for mode II loading it assumed the condition presented by equation (12):

$$K_{ss} = K_{tt} = 3\left(\frac{G_a}{T_0}\right)^{0.65}, \tag{12}$$

Where E_a is the elastic modulus of adhesive, T_0 is the original thickness of the adhesive layer and G_a is the shear modulus of the adhesive.

The main damage criterion is that the damage starts at the point with τ_{max} and δ_0 – in other words, it is assumed that the damage occurs when the maximum nominal stress ratio reaches one. This criterion can be presented by equation (13)

$$max\left\{\frac{\langle t_n\rangle}{t_n^0}, \frac{t_s}{t_s^0}, \frac{t_t}{t_t^0}\right\} = 1. \tag{13}$$

4. Results and analysis

Figure 6 present the result from the experimental and FE simulation. "J2", "J4", "J6" and "J8" are the data from the experimental results. "J2Num", "J4 Num", "J6 Num" and "J8 Num" are the data obtained from the FE analysis results. The results show, that the FE simulation with a good agreement represents the mechanical behavior. Thus, the FE model can be validated with the obtained result.

Table 2. Study of the influence of the number of layers

Number of layers of CFRF	F_u Numerical	F_u/F_0x100 [%]	Experimental	%
0	260kN	100%	259kN	0.364857
1	468kN	180%	-	-
2	514kN	197%	-	-
3	528kN	203%	538kN	1.782156
4	568kN	218%	-	-
5	688kN	264%	-	-

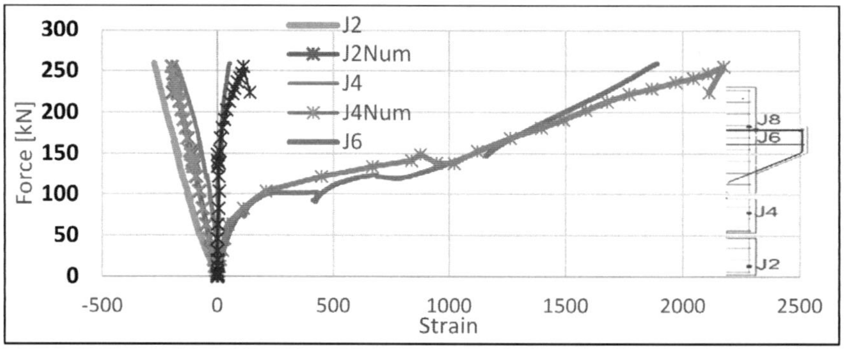

Figure 6. Force-Strain curve comparing the experimental and numerical data non-strengthened assembly.

Table 2 shows the study performed for the structure strengthened with different numbers of the layer of CFRF, where *Fu* is the obtained ultimate force and F_0 is the ultimate force for the structure without strengthening. The optimal strengthening is by bonding with three layers of CFRF, because the obtained gain for the load-bearing capacity compared to the invested materials are more economically appropriate.

(a)  (b) 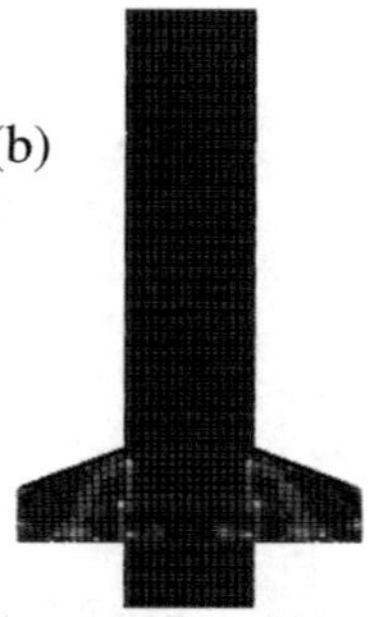

Figure 7 comparing the damages obtained by the numerical and the experimental study.

Figure 7 represents the crack pattern of the damaged structure "a)" and those from the FE model. Once the model is validated with the experimental result, obtaining the bearing capacity, mechanical behavior of the steel reinforcements, and the crack pattern, it was possible to investigate the influence of the different number of layers. The result is presented in ***Table 2***.

By analyzing the curves given by the strain gauges, the ultimate loads, and the failure modes, it can conclude that when an (RC) column-short corbel assembly, is subjected to the increasing load, the behavior of the reinforced structure can distinguish three different phases:

- Elastic phase

In this phase, the concrete, CFRF, and steel reinforcements are deforming in the same way, depending on the applied load. The appearance of cracks in the area of the tensile concrete due to bending is the end of this phase.

- Crack propagation

When the tensile stress on the lower fiber reaches its limit, the concrete cracks. However, the value of the stress remains quite low. After concrete cracking, the compressive concrete, steel, and composite plate continue to behave elastically with increasing load.

It has to be noted that, the increase in the load leads to the opening of the crack due to bending. In this phase, the contribution of CFRF to strength is greater than that of steel.

This phase ends with the appearance of the diagonal crack due to the shear forces. The CFRF therefore takes up all the efforts.

- Unstable phase

In this phase a diagonal cracks is opening. After the strengthening, the structure became more rigid. However, the fracture is characterized by a diagonal crack, opens progressively with increasing load.

5. Conclusions

Finite element models were developed to simulate the mechanical behavior of reinforced concrete (RC) column-short corbel assembly strengthened by bonding carbon fiber reinforced fabrics (CFRF). Numerical analyses were carried out and validated using experimental results. From the experimental-numerical comparisons, the following considerations were drawn:

1) The finite element procedure yields predictions that are in good agreement with the experimental results when using accurate non-linear material models. In addition, reasonable results were obtained when using the CDP model for the concrete, the simplified model for steel, making it suitable for design purposes, and Hashin Damage for the CFRF fabrics.

2) Experimental tests and numerical modeling demonstrate the efficiency of the proposed system to strengthen the RC corbels. The overall structural behavior is very satisfying, exhibiting a ductile deformation up to failure as the anchorage system changed the overall behavior of the CFRF-strengthened RC column-short corbel assembly to a composite one.

In our first experiments on our reinforced concrete column - short corbel reinforced with carbon fiber composite fabrics, we observed a marked improvement in the ultimate load-bearing capacity which could reach twice the value obtained by an unreinforced structure (207% i.e. + 107%) and an increase of a third of the rigidity of the corbel strengthened by wrapping.

3) The developed model show with a good agreement the crack pattern and the behavior of the main reinforcement of short corbel, obtained from the experimental study.

In the present case, we have observed different modes of failure:

- Failure by bending - shearing,
- Shear failure,
- Compression failure of concrete,

The damage remarkably reduces the elastic phase of the structure.

The bearing capacity, of the assembly, is also obtained with a difference of less than one percent. Thus, the model is validated and a parametrical study is performed, strengthening by wrapping the RC structures with different numbers of CFRF layers. Parametrical studies show, that the more economical solution is to strengthen the considered structure by wrapping it with two layers of CFRF

6. References

[1] Ivanova I 2013 Comportement mécanique de console courte en béton armé renforcée ou réparée par collage des matériaux composites *PhD Thesis URCA Reims France*

[2] Ivanova I, Assih J, Li A, Dontchev D and Delmas Y 2015 Experimental investigation into strengthened short reinforced concrete corbels by bonding carbon fiber fabrics *JAST* **vol 29 (20)** pp 2176-89

[3] Ivanov G 2018 Basic methods for strengthening of reinforced concrete floor structures *XVIII International Scientific Conference VSU ed.* Sofia, Bulgaria pp 621-26.

[4] Khadraoui A 1998 Analyse et modélisation du comportement mécanique des consoles courtes en béton armé *PhD Thesis Reims France* UFR Sciences de Reims

[5] Rafati A and Razavi S 2016 Finite element analysis of reinforced rc bracket using CFRF plates, JSM **vol 8 (4)** pp 734-743.

[6] Nguyen K 2015 Contribution a la compréhension du comportement des structures renforcées par FRP sous séismes *PhD Thesis INSA de Lyon*

[7] Abaqus 6.13 Simulia abaqus analysis user's guide

[8] Ben Ftima M 2005 Comparaison des trois logiciels d'éléments finis: Ansys, Adina et Abaqus, Rapport interne, École Polytechnique de Montréal

[9] Lubliner J, Oliver J, Oller S and Oñate E 1989 A Plastic-Damage Model for Concrete *Int. J. Solids Struct.* **vol 25** pp 299–329

[10] Lee J and Fenves G Plastic-Damage Model for Cyclic Loading of Concrete Structures 1998 *J. Eng. Mech.* **vol 12 no 8** pp 892–900

[11] Yasmeen T 2011 Structural retrofitting of concrete beams using FRP - Debonding Issues, Lund University *PhD Thesis*, Division of Structural Mechanics

[12] Borowski E, Soliman E, Kandil U and Taha M 2015 Interlaminar Fracture Toughness of CFRF Laminates Incorporating Multi-Walled Carbon Nanotubes *Polymers* **vol 7** pp 1020-45

[13] Kirn J 2016 Investigation of a shape adaptive aircraft wing leading edge with pneumatic actuation Munich : Technischen Universität München

[14] Hashin Z 1980 Failure criteria for unidirectional fibre composites *J. Appl. Mech* pp 329-334

[15] Kmiecik P and Kamiński M 2011 Modelling of reinforced concrete structures and composite structures with concrete strength degradation taken into consideration *Arch. Civ. Mech. Eng.* **vol XI** pp 623-636

[16] Carreira D and Chu K 1985 Stress-Strain Relationship for plain concrete in compression *J. Am. Concr. Inst.* **vol 82 no 6** pp 797-804

[17] Alfarah F, López-Almansa F and Oller S 2017 New methodology for calculating damage variables evolution in Plastic Damage Model for RC structures, *Eng. Struct.* **vol 132** 70–86

[18] Wahalathantri B, Thambiratnam D, Chan T and Fawzia S 2011 A material model for flexural crack simulation Brisbane: Proceedings of the First International Conference on Engineering, Designing and Developing the Built Environment for Sustainable Wellbeing, Queensland University of Technology pp 260-264

[19] Aslani F and Jowkarmeimandi R 2012 Stress-strain model for concrete under cyclic loading Mag. Concr. Res. **vol 64 (8)** pp 673-685

[20] Bazant Z and Oh B 1983 Crack band theory for fracture of concrete *Materiaux et Constructions* **vol 16(93)** pp 155-177

[21] Hordijk D 1992 Tensile and tensile fatigue behaviour of concrete; experiments, modelling and analyses *Heron* **vol 37(1)** pp 1-37

[22] CEB-FIP 2010 Model Code London: Thomas Telford,.

[23] Jones R 1975 Mechanics of composite materials *New York: Hemisphere Publishing*

[24] Al-Zand A, Wan Badaruzzaman W, Mutalib A and Hilo S 2017 Modelling the delamination failure along the CFRF-cfst beam interaction surface using different finite element techniques *J. Eng. Sci. Technol.* **vol 12(1)** pp 214-228

[25] Al-Zubaidy H, Al-Mahaidi R and Zhao X 2013 Finite element modelling of CFRF/steel double strap joints subjected to dynamic tensile loadings *Compos. Struct.* **vol 99** pp 48–61

[26] Fawzia S, Al-Mahaidi R and Zhao X 2006 Experimental and finite element analysis of a double strap joint between steel plates and normal modulus *Compos. Struct.* **vol 75(1-4)** pp 156-162

[27] Camanho P and Davila C 2002 Mixed-Mode Decohesion Finite Elements for the Simulation of Delamination in Composite Materials *NASA/TM* **2002–211737** pp 1 –37

Study of the behavior of a experimental reinforced concrete slab at different stages of loading from the production of the reinforced concrete slab over it

S Bakardzhiev[1]
[1]University of Structural Engineering and Architecture (VSU) "Lyuben Karavelov", Bulgaria, Sofia, "Suhodolska" str. 175
E mail: st.bakardzhiev@mail.bg

Abstract. This report examines the behavior of a reinforced concrete slab, part of an experimental reinforced concrete structure, under loads derived from the casting of the formwork system of the subsequent plate and from the laid concrete mix in the slab above it. The deflections in the individual stages were measured with the help of dial gauges. The full strength and deformation characteristics of the concretes used in the construction were researched. The behavior at all stages of loading of the slab was studied, including each phase of formwork of the slab above it and the laying of the concrete in the formwork itself. Conclusions have been made about the behavior of the slabs during production using the method for Early striking of reinforced concrete slabs.

1. Introduction

In the building industry, ways are always being sought for optimal use of formwork systems in the making of reinforced concrete slabs. Various combinations of casting, striking, propping and re-propping are used, as well as the Early Striking method. The early striking is based on the technology of proping and re-propping a field of slab with props and it makes more smaller fields, with smaller bending moments and shear forces in them [1], [2]. In order for formwork systems to be used as rationally and safely as possible, it is necessary to know well the behavior of slabs in the process of their casting and striking. There are studies that clarify questions about the load transfer of reinforced concrete slabs through the props of slabs below them at different ages and in different propping combinations [3], [4], [5]. A big part of the reports are summarized in [6]. In other studies, it is based on the use of the method for Early striking [7], [8], [9], [10],[11], [12], [13], [14], [15]. This study is an extension of research from a report by [1], focusing on the behavior of one of the first slab at a certain point in the production of the structure including two reinforced concrete slabs, when made it with an Early striking formwork system. This report studies the behavior of a experimental reinforced concrete slab at various stages of loading from the making of the slab above it - in particular the formwork and laying of the concrete for the top slab.

2. Purpose and main tasks in the report

2.1. Purpose

Purpose in the study is to experimentally examines the behavior of an experimental reinforced concrete slab loaded to the loads imposed on it by the construction of a slab above it, manufactured with a formwork system with specific drop heads suitable for application the Early striking method. To study in detail two of the main stages of loading, namely the behavior of the first slab during formwork and the behavior of the slab during the caasting of the concrete mixture in the slab above it

2.2. Main tasks

The main task is to measuring the deflection of the considered slab (level +1.21m) from the experimental construction in the different loading stages and comparing the real deflection in the slab with the elastic diflections obtained from a numerical model of the structure. Analyzing the results.

3. Construction

The dimensions of the construction is taken from [1]. The dimensions of the plates are given in Figure 1. The shors used between the two slabs are made for small floor heights [1]. For clearer results when considering the deflections, the reinforcement inserted in the slabs is very close to the minimum reinforcement coefficient [1]. The loads on the slab under consideration are from its own weight, the load from the formwork system and the concrete mix for the slab above it. With help of dial gauges are measured the deflections. The gauges are attached to a metal box frame between the slabs. The metal frame is separated from the construction.

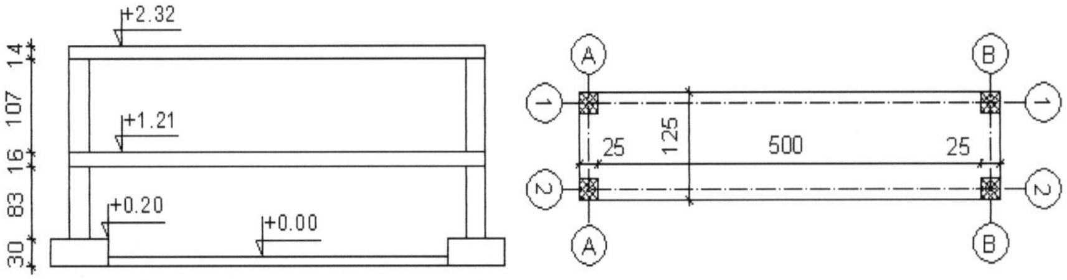

Figure 1. Drawing plan of the construction [1].

Figure 2. Stage of construction of the formwork to make the second slab.

Figure 3. Stage of loading the structure after laying concrete mixture in the formwork of the top slab.

4. FEM of the construction

The numerical model of the construction is based on FEM. The slabs were modeled with plate two-dimensional finite elements. The reinforced concrete columns and the metal shores were modeled with beam finite elements [1],[16]. The shores are modeled as with two pipe sections with a length of 50cm. The first section is with a diameter of 60mm and a thickness of 2mm. The upper section is with a diameter of 50mm and a thickness of 1.7mm. It should be noted that the prop height adjustment mechanism is not considered. Column-slab and shore-slab nodes are modeled with simple joint in the nodes. For each of steps of conducting the experiment, the real values of the modulus of the elasticity obtained during laboratory tests were entered into the numerical model.

5. Experimental research

In the report, there are tracked of the deflections and crack formation in the plates. The study was carried out in 10 steps in Figure 5. These are the steps that lie between step 2 and step 4 (inclusive) of the report of Figure 4[1]. Each step of this report shows the level of casting and striking, the age of the first slab concrete and the modulus of elasticity of the first slab concrete. Three stages out of the nine stages of the report [1] are considered, with the steps included in the individual stages indicated in Table 1. In this report, we look at stages 1, 2 and 3.

5.1. Steps from the report [1] on which the research is focused

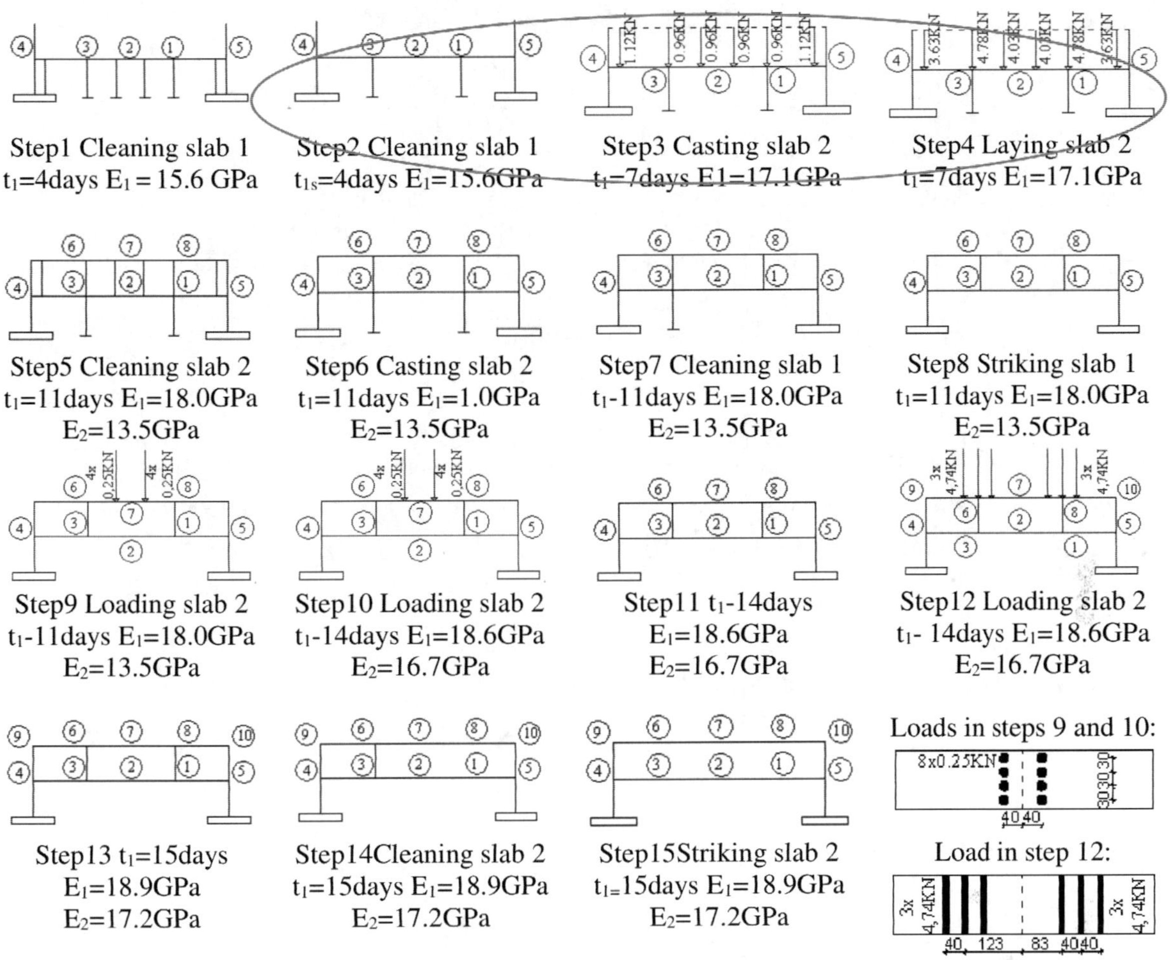

Figure 4. Construction process.

5.2. Stages from the report [1] on which the research is focused

Table 1. Stages of the constructive process in [1].

Stage	Step	Period (days)	Time (days)	Actions
1	1-2	4th	1	Cleaning 1st slab
2	3	4th – 7th	3	Casting 2nd slab
3	4	7th	1	Laying 2nd slab
4	4-5	7th – 11th	4	-
5	5-8	11th	0,5	Cleaning 2nd slab and striking 1st slab
6	9-10	11th – 14th	3,5	Loading 2nd slab with temporary load
7	11	14th	0,5	Unloading the temporary additional load
8	12	14th	0,5	Loading 2nd slab with temporary load
9	13-15	15	1	Removing the additional load and striking 2nd slab

doi:10.1088/1757-899X/1276/1/012024

5.3. Steps in this report between step 2 and 4 of the report [1]

Figure 5 shows the loading situations in each of the steps of the present study.

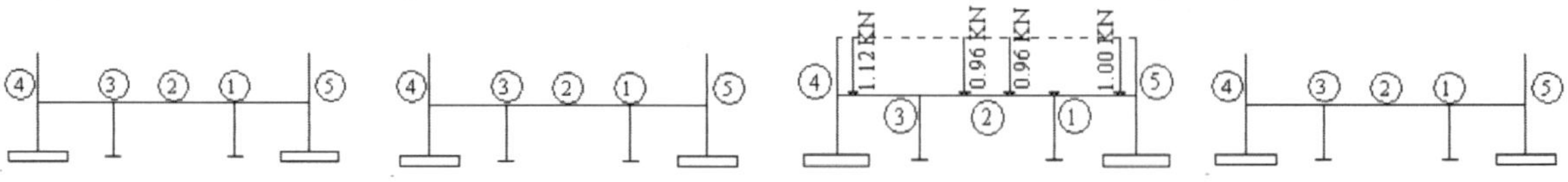

Step2 Partial striked level 1 t_{ls}=4days E_l=15.6GPa | Step3.1 Partial striked level 1 t_l=5days E1=16.8GPa | Step3.2 Loads from workers level 1 t_l=5days E1=16.8GPa | Step3.3 Partial striked level 1 t_l=5days E1=16.8GPa

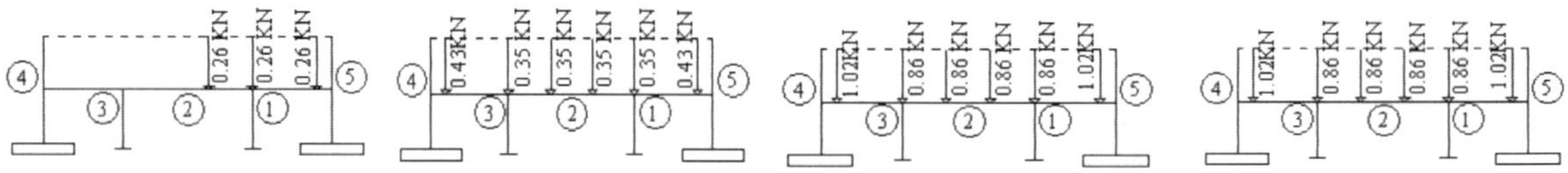

Step3.4 Partial casted level 2 t_l=5days E1=16.8GPa | Step3.5 Partial casted level 2 t_l=5days E1=16.8GPa | Step3.6 Partial casted level 2t_l=5days E1=16.8GPa | Step3.7 Partial casted level 2t_l=6days E1=17.0GPa

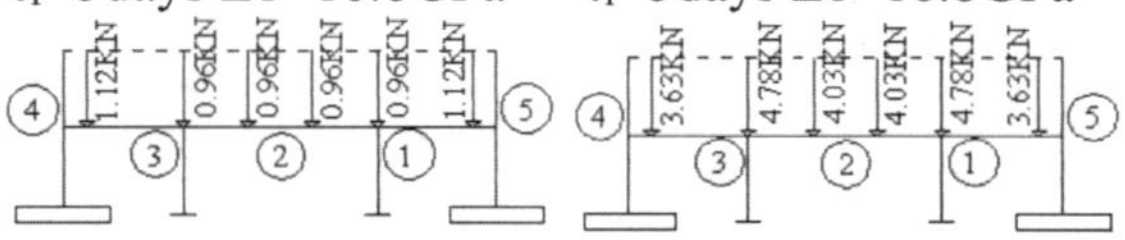

Step3.8 Casted level 2 t_l=7days E1=17.1GPa | Step4 Laying level 2 t_l=7days E_l=17.1GPa

Figure 5. Construction process in the study.

5.4. Load situations in photos

Figure 6 depicts each load situation from the current study

Step2 Partial striked level 1 t_{ls}=4days E_l=15.6GPa | Step3.1 Partial striked level 1 t_l=5days E1=16.8GPa | Step3.2 Loads from workers in level 1 t_l=5days E1=16.8GPa

Step3.3 Partial striked level 1 t_l=5days E1=16.8GPa | Step3.4 Partial casted level 2 t_l=5days E1=16.8GPa | Step3.5 Partial casted level 2 t_l=5days E1=16.8GPa

Step3.6
Partial casted level 2 t_l=5days
E1=16.8GPa

Step3.7
Partial casted level 2 t_l=6days
E1=17.0GPa

Step3.8 Casted level 2
t_l=7days E1=17.1GPa

Step4 Laying level 2
t_l=7days E_l=17.1GPa

Figure 6. Casting situations in photos.

6. Results and consideration

The elastic deflections in the slab (level +1.21m) are shown in Table 2. The measured real deflections in the slab (level +1.21m) are shown in Table 4. The real measured cracks are shown in Table 3.

6.1. Elastic deflections from the numerical model (mm)

Table 2. Elastic deflections.

Day	Point									
	1	2	3	4	5	6	7	8	9	10
4th (2) day	0.15	0.31	0.15	0.00	0.00	0	0	0	0.00	0.00
5th (1)day	0.15	0.31	0.15	0.00	0.00	0	0	0	0.00	0.00
5th (2)day	0.17	0.36	0.17	0.00	0.00	0	0	0	0.00	0.00
5th (3)day	0.15	0.30	0.15	0.00	0.00	0	0	0	0.00	0.00
5th (4)day	0.16	0.31	0.16	0.00	0.00	0	0	0	0.00	0.00
5th (5)day	0.16	0.33	0.16	0.00	0.00	0	0	0	0.00	0.00
5th (6)day	0.18	0.38	0.18	0.00	0.00	0	0	0	0.00	0.00
6^{h} (1)day	0.18	0.37	0.18	0.00	0.00	0	0	0	0.00	0.00
7th (1)day	0.19	0.38	0.19	0.00	0.00	0	0	0	0.00	0.00
7th (2)day	0.32	0.67	0.32	0.00	0.00	0	0	0	0.00	0.00

6.2. Results of measuring the cracks (mm)

Table 3. Real measures of cracks in the slabs (mm).

Period	Point									
	1	2	3	4	5	6	7	8	9	10
Stage 1	0.00	0.00	0.00	0.00	0.00	0.00	0.00	0.00	0.00	0.00
Stage 2	0.00	0.00	0.00	0.00	0.00	0.00	0.00	0.00	0.00	0.00
Stage 3	0.00	0.00	0.00	0.00	0.00	0.00	0.00	0.00	0.00	0.00

6.3. Experimental results of deflections (mm)

Table 4. Measured real deflections in each of the points.

Day	Point									
	1	2	3	4	5	6	7	8	9	10
4th (2) day	0.30	0.48	0.27	0.00	0.00	0	0	0	0.00	0.00
5th (1)day	0.33	0.49	0.29	0.00	0.00	0	0	0	0.00	0.00
5th (2)day	0.37	0.70	0.40	0.00	0.00	0	0	0	0.00	0.00
5th (3)day	0.30	0.60	0.41	0.00	0.00	0	0	0	0.00	0.00
5th (4)day	0.39	0.74	0.46	0.00	0.00	0	0	0	0.00	0.00
5th (5)day	0.49	0.81	0.51	0.00	0.00	0	0	0	0.00	0.00
5th (6)day	0.69	1.25	0.63	0.00	0.00	0	0	0	0.00	0.00
6h (1)day	0.73	1.24	0.84	0.00	0.00	0	0	0	0.00	0.00
7th (1)day	0.53	1.01	0.65	0.00	0.00	0	0	0	0.00	0.00
7th (2)day	0.99	1.66	1.06	0.00	0.00	0	0	0	0.00	0.00

6.4. Comparisons

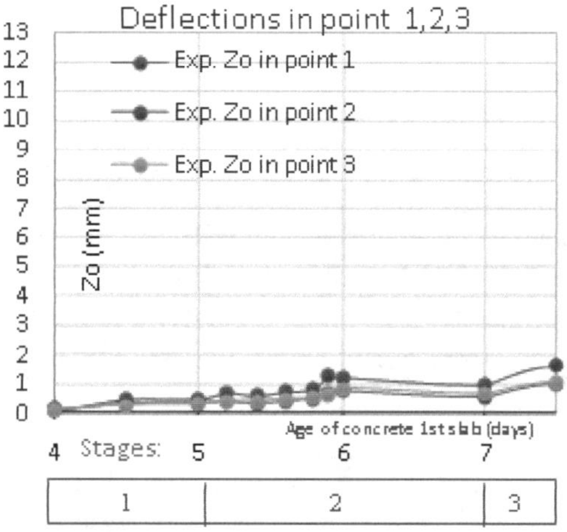

Figure 7. Point 1, 2 and 3.

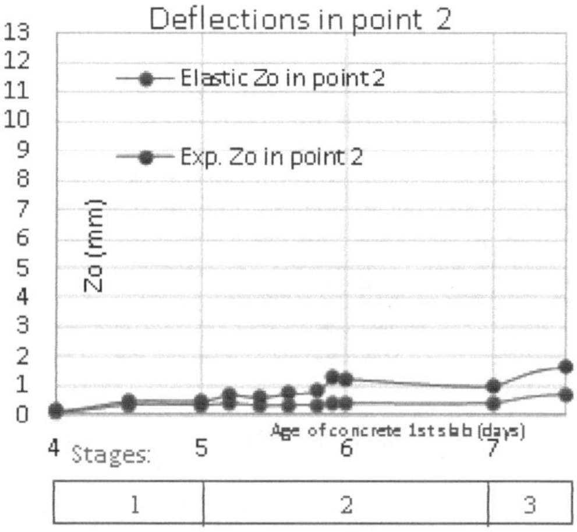

Figure 8. Point 2.

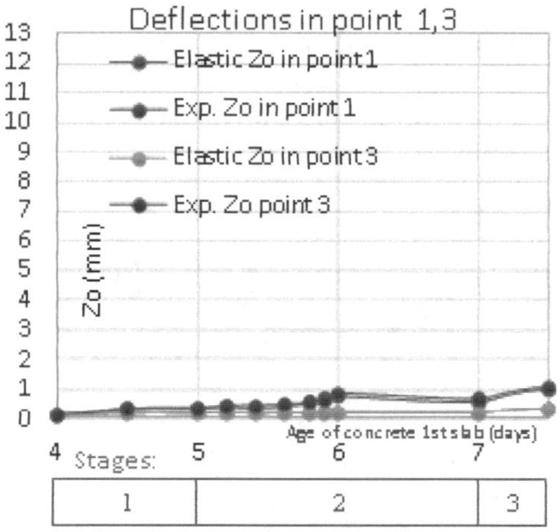

Figure 9. Point 1 and 3.

IOP Conf. Series: Materials Science and Engineering 1276 (2023) 012024 doi:10.1088/1757-899X/1276/1/012024

6.5. Observations on the deflections in the shores and mid-field areas from the first of slabs experimental measurements.

- In the zone from stage 2 to stage 3 of Figure 7, is observed more increase in the deflections at node 2 than these at nodes 1 and 3. The reason for this is non-centrically placed shores with loads over them from the formwork and the concrete of the slab over considering slab.

- At the beginning of stage 2 of Figure 7 with step 3, there is an increase in the elastic and experimental sags in nodes 1, 2 and 3, and this is due to the workers who are on the slab Figure 15. The people are arranged next to each other statically , with their own weight accounted for in the model as concentrated forces, in order to simulate the load of workers on the slab.

Figure 10. Step3.2 of loading with a simulated human load on the slab.

- At the end of stage 2, despite the increase in the loads from the formwork system, due to the complete and implementation, namely placing all horizontal frame panels, edging of the slab formwork and reinforcement - the deflections decrease. This is due to the fact that in this case, while the formwork system is not edged with formwork strips and a wooden frame behind them, the individual frame panels work as simple beams, and when we edge the formwork system, the sides of the formwork and begin to act as a frame, as this stiffens the bottom of the plate and the loads are distributed more evenly along the shores which transmit the loads to the first slab Figure 11 and Figure 12. The skirting and the wooden beam are clamped to the formwork panels.

Figure 11. Stage of partial construction of the formwork for the next slab.

Figure 12. Stage of full casting of level 2 with formwork strips. The skirting and the wooden beam are clamped to the formwork panels.

6.6. Comparison of elastic and experimental deflections in the considered slab

- In the zone from stage 1 to stage 3 of Figure 9, the experimental there are more deflections than the elastic ones. This is because the mechanism made for adjusting the height of the shores was not taken into account in the numerical model. Therefore shores are with bigger stiffness in the numerical model.

- In step 4 of stage 2, a return of the elastic deflections to the values of step 2 of the same stage was observed, but this was not observed in the experimental ones, which, in turn, speaks of the appearance of plastic deformations in the plate after unloading the load by workers on the plate. This is another reason for the bigger deflections than the theoretically calculated ones.

7. Conclusions

The following conclusions and recommendations are reached from the analysis of the results.

- Bigger deflections than calculated were observed due to two main reasons, that was discussed in point 6.6. That means that designers (structural engineers) should keep in mind this and must use adequate theoretical models for more precise calculations.

- Since the plates of the experimental structure work together, and the transmission of the loads from the first slab to the second slab is done by means of the telescopic supports in each of the stages, we come to the conclusion that each subsequent plate accepts part of the sagging of the plate below it from the formwork and this in some cases it is necessary to take into account.

- It is recommended that the telescopic supports be maximally arranged one below the other and to avoid their non-centric arrangement on the floors.

- Due to the fact that the telescopic struts directly affect the deformations in the slab, it follows that the more detailed the telescopic struts and their mechanisms for adjusting their height are modeled, the more precise results will be recieved in the inspections in the slabs by using the in the model. Not accounting for the telescoping strut height adjuster in the model would make the shore numerical model a stiffer element than it actually is.

Acknowledgments

Acknowledgments to companies "Tehstroy" Ltd. and "doka" Bulgaria, and eng. Radoslav Stafunski.

8. References

[1] Bakardzhiev S, 2020 *Experimental study of flat ,reinforced concrete slabs created in the method of early striking*, XI ISC „Civil Engineering Design and Construction", Sept. 10-12, Sept. 10-12, 2020, Varna, Bulgaria ISSN 2603-4255 (CD-ROM); ISSN 2683-071X (online),doi:10.1088/1757-899X/951/1/012010

[2] Haidarzhynska M., Hrischev L., Traykova M., Ivanov G., 2020 *Analysis of reinforced concrete slabs using formwork with dropheads during early striking*, Conference Proceedings XI International Scientific Conference „Civil Engineering Design and Construction", Sept. 10 - 12, Varna, Bulgaria ISSN 2603-4255 (CD-ROM); ISSN 2683-071X (online), pp. 312-324.

[3] Alvorado Y, Calderon P, Gasch I and Adam J, 2010 *Experimental and numerical study during construction of multistory buildings using clearing or partial striking*, (XXXVII IAHS World Congress on Housing)

[4] Gasch I, Alvorado Y, Calderon P and Ivorra S, 2013 *Construction loads using a shoring–clearing striking process*, (Structures and Buildings), Volume 167 Issue SB4, p 217-229

[5] Buitrago M, Alvorado Y, Adam J and Calderon P, 2016 *Maximum loads on shores during the construction of buildings*, (Article in ICE Proceedings Structures and Buildings)

[6] Buitrago M, Adam J, Moragues J and Calderon P, 2018 Load transmission between slabs and shores during the construction of RC building structures – A review, (Engineering Structures 173 (2018)) p 951–959

[7] Pallet P, 2019 *Backpropping flat slabs*, (Pallet Temporary Work), PFP / 136 F

[8] Zigmunt O, 2010 *The analysis of slab resistance during early striking exemplified by Gridflex Peri formwork system*, (Vilnius, Lithuania The 10th International Conference)

[9] Hrischev L, Tsvetkov S and Nedelchev N, 2015 *Early striking calculations of formwork for reinforced concrete slabs according to Eurocode 2*, (VII International scientific conference architecture, civil engineering – modernity Varna Bulgaria)

[10] Buitrago M, Adam J and Calderon P, 2018 *Structural failures in cast-in-place RC building structures under construction*, (ICITECH, Universitat Politècnica de València. Camino de Vera s/n, 46022 Valencia, Spain)

[11] Ivanchev I and Slavchev V, 2019 *About the Possible Limitations in the Usage of the Non-Destructive Ultrasonic Pulse Velocity Method for Assessment of Cracks in Reinforced Concrete Structures*, Subjected to Direct Environmental Exposure, Buildings, Volume 9, Issue 9, 202, http://dx.doi.org/10.3390/buildings9090202

[12] Bakardzhiev S and Lazova O, 2019 *Research of the strenght characteristics of concretes in the quality receiving period*, (XIX International scientific conference by construction and architecture VSU'2019)

[13] Ivanchev I, Slavchev V, Dobreva I and Daalov B 2018 *Experimental determination of modulus of elasticity and stress-strain diagrams of concrete*, (XVIII Anniversary international scientific conference by construction and architecture VSU'2018)

[14] Ivanchev I, Slavchev V, 2019 *Probable compressive strength and modulus of elasticity of concrete, determined by non-destructive ultrasonic pulse velocity method (NDUPVM) in different ways of measuring transducers placement*, Conference: XXIX International Scientific Symposium 'Metrology and Metrology Assurance'At: Sozopol, Bulgaria, September 6-10, DOI: 10.1109/MMA.2019.8936002

[15] Hrischev L, Rostovsky I, Conev I, Nikolov V, 2020 *Establishing the strength-maturity relationship of the concrete trough laboratory tests* – pioneer study in Bulgaria, Conference Proceedings XI International Scientific Conference „Civil Engineering Design and Construction", Sept. 10 - 12, Varna, Bulgaria ISSN 2603-4255 (CD-ROM); ISSN 2683-071X (online), pp. 338-349.

[16] Kazakov K, 2010 *Finite Element Method for structural modeling*, (Prof. Marin Drinov Academic Publishing House), Sofia, Bulgaria

Monitoring of road infrastructure operation indicators using mobile laser scanning

E Ivanova[1], S Gospodinov and D Boyadzhieva

University of Civil Engineering, Architecture and Geodesy, Faculty of Transportation Engineering,
1 Hr. Smirnensky blvd., 1000 Sofia, Bulgaria

E-mail: [1] altera1979@abv.bg

Abstract. The main problems of the national and local road infrastructure in the Republic of Bulgaria are the significant percentage of routes with poor and unsatisfactory operation indicators, and the significant number of road traffic accidents annually. This paper provides a summary assessment of Mobile Laser Scanning (MLS) as a means providing a package of activities that improve roads' technical and operational parameters and increase road's safety. It is noted that the use of MLS technology provides high-information value results, and surveying process is significantly faster and safer than that of conventional geodetic methods.

1. Main groups of problems of the road infrastructure

The national road network (NRN) of the Republic of Bulgaria includes motorways, expressways, and 1st, 2nd, and 3rd class roads. They provide transport connectivity of national importance and make up the state road network. Local roads are municipal and private. They are designed for public use and constitute the transport links of local importance.

The main problems of the road infrastructure in Bulgaria are the routes that are in poor and unsatisfactory condition and the high number of road traffic accidents annually.

The aim of this analysis is to accelerate the processes of condition assessment and to increase the options for higher roads' durability and operability in parallel with prevention and reduction of traffic accidents.

2. Necessary activities for road infrastructure maintenance

Package of construction and operation activities envisaged for building and maintenance of the national road network may include: new building, reconstruction, major overhauls (rehabilitation), maintenance (preventive, ongoing, emergency, and winter maintenance).

2.1. New construction

It includes the activities to build new roads respectively road sections, road links, utility facilities and appurtenances.

2.2. Road reconstruction

Activities to reconstruct road elements without altering the main direction of an existing route. It involves changes to structural elements, main parts, facilities and engineering networks. Adjustments to the road axis, changes to geometric elements of the road (route, dimensions, horizontal and vertical curves radius, longitudinal and transverse gradients, etc.) are also possible.

2.3. Major overhaul (rehabilitation)

Major overhaul includes a complex of repair works intended to restore road to operational condition. Its implementation is aimed to ensure traffic safety and extend road's service life. The activities envisage partial rehabilitation and/or partial replacement of depreciated materials, structures and structural elements. During the overhaul, level and partial plan changes of the route, roadway widening, etc. may be carried out within the restrictive strip - Figure 1 [1].

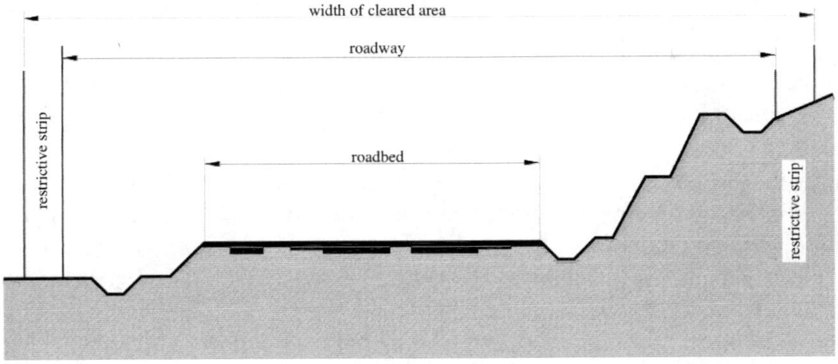

Fig. 1. Schematic cross-section of a roadway [2].

2.4. Road maintenance

This is a set of activities to ensure continuous, safe and comfortable road traffic, as well as increase of road durability leading to improved service life. It is subdivided into preventive maintenance, ongoing maintenance, emergency rehabilitation, winter maintenance.

3. Measures ensuring package of activities improving roads' technical and operational parameters and increasing road safety level

3.1. Survey of current road condition

Inspection of actual condition of existing roads is carried out to implement reconstruction, rehabilitation, and maintenance activities (preventive, ongoing) [3, 4], as well as to assess the whole impact of this parameter on road safety.

3.2. Monitoring of operational condition of transport facilities – bridges, tunnels, culverts, pedestrian links, etc. [5].

Surveying the current condition of bridges, tunnels, etc. includes a number of parameters. Capture specifics and summary results for this application will be presented in a subsequent publication in view of the many features.

3.3. Vertical planning and permanent road surfaces of urban transport communications, port and airport complexes

Surveying the current condition of urban transport communications, port and airport complexes includes a number of parameters. Capture specifics and summary results for this application will be presented in a subsequent publication in view of the many features.

3.4. Assessment of road surface condition (respectively of road's condition) [3]

Road surface condition assessment is carried out considering the condition of its main transport and operational indicators:

- Road surface bearing capacity
- Road surface levelness [4]

- Road surface slip resistance /roughness/
- Road surface damages – presence of damages and deformations [3, 4]

3.5. Winter maintenance of Bulgarian NRN with focus on snowdrift prevention [3, 6, 7, 8]
Snowdrift prevention includes a set of measures outlined in [3]. They have a number of weaknesses and do not lead to a lasting solution to the problem of NRN sections closure during snowfall combined with wind. In this regard, [6, 7, 8] propose an approach and methodological sequence to increase the effectiveness of the proposed protection. It recommends an approach for selecting a suitable snowdrift barrier through a common assessment methodology, which must be applied individually for assessment of each project assignment.

3.6. Road safety management and registering the causes for traffic accidents [1, 2, 9]
For each road class, specific road sections are analysed and categorised according to safety factors such as accident concentration, traffic volume and typology, etc. The package of activities can be summarized in three main groups:

- Visibility assessment to ensure traffic continuity and road safety [2].
- Survey routes' geometry and take into consideration their actual design parameters in operation which directly affect the mileage and level of road safety achieved.
- Survey the road and traffic situation in the event of traffic accidents or incidents.

3.7. Geological risk control and improvement of statistical condition assessment of terrains adjacent to roads and railway tracks with preconditions for forming rockfall and landslide areas in their side-slopes [4, 9, 10].
Surveying for this application includes a number of parameters. Capture specifics and summary results will be presented in a subsequent publication in view of the many features.

4. Rationale of the study

A major problem of Bulgarian road infrastructure is the large percentage of routes with poor and/or unsatisfactory performance. This is a prerequisite for reduced capacity, poor connectivity and serious combination of risk factors that lead to traffic accidents (largely with severe trauma). The results of survey measurements, allow to proposal of initial parameters to support a conceptual methodology for qualitative assessment of the condition of road elements and structures in service, including after repair and rehabilitation works. Based on this, a set of practical procedure sequences including modern "contactless" surveying methods in investment processes to ensure operability, traffic continuity and road safety can be sought. The implementation of modern technologies for surveying the transport infrastructure elements and their condition provides an opportunity to complement the existing procedural rules for road prevention and maintenance, which are adopted and regulated by the Ministry of Regional Development and Public Works, Ministry of Transport, Information Technology and Communications, Agency "Road infrastructure" (ARI), Road and Bridge Institute (RBI), and others. As a result, the assessment will be obtained with the involvement of minimal human resources with a high degree of survey precision (visual and geodetic) while maintaining continuity of traffic and safety of road users. All of this will reduce the cost of the investment process for road sections monitoring.

5. Scope of study and parameter selection of suitable road sections

Roads in Bulgaria that are subject of the study are the national and municipal roads. From the road infrastructure in or outside the settlements were chosen those sections that are not closed for repairs and have a variety of design geometric parameters in ground plan, longitudinal and transverse section.
A verification criterion was also the requirement for recorded damages, defects, local failures, etc. within the road section. For this purpose, we relied on a list of NRN road sections that are in poor condition and scheduled for repairs according to ARI's data for 2021 [https://api.bg/bg/1642084331.html]. The final selection was made after a walk-through and site inspection of selected routes from the list.

A condition that limited the selection mainly within the national road network was ensuring the continuity and uniformity of the mobile survey. This excluded roads in urbanised areas where road safety precautions have been implemented and unauthorised pedestrian crossing is possible.

6. Main research methods, testing and processing of results. Work programme stages.

- Direct measurements with a mobile laser system and rigorous processing using the least squares method.
- Direct survey measurements with rigorous least squares processing.
- Structural analysis and probabilistic-statistical methods.
- Modelling and simulations in BIM software.
- Application of a systematic approach to create synergy of the subject area.
- Evaluation and analysis of MLS results (through internal and external convergence in the country's coordinate and vertical systems and associated geodetic projections).

The study included the following work stages:

- Theoretical justification and identification of problem areas.
- Field work – surveying problem areas using MLS technology in the following stages: survey, preliminary preparation, field work. Creation of a data cloud.
- Processing of field work data – survey results values: trajectory, point cloud, alignment, georeferencing, colouring, filtering, export, obtaining of results in digital format.
- Processing of survey measurement results (scanning). Obtaining of graphical results – 3D models describing the route, facility or both.
- Final assessment of technological condition and compilation of a road's technical characteristics. Derivation of application principles for monitoring road infrastructure operation parameters using MLS.

7. Research results – evaluation of MLS to secure measures for package of activities, which improve roads' operational and technical parameters and increase road's safety

7.1. Surveying of current roads' condition

The study included the following activities:

- Surveying the current position of the existing routes and establishing the actual condition of the existing elements of road sections according to the scope shown in Figure 1 [2].
- Vertical planning and condition of the existing permanent road surface.
- Subsequent assessment of repair and rehabilitation works of road surface, shoulders/pavements and appurtenances, which took place during the year.

7.2 Monitoring the operational condition of transport facilities - bridges, tunnels, culverts, pedestrian links, etc.

Surveying the current condition of bridges, tunnels, etc. includes a number of parameters. Capture specifics and summary results for this application will be presented in a subsequent publication in view of the many features.

7.3. Assessment of road surface condition (respectively of the road) [3] - Figure 2

The assessment of road surface condition was made by surveying the current state of its main traffic and operation indicators:

- Levelness of road surface, respectively the possibility to determine slip resistance of the road surface.
- Presence and characteristics of existing road surface damage (failures and deformations).
- Subsequent assessment of repair and rehabilitation works carried out on road surfaces, shoulders/culverts and appurtenances, which took place during the year.

		Depending on weather conditions	Safety of the teams carrying out the survey	Human and traffic conflict-free points	Ensuring the continuity of traffic	Influence of the human factor	Possibility to set additional points every 10 m and the continuity of the survey along the entire length	Need for supplementation through stationary survey	Simultaneous geodetic and photographic survey	Accuracy of the results (for the needs of the standardized inspection)	Ability to obtain a 3D models	Survey speed and continuity
Surveying of current roads' condition	MLS	✓	✓	✓	✓	✓	✓	✗	✓	✓	✓	✓
	conventional geodetic methods	✗	✗	✗	✗	✗	✓	✗	✗	✓	✗	✗
Monitoring the operational condition of transport facilities	MLS	✓	✓	✓	✓	✓	✓	✗	✓	✓	✓	✓
	conventional geodetic methods	✗	✗	✗	✗	✗	✓	✗	✗	✓	✗	✗
Assessment of road surface condition	MLS	✓	✓	✓	✓	✓	✓	✗	✓	✓	✓	✓
	conventional geodetic methods	✗	✗	✗	✗	✗	✓	✗	✗	✓	✗	✗

Fig. 2. Assessment of road surface condition.

7.4 Winter maintenance of Bulgarian NRN with focus on snowdrift prevention - Figure 3

In order to improve snowdrift prevention of the NRN and municipal road network, an assessment procedure applicable to each specific road section should be used. For this purpose, a research model can be applied using numerical modelling based on computational fluid mechanics. This will significantly improve the applicability of the theoretical protective measures described in [3].

- An example procedure is presented in [6, 7, 8]. In its basic package of activities, it includes:
- Identification of road sections with high susceptibility to snowdrifts on the territory of the country.
- Identification of road sections according to geometric design parameters of the routes.
- Identification of road sections according to the landscape susceptibility of terrains outside the roadway.
- On-site investigation and survey of selected road section.
- Creation of a three-dimensional numerical model of terrain surface of the surveyed road section.
- Final three-dimensional numerical model of the terrain surface and road alignment.
- Conceptual options for snowdrift protection facilities.
- Efficiency evaluation and final selection of snow protection measures (facilities).
- In order to provide such an algorithm, precise surveying of the road within its range and of adjacent terrain must be performed to create a three-dimensional digital terrain surface within which snowdrift is investigated and subsequent generation of a digital three-dimensional terrain surface model that can be monitored for snowdrift situations using computational fluid mechanics software.

		Depending on weather conditions	Safety of the teams carrying out the survey	Human and traffic conflict-free points	Ensuring the continuity of traffic	Influence of the human factor	Possibility for continuity of the photo along the entire length	Need for supplementation through stationary survey	Simultaneous geodetic and photographic survey	Accuracy of the results (for the needs of the 3D models for Fluid Mechanics) **	Ability to obtain a 3D models	Survey speed and continuity
Winter maintenance of Bulgarian NRN with focus on snowdrift prevention	MLS	✓	✓	✓	✓	✓	✓	✓	✓	✓	✓	✓
	conventional geodetic methods	✗	✗	✗	✗	✗	✗	✗	✗	✓	✗	✗

Fig. 3. Winter maintenance of Bulgarian NRN with focus on snowdrift prevention.

7.5 Road safety management and registering of traffic accidents' causes – Figure 4
The assessment includes the following summary set of activities:

- Survey the current situation of existing routes and visibility to ensure traffic continuity and road safety [2] – as the analysis in Figure 3.
- Survey the route geometry and estimate their actual design parameters in operation, which directly affect the mileage and level of road safety achieved (item 3) – as the analysis in Table 1.
- Survey the actual traffic situation when accidents and/or incidents occur on the road. This includes road's design operational components, the presence of obstructions impairing the visibility, the actual location of participants in the traffic accident.

		Depending on weather conditions	Safety of the teams carrying out the survey	Human and traffic conflict-free points	Faster recovery of traffic in the area with a traffic accident	Influence of the human factor	Possibility for continuity of the photo along the entire length	Need for supplementation through stationary survey	Simultaneous geodetic and photographic survey	Accuracy of the results (for the needs of the investigation)	Ability to obtain a 3D models	Survey speed and continuity
Road safety management and registration of the causes of traffic accidents	MLS	☑	☑	☑	☑	☑	☑	☒	☑	☑	☑	☑
	conventional geodetic methods	☒	☒	☒	☒	☒	☒	☒	☒	☒	☒	☒

Fig. 4. Road safety management and registering of traffic accidents' causes.

8. Summary and applicability recommendations

Geodetic survey using the MLS method to secure transport infrastructure maintenance procedures in Bulgaria – summarized advantages of all application cases:

- Millimetre accuracy
- Image detail
- Minimum manpower
- Automated post-processing
- Minimum subjective errors
- Increased productivity
- Cost reduction

The study confirmed these significant advantages over conventional surveying methods, as its use leads to the following direct benefits:

- Increased safety of the teams carrying out the survey. Minimized human and traffic conflict points.
- Supplementing and refining the information from captured road damage and deformations such as cracks, ruts, drainage elements, etc.
  Respectively, a comparison can be made with the visual inspection at the date of its completion.
- Shortening survey time and meeting the project deadline. This is an element of the process that is essential to save financial resources for repair and restoration procedures.
- Increasing the level of safety by surveying elements that obstruct visibility (for example, in complex geometric and terrain conditions, in areas of road connections, etc.) and evaluating implemented protection systems for prevention of traffic accidents.

9. Conclusions

Survey works are main package of activities in the investment cycle of road network maintenance in Bulgaria. Respectively, their implementation is an essential part of the time and cost framework of proposals that appear in the projects to ensure road infrastructure operability.

The main conclusion of the study is that modern surveying methods can be a basis for optimizing maintenance procedures. This is evident from the analysis carried out along the main lines, as they illustrate that using MLS, can support rapid data collection on the actual condition of all NRN elements.
It can be expected that the application of modern surveying methods will support road infrastructure management processes in Bulgaria. This will finally enable us to respond adequately to the demand for compiling a detailed information database on:

- All sections in poor and unsatisfactory condition – this will enable proper targeting of maintenance funds from ARI's budget.
- All sections where a combination of design parameters, landscape and weather conditions require snowdrift prevention.
- All route parameters in locations with concentrations of traffic accidents (according to the Ministry of Interior) and those whose condition suggests possibility of severe traffic accidents. This will enable taking of measures to minimise the preconditions for accidents.

The expected result is that the widespread use of MLS-type methods will lead to a highly efficient road network that will meet the need for increased travel comfort (reduced travel times) and increased capacity of the infrastructure (throughput, continuity of traffic and opportunities for traffic growth, improved road safety, respectively reduced traffic accidents).

References
[1] Ministry of Regional Development and Public Works, ARI, Guidelines for Road Infrastructure Safety Management.
[2] Regulation No. RD-02-20-14 of 29 September 2011 on the scope and content of the road safety impact assessment and road safety audit, the conditions and procedure for their performance and for the acquisition and recognition of the professional qualification "Road Safety Auditor".
[3] National Agency "Road Infrastructure", Central Laboratory of Roads and Bridges, Technical rules and requirements for road maintenance, 2009.
[4] Shtereva K 2021*Specific applications in terrestrial laser scanning* (Sofia: Military Geographical Service).
[5] Postolovski Al, Dimitrov M, Markov M 2017 Application of MTLS – Technology in Tunnel Construction Audit 207-212.
[6] Ivanova E, Sulay I, Kacarova V, Miteva D, Stajkov I, Ivanova S,. Gospodinov S., Boyadzhieva D and Boyadzhiev S 2019 Investigation of the problem of snowdrifts on automobile roads in Bulgaria. Evaluation of the effectiveness of variant solutions of protective constructions with installation of PV panels *Annual of the UACEG* 52(3) 719-737.
[7] Ivanova S and Miteva D 2020 Investigation of the problem with the snowdrifts on the roads in Bulgaria. Evaluation of the efficiency of installation of PV panels in variant solutions of protective non-movable facilities *Annual of the UACEG* 53(3) 715-731.
[8] Ivanova E, Markov D and Mijorski S 2020 Investigation of the problem of snowdrifts on automobile roads in Bulgaria. Evaluation of the effectiveness against snowing of variant solutions of protective constructions with installation of PV panels *Annual of the UACEG* 53(3) 733-753.
[9] DIRECTIVE No. 5 of 23 September 2003 On the identification and securing of sections with concentration of traffic accidents and on road safety categorisation.
[10] Altyntsev M 2016 Automated detection of characteristic road lines from mobile laser scanning data, UDK 528.4.

Acknowledgments
Acknowledgements to the Research, consultancy and design centre at the University of Architecture, Civil Engineering and Geodesy which financed the project "Principal Methodology for Monitoring of Road Infrastructure Operational Indicators through Implementation of Modern Geodetic Technologies" - BN 256/21.

IOP Conf. Series: Materials Science and Engineering 1276 (2023) 012026 doi:10.1088/1757-899X/1276/1/012026

Water tanks - checks of tensile internal forces in the walls of cylindrical tanks

S Tsvetkov

University of Structural Engineering and Architecture "Lyuben Karavelov" Bulgaria, Faculty of Construction, Department "Building Structures", 1373 Sofia, 175 Suhodolska str.

E-mail: st.cvetkov@vsu.bg

Abstract. The report examines parametric solutions of the facilities - by changing the elements of the load-bearing structure of reinforced concrete tanks. The created computational models have specialized software and seismic analysis included. The hydrodynamic impact of water during an earthquake taken into account. The annular efforts in the walls were checked to determine the double (symmetrical) reinforcement in the radial direction. Comparisons are given for the individual solutions. The construction of the wall has an effect on the internal forces. Envelope combinations of static and seismic impacts must be made to provide the elements for peak values not taken into account in the analysis.

1.Introduction

Water tanks are important facilities for human life. The structures of the tanks are often reinforced concrete. Several computational models without roof reinforced concrete structures can be seen in the report, with monitoring of the value of the annular effort. The most labor-intensive are seismic analyzes, by far. They should be especially careful conducted because they use formulas from Dynamics of Building Structures, Eurocode 8-4 [1] and instructions (skills) when working with specialized software.

The solutions shown below are for open tanks, widely used in sewage treatment plants and industrial plants. The facilities must be filled with water during longer service life and out of service to protect it from impacts caused by rising water (empty tank loaded with pressure back to gravity) at high groundwater levels. It is no coincidence that, for example, radial precipitators (cylindrical vessels of small height) in treatment plants remain full even when not in use. Successful solutions to such problems are drilling with direct water abstraction.

2.General information about the objects of the study

2.1. Geometry and impacts

In figure 2 shows the formwork schemes of some of the studied tanks. They are grouped by models described in table 1. In figure 3 shows the diagrams of the obtained profiles of the walls.

Material for the construction: For the decisions it is accepted: concrete grade C25/30 - for structural elements.

Main dimensions of the facilities:

• radius of the wall $R_{cyl}=7,75m$;

• the cylindrical walls are with wall thickness $\delta_{cyl}=18cm$;
• the upper support rings are with sizes $b_r/h_r=35/50cm$;
• ring-thickening along the height of the wall - $b_{local}/h_{local}=36/60cm$;
• the annular foundation has a width of $a_{found.}=200/h_{found.}=40cm$;
• the bottom plate is 25 *cm* thick;
• the piles are with 30 *cm* diameter and 6 *m* long (see figure 1);

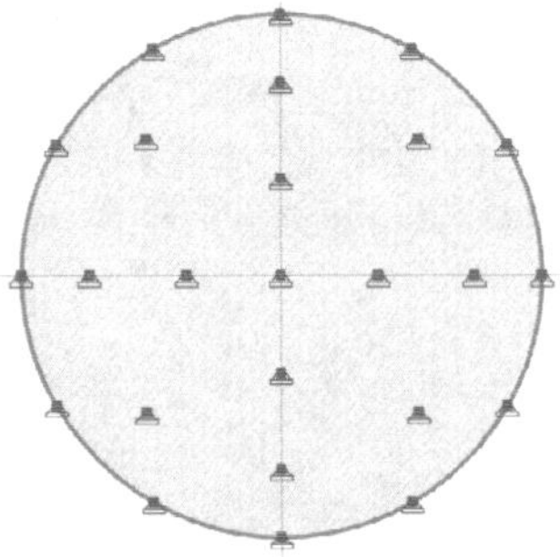

A total of 25 pilots located through 4 *m* along the contour of the bottom, as well as along the main coordinate axes through 2,90 *m*.

Figure 1. Plan of piles - picture from software.

• "hidden" columns in the contour of the wall - with dimensions 18/18 *cm*, distributed through 3,50 *m* along the cylindrical wall.

Fluid storage in tanks - data:
• height of the stationary water level $H_w=4,80m$;
• light radius of water $R_w=7,66m$.

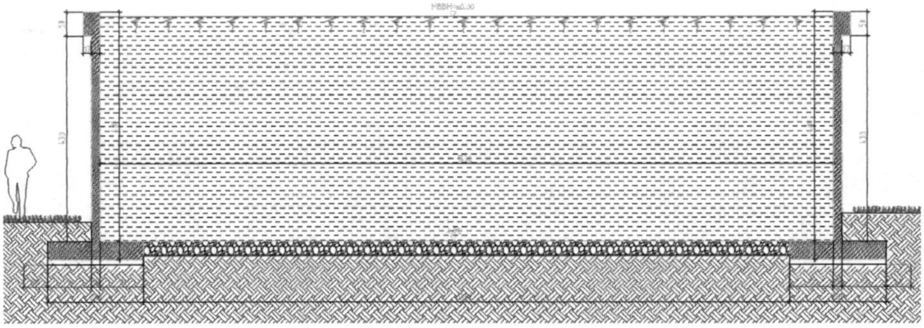

a). Model 1.

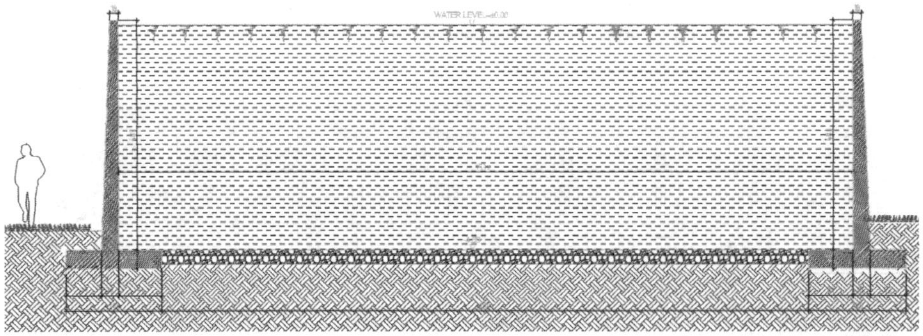

b). Model 4.

Figure 2. Fragments of formwork plans.

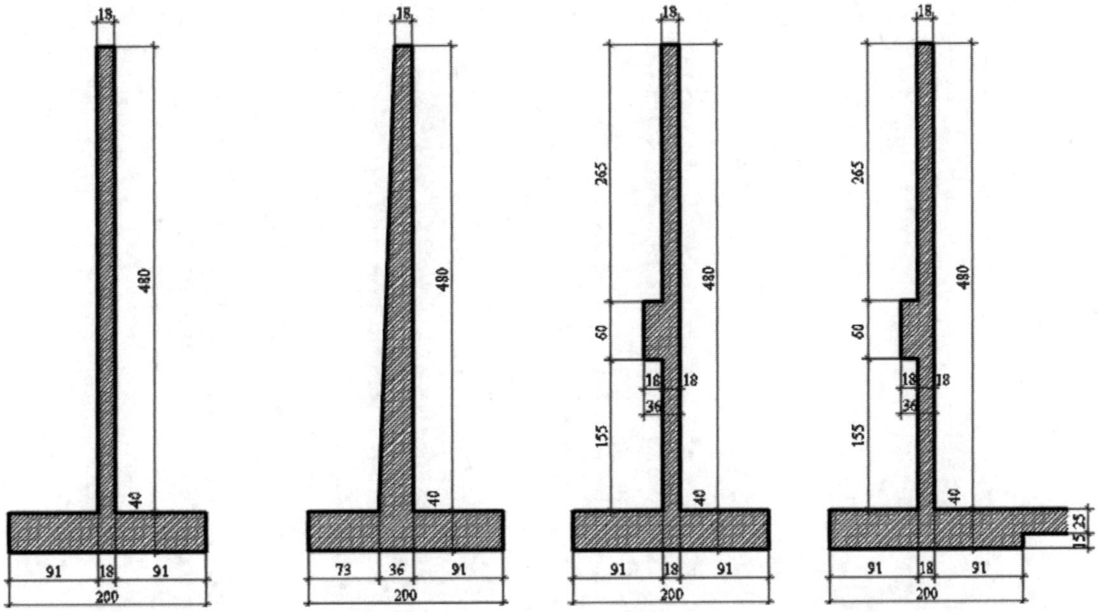

Figure 3. Transverse profiles of the walls: Model 2; Model 4; Model 6 and Model 7 – fragment.

In the table 1, the models M1 ÷ M9 are systematized, with a description of the study of the structure. Considered models, represent: models with annular foundations (under the walls) and models with a bottom plate - with piles (see figure 1).

Table 1. Characteristics of computational models

Model – M...	Wall type
M1	with support ring (at the upper end of the wall)
M2	similar to the M2 model, but without a support ring
M3	with a variable thickness on the wall and with a support ring
M4	similar to the M3 model, but without a support ring
M5	similar to the M1 model with a support upper ring, but also with thickened ring along the height of the wall
M6	similar to the M5 model, but without a support ring
M7	with modeled: local ring in the wall and bottom plate with piles
M8	similar to the M7 model - with a bottom plate and piles, but without a local ring
M9	with "hidden" columns in the contour of the wall

The considered models with annular foundations (under the walls) – models M1 ÷ M6 and M9, as well as with a bottom plate - with piles (models M7 and M8), have the following charasteristics for the soil:

- Winkler constant: $\left| \begin{array}{l} K_{Winkler}^{static\ loads} = 30\,MPa/m' \\ K_{Winkler}^{seismic\ loads} = 70\,MPa/m' \end{array} \right.$;
- coefficients of proportionality in height of the frame elements for piles: $K, R_{x/y/z}$ (see table 2).

Table 2. Coefficients of proportionality for characteristic points entered

	Stiffness		
№	K,R1	K,R2	K,R3
1	7.000e+4	7.000e+4	7.000e+4
2	2.375e+2	5.938e+2	5.938e+2
3	7.125e+2	1.781e+3	1.781e+3
4	1.188e+3	2.969e+3	2.969e+3
5	1.663e+3	4.156e+3	4.156e+3
6	2.138e+3	5.344e+3	5.344e+3
7	2.613e+3	6.531e+3	6.531e+3
8	3.088e+3	7.719e+3	7.719e+3
9	3.563e+3	8.906e+3	8.906e+3
10	4.038e+3	1.009e+4	1.009e+4
11	4.513e+3	1.128e+4	1.128e+4
12	4.988e+3	1.247e+4	1.247e+4
13	5.463e+3	1.366e+4	1.366e+4
14	5.938e+3	1.484e+4	1.484e+4
15	6.413e+3	1.603e+4	1.603e+4
16	6.888e+3	1.722e+4	1.722e+4
17	7.363e+3	1.841e+4	1.841e+4
18	7.838e+3	1.959e+4	1.959e+4
19	8.313e+3	2.078e+4	2.078e+4
20	8.788e+3	2.197e+4	2.197e+4
21	9.263e+3	2.316e+4	2.316e+4
22	9.738e+3	2.434e+4	2.434e+4
23	1.021e+4	2.553e+4	2.553e+4
24	1.069e+4	2.672e+4	2.672e+4
25	1.116e+4	2.791e+4	2.791e+4

The permanent impacts on the wall are accepted from: waterproofing and double-sided screed. Hydrostatic pressure is a basic load, by law of triangle: $w_{k,max}=\gamma_w.H_w$. In the models with bottom, the hydrostatic pressure loads with a constant maximum value over the entire surface.

2.2. Modeling [2]
For the models, the so-called "infinitely rigid zones" in the walls, in the areas of the upper support rings have not been introduced. This is a solution, as roof structures are missing, and efforts are also sought that are not laden with additional prerequisites for analyzes and coefficients.

The finite elements are 20 *cm* in size.

2.3. Analysis [2]
The combination at full to maximum stationary water level tank is considered.

The decisions were performed with Radimpex Tower 7 software product, based on work on the Finite Element Method.

Static analysis: The ULS combination includes the impacts: constant, multiplied by **1,35** and hydrostatic pressure multiplied by **1,2**.

Seismic analysis:
Analyzes are spectral with preliminary modal analysis.

The studies were performed for soil type B (according to Eurocode 8) and reference acceleration $a_{gR}=0,23.g$ for the location of the site on the territory of Bulgaria.

The coefficients of: importance γ_I and behavior q - for fluid (for impulsive and convective hydrodynamic components) and structure of the tanks, are accepted: **1**.

The nature of the masses (see figure 4):
→ impulsive mass m_i: its oscillation frequency corresponds to that of the tank and has a maximum value at the base.
→ convective mass m_c: oscillates independently, creating hydrodynamic waves and has a maximum value in the upper part of the water.

The maximum value of the hydrodynamic wave d_{max} is not monitored in the investigated tanks, as lids will be provided.

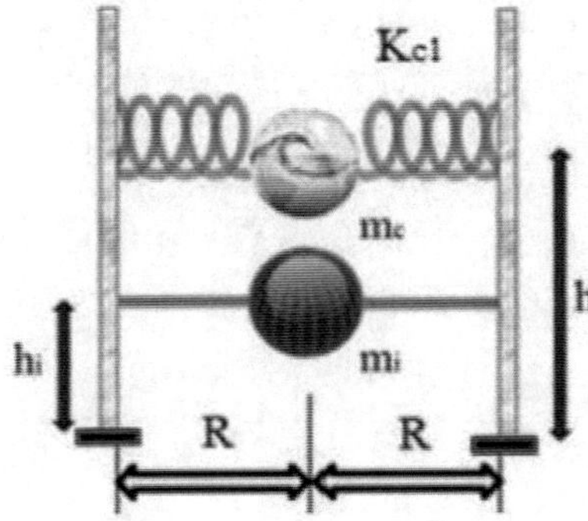

a). Mechanical-mathematical model - model of G. W. Housner [3].

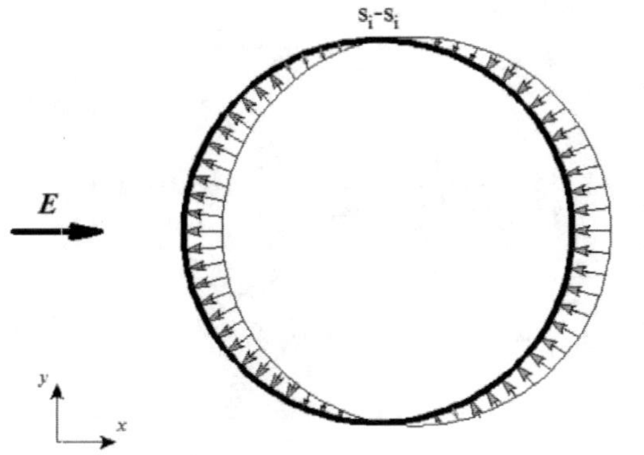

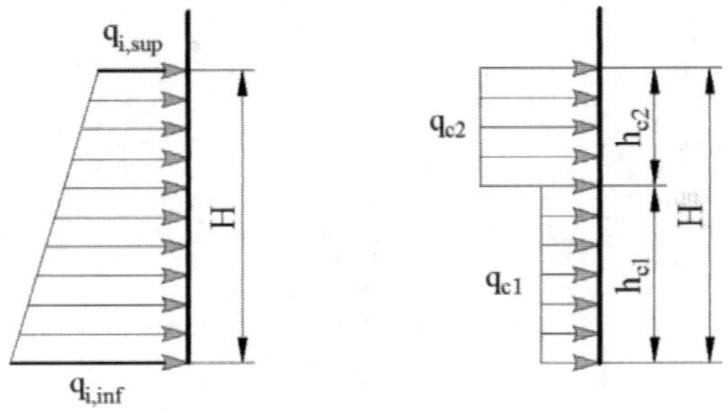

b). Pressures in cross section c). Equivalent impulsive and convective loads for the wall.

Figure 4. Hydrodynamic loads - "transformed loads" from m_i and m_c [4], [5], [6], [7]

Loads on the wall:

$$\left| \begin{array}{l} q_{i,sup}=11,5\,kN/m^2 \\ q_{i,inf}=45\,kN/m^2 \end{array} \right.$$

$$\left| \begin{array}{l} q_{c2}=21,1\,kN/m^2 \leftarrow h_{c2}=2,11\,m \\ q_{c1}=16,6\,kN/m^2 \leftarrow h_{c1}=2,69\,m \end{array} \right.$$

The seismic combination ALS includes the loads: permanent, hydrostatic and hydrodynamic pressure, incl. seismicity, with coefficients **1**.

3. Results

In the table 3 the values of the observed internal forces are systematized, in the combination - full tank, the maximum values are indicated in red.

Table 3. Internal forces N_x, *[kN/m']*, in the models

Combinations	ULS	upper end	-18,02	-3,18	-10,60	11,66	-13,78	7,42	-20,14	-12,72	-16,96
		middle	203,52	202,46	199,28	198,22	**403,86**	401,74	231,08	230,02	212,00
		lower end	290,44	290,44	331,78	331,78	275,60	276,66	2,12	4,24	287,26
	ALS	upper end	167,48	219,42	189,74	424,00	107,06	250,16	150,52	332,84	108,12
		middle	584,06	564,98	486,54	470,64	**827,86**	802,42	588,30	577,70	574,52
		lower end	-108,12	-112,36	215,18	246,98	-124,02	-108,12	-119,78	-115,54	-111,30
Model			M1	M2	M3	M4	M5	M6	M7	M8	M9

In figure 5 are shown diagrams by isolines of the internal forces for models M3 and M4, showing that in ULS combinations the maximums are at the lower end, and in ALS - in the middle.

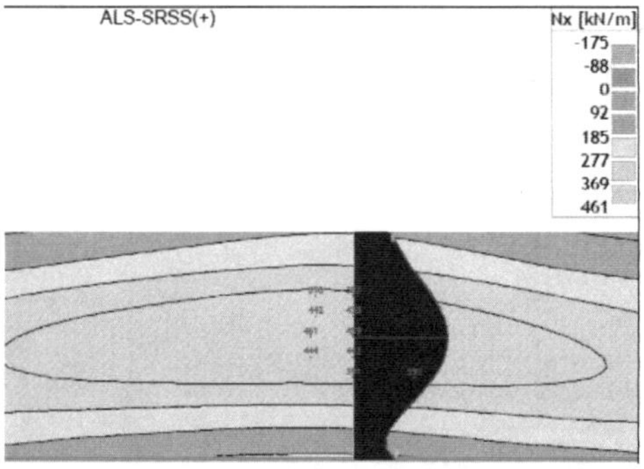

a) . Model M3.

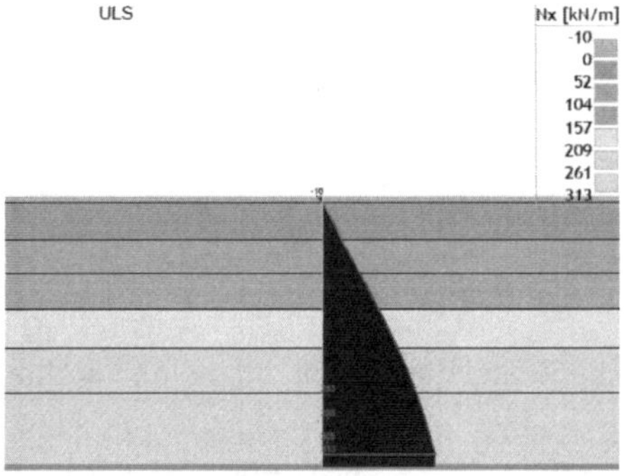

b). Model M3.

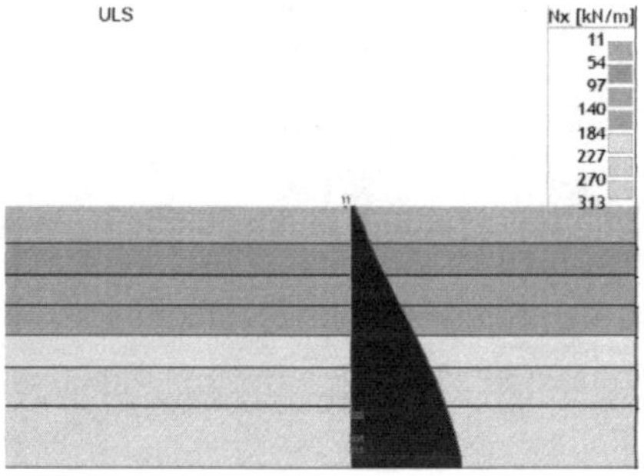

c). Model M4.

Figure 5. Diagrams with isolines for M3 and M4.

This shows that neglecting the hydrostatic pressure in the connection "wall-foundation" can compromise the part. Conversely, if we consider the hydrodynamic pressure, which is combined with the hydrostatic - the hydrodynamic pressure creates in the wall tensile forces in its middle part. This, in turn, is a prerequisite for changing the load-bearing reinforcements in the annular direction (compared to the upper and middle section) and reinforcement bars are without change at the middle and lower end. The ALS combination shows the maximum values of the effort with which the shell must actually be reinforced. All this can be seen from table 3.

In computational models, with realized stiffening between wall and foundation, there is an "upload" of the maximums in wall height at ULS - combination. Also, "the hardening", from the rings (supports or foundations) in many cases turn the diagrams of the effort in their field - this can be used by engineers as a technique to reduce peak values.

Two more models (M10 and M11) were made for verification (see table 4). In the table 5 the values of the observed internal forces are systematized, in the combination - full tank, the maximum values are indicated in red. In figure 6 are results for M10 and M11.

Table 4. Characteristics of computational models - new

Model – M...	Wall type
M10	straight wall and is fixed at the lower end (without foundation)
M11	straight wall, but with modeled foundation and bottom plate, on an elastic base

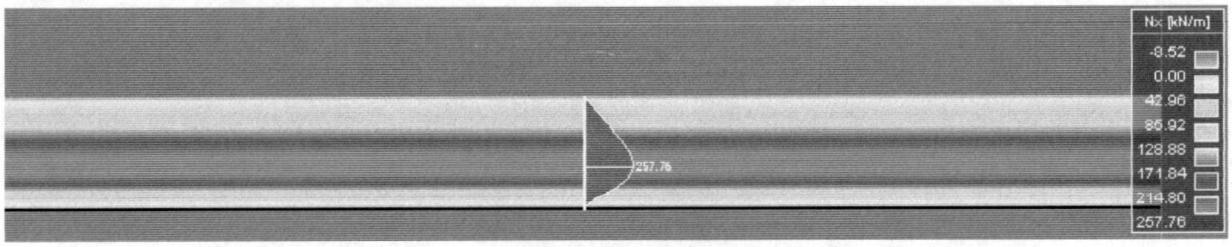

a). Model M10.

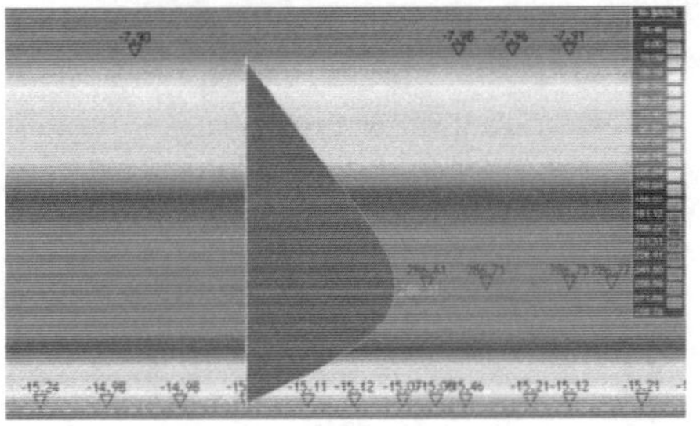

b). Model M11.

Figure 6. Diagrams with isolines for M10 and M11.

Table 5. Internal forces N_x, *[kN/m']*, in the new models: a). ULS; b). Seismic - *quasi* increase

ULS	upper end	-7,39	-7,90
	middle	257,76	286,51
	lower end	-8,52	-15,24
Model		M10	M11

ALS	upper end	-9,98	-10,67
	middle	347,98	386,79
	lower end	-11,50	-20,57
Model		M10	M11

The multiplication factor MF=**1,35** - with which the internal forces from table 5 – a) turn into internal forces from table 5 – b), without direct seismic computational analysis [4].

4. Conclusions

After analysis, it was found that the computational model M5 shows the largest ring internal forces with a positive sign, of the two combinations - static (ULS) and seismic (ALS). The characteristic of this model is that the structure has an inserted support ring along the height of the wall. In this situation, part of the annular reinforcement can be concentrated in the local ring, hardening the wall. The difference in results is 51,22% between the two different combinations. This shows that seismic analysis of the facility is mandatory and neglecting it would lead to dangerous consequences.

In the middle sections of the walls, from ULS and ALS combinations model M4 has a minimum value of tensile forces. This shows that in addition with the economy of the material (by forming a wall sloping from the outer edge of the water), as well as this is model without supporting upper ring compensated by weight from the wall itself. For the middle of the wall, the difference in effort between the two combinations is 57,88%.

The models for verification – M10 and M11 there are small values of internal forces (in the combinations).

All this shows that it is necessary to make enveloping combinations of loads and results, for the wall.

Can be used successfully the postulates from [8] and [9].

Figure 7 shows the envelope of forces N_x diagram by which the reinforcements are selected. As shown, the graph coverage is smooth - by zones.

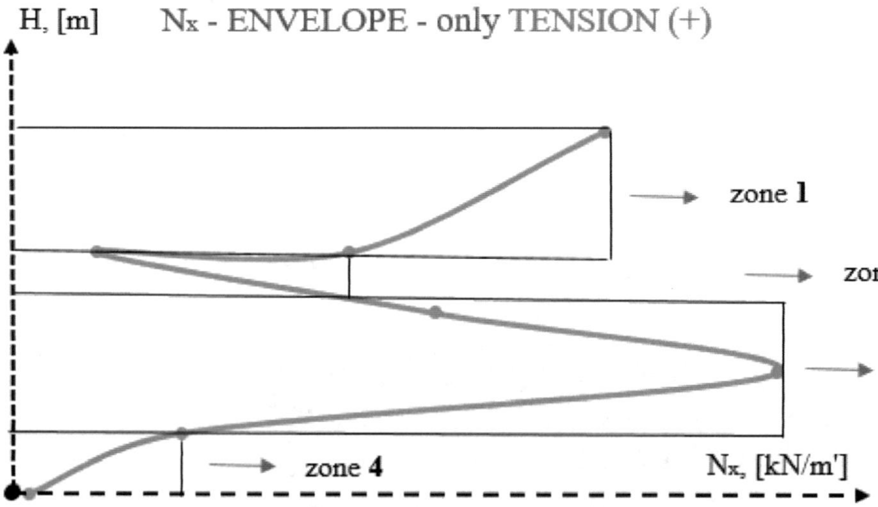

Figure 7. Diagram – ENVELOPE

With already found and localized values of the annular forces, the analysis of the reinforcements in this direction is quick and easy. It is based on the formulas in the stress state of pure tension:

$$A_{sl}^{tot.} = \frac{\left(N_{x,max,i} \cdot \ell_i\right)}{f_{yd}}, \left[cm^2/m'\right] \quad (1)$$

according to which: the wall is divided into strip i with widths ℓ_i, and in each strip for the action maximum internal forces in the strip $N_{x,max,i} \equiv N_{2,max,i}$, is found the value of the total area of annular reinforcement for the group (on both sides of the wall) bearing reinforcement bars (in tension).

Figure 8 shows the arrangement of the meridian and ring reinforcing bars. In order not to overload the concrete cover on the inside and not break away, the bars should be of small diameters and distances between them, as well as on the inside of the grid - not directly on the water side.

It was found that up to a height of about 2,40 m, the horizontal tension in the region of the small annular forces in the foundation region, as a result of the limited drying deformations, is taken up by extending the reinforcement from the adjacent zone (see zones 3 and 4 of figure 7).

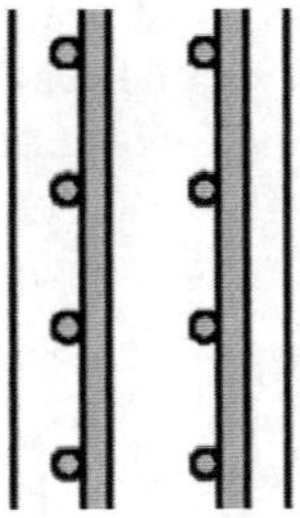

in – side of the water

Symmetrical – for seismic impact.

Figure 8. Scheme of installation

Figure 9 shows construction (fragment of the wall) of the reinforcement bars.
Figure 10 shows a stage of construction activities of a facility.

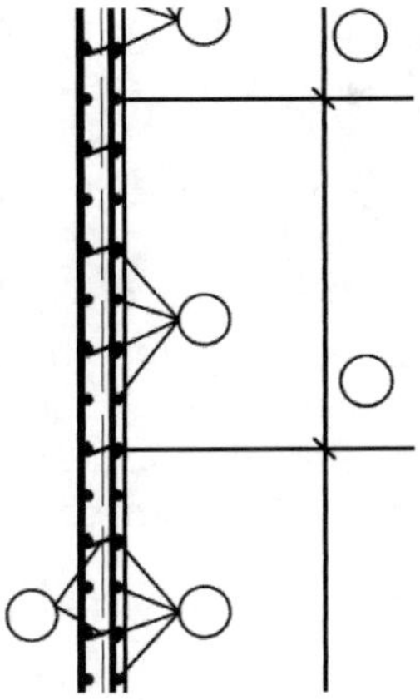

2x8N10/20 - ℓ=140 cm

(2x8N10)x6x600+(2x8N10)x1x100

Figure 9. Basic labels – partially (*for example*)

Figure 10. Object in implementation [10]

Hydrotechnic facilities must be checked for crack opening, especially the walls in radially (with accepted reinforcement, for internal force $N^{SLS}_{x,max}$) and meridionally (with reduce/neto characteristics, with guaranteed height of the compression zone, for internal force $M^{SLS}_{y,max}$) directions, mandatory - at fixed limiting values of crack-forming maximums.

In: [11], [12] and [13] can view information for other constructions and their solutions.

IOP Conf. Series: Materials Science and Engineering 1276 (2023) 012026 doi:10.1088/1757-899X/1276/1/012026

The considered developed examples show possible constructions of the wall, in which the effort can be "controlled" - as a value and location along the height of the wall. It was found that the hydrodynamic pressure leads to a significant increase in the internal forces in the seismic combination. Ignoring this combination can lead to a significant "unloading" of the structure, and hence to collapse (with unsecured for the maximum possible impact walls and lack of reinforcement bars).

Acknowledgements
This paper is financially supported by USEA (VSU) “Lyuben Karavelov”, Sofia, project 04/2022 – “Analysis of speciality reinforced concrete structures”.

References

[1] BDS EN 1998-4: 2006 Eurocode 8: Design of seismic structures impacts Part 4: Silos, tanks and pipelines
[2] Georgiev G and Tsvetkov St 2014 Guide for design of Special Reinforced Concrete Structures according Part 1 Chapter 2 – Development of course project (Design of water tanks)” *VSU Publishing house Sofia*
[3] Housner G W 1963 The dynamic behavior of water tanks. Bulletin of Seismological Society of America **53** pp 381-387
[4] Tsvetkov St 2018 Antiseismic provision of special R.C. structures USEA L. Karavelov – Sofia *Dissertation for the award of educational and scientific degree PhD*
[5] Tsvetkov St 2018 R. C. tanks for water storage – seismic analysis *Proceedings of the 16th International Conference on New Trends in Statics and Dynamics of Buildings* October 18-19 2018 Bratislava Slovakia Faculty of Civil Engineering STU Bratislava Slovak Society of Mechanics SAS pp 85-94
[6] Tsvetkov St and Nedelchev N 2020 Modeling and construction of R.C. water tower *Academic journal „Mechanics, Transport, Communications*” (MTC AJ) Scientific paper ID 1913: 2020/1
[7] Tsvetkov St 2020 Hydrotechnical facility – R.C. radial precipitator. features in the structural design *XX International Scientific Conference by Construction and Arhitecture VSU'2020* pp 331-337
[8] *ACI Committee 350.3-06* 2006 Seismic Design of Liquid-Containing Concrete Structures and Commentary
[9] Newmark N. M. and Hall W. J. 1982 Earthquake spectra and design M. S. Agbatian California: *Earthquake Engineering Research Institute* pp 33-37
[10] https://structville.com/2021/11/design-of-underground-rc-circular-water-tanks.html (Last visited *July 29, 2022 for Internet – link*)
[11] Georgiev T 2018 Original structural decision for steel bunkers in order to speed up construction *Proc. in XVIII Anniversary International Scientific Conference by Construction and Arhitecture VSU'2018* pp 615-620
[12] Handruleva An 2020 Mechanisms of destruction of spatial frame structure VSU Sofia
[13] Stoynova Il and Hristov Ts 2019 Compatative analysis between by classic solutions methods of flat slab and numerical solutions. Verification of finite elements *18-th International Symposium MASE 2019 Ohrid*

IOP Conf. Series: Materials Science and Engineering 1276 (2023) 012027 doi:10.1088/1757-899X/1276/1/012027

The Septemvri-Dobrinishte narrow-gauge railway line – a unique project for Balkans.

M. G. Lepoev[1]

1. Prof. D-r Eng. Department of Railway construction, Faculty of Transport Engineering, University of Architecture, Civil Engineering and Geodesy, Sofia, Bulgaria

mlepoev_fte@uacg.bg

Abstract. The first section of the narrow-gauge railway line Septemvri-Dobrinishte was put into operation in the distant 1926, and the last sixth stage of it was completed in 1945. It is single line, non-electrified, with a track gauge of 760mm, which makes it unique for Bulgaria and the Balkans. Apart from its track gauge, the line impresses with its spiral ascents, maximum longitudinal slopes - up to 32 ‰, minimum radii of horizontal curves of 50m and the highest station on the Balkan Peninsula, Avramovo 1267.4m above sea level. Within a total length of 125 km, 35 tunnels, 23 bridges, retaining walls and numerous culverts have been built. The article presents the current state of the major engineering facilities along the line - bridges and tunnels, as well as the state of the superstructure of the railway. Its technical characteristics, which makes it a unique achievement, both in terms of transport structural facilities and for railway construction in general, are also commented.

1. Introduction

The railway line Septemvri - Dobrinishte narrow-gauge railway line is the 16-th railway line in Republic of Bulgaria in exploitation from 1926 with track gauge 760mm (Bosnian track gauge as it is known) [1], [2]. The last sixth stage of it was completed in 1945. This railway crosses difficult mountainous terrain and for this reason a total length of 125 km includes 35 tunnels, 23 bridges, retaining walls and numerous culverts. The plan of the railway near highest railway station in Bulgaria - Avramovo with 1267m altitude is presented on Fig.1.

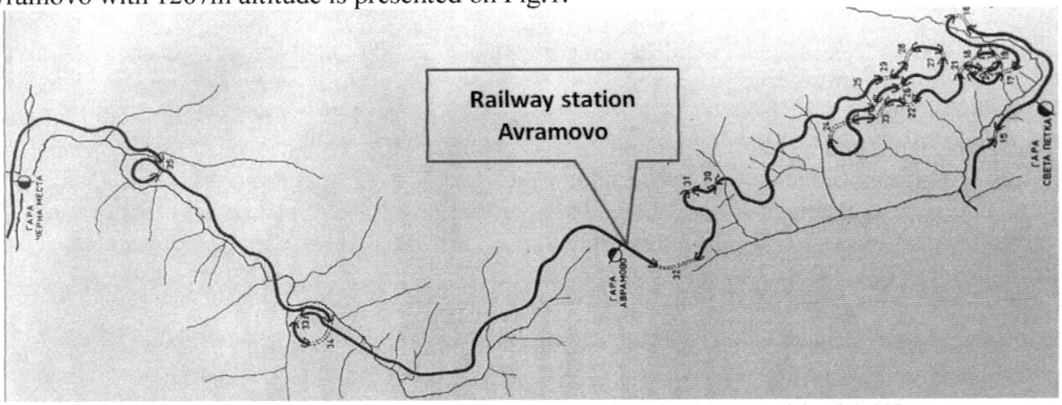

Figure 1. Complex railway near Avramovo station (at 1267m altitude) [2]

During its exploitation a serious conceptual and detailed alterations for track design, construction gauge allowing the movement of a railway vehicle and exploitation is realized [7]. During its exploitation a serious conceptual and detailed alterations of the design standards for structural constructions in respect to transport structural facilities are realized [3], [4], [5], [9], [10], [11]. In such cases condition analysis is needed to evaluate the safety and conformity of the described structures concerning present demands [6].

The railway track is constructed as a narrow track-gauge 760mm with timber sleepers as it is shown on Fig.2. The maximum longitudinal slopes along the railway line are up to 32 ‰. The highest railway station Avramovo with 1267m altitude is presented on Fig.3.

Figure 2. Railway track.

Figure 3. Railway station Avramovo [12].

2. Description of bridge types along the railway line.

There are several types of bridge superstructures along the railway line:

- ✓ Steel riveted plate girder bridges
- ✓ Steel riveted truss bridge over river Elidere with 30m span – Fig.10 [2]
- ✓ Concrete arc bridges
- ✓ Reinforced concrete girder bridges

The main part of those bridges are simply supported with respect to support conditions of the superstructure. Seven of them are with steel superstructure and 16 with masonry and concrete superstructures. The steel bridges are with open type deck and as it is discussed in other investigations problems with drainage [8] and the longitudinal inertia forces from the traffic can be expected [3], [4], [5].

There is a new continuous plate girder steel welded railway bridge with 19,4m span at km. 006 + 783 (Varvara Dolene) – Fig.4. Superstructure is supported on steel tube piers and on a reinforced concrete girder supported on the existing piers from approaching arc span. The bridge deck is standard of open type with longitudinal girders and cross girders – Fig.5. It can be noticed the absence of continuous horizontal bracing between main girders. The railway track supported directly on longitudinal girders is presented on Fig.6 and the connection between steel superstructure and the steel tube pier is presented on Fig.7. This new bridge replaces an old steel riveted plate girder superstructure – Fig.8. Problems with clearance with approaching arc spans of the old bridge superstructure are presented on Fig.9 and it is obvious the risk for the road traffic.

An example of a concrete three hinge arc bridge with 12,1m spans is presented on Fig.11.

Another concrete arc bridge with 15,6m span at km107+503 is presented of Fig.12 and Fig.13.

Figure 4. Bridge at km. 006+783.

Figure 5. Bridge at km. 006+783.

Figure 6. Bridge at km. 006+783.

Figure 7. Connection at steel pier.

Figure 8. Old bridge at km. 006+783.

Figure 9. Old bridge at km. 006+783.

Figure 10. Bridge over river Elidere [2].

Figure 11. Arc bridge km118+205.

Figure 12. Bridge at km107+503.

Figure 13. Bridge at km107+503.

3. Description of tunnels.

There are 35 tunnels along Septemvri - Dobrinishte narrow-gauge railway line. Most of the tunnels are standard structures built with new Austrian method. At Fig.14 the Portal of tunnel 9 is presented (between km14+788 and 14+863). The railway track in tunnels is ballasted – Fig.15.

Figure 14. Tunnel 9.

Figure 15. Railway track in the tunnel.

Figure 16. Protective structure – Tunnel 8.

Figure 17. Slope tunnel.

The complex structural conditions in the Rhodopi mountain requires also to build protective structures for falling rocks – Fig16 (at the entrance of Tunnel 8 between km13+397 and 13+426). Also slope tunnels are built – Fig17 (tunnel between km13+377 and 13+397). As it can be seen from Fig.17 the main maintenance problem for most of the tunnels is connected with water.

4. Summary and Conclusions

The Septemvri-Dobrinishte narrow-gauge railway line is a unique project for Balkans. During its building different type of transport structural facilities and execution technologies were implemented. This railway line is of great importance for connected villages and towns especially at winter when the road traffic is difficult as a function of the snow cover. Also during summer due to beautiful mountain landscape great numbers of tourists are using the line. The maintenance of the railway line and structures is realised by the NRIC in Republic of Bulgaria. It is important to provide additional activities to evaluate the condition of the railway track, bridges and tunnels along the described railway line in respect to present design guides and exploitation standards in order to investigate the possibilities to increase the level of safety and comfort.

References

[1] https://bg.wikipedia.org/wiki/%D0%A2%D0%B5%D1%81%D0%BD%D0%BE%D0%BB%D0%B8%D0%BD%D0%B5%D0%B9%D0%BA%D0%B0_%D0%A1%D0%B5%D0%BF%D1%82%D0%B5%D0%BC%D0%B2%D1%80%D0%B8_%E2%80%93_%D0%94%D0%BE%D0%B1%D1%80%D0%B8%D0%BD%D0%B8%D1%89%D0%B5, 06.2022

[2] Dimitar Deyanov, The railway network in Bulgaria 1866-1975, ISBN: 954-9363-05-4, Stara Zagora, 2005 (in bulgarian)

[3] Georgiev, L.D., Ivanov, S.D., Retrofitting Old Steel Riveted Railway Truss Bridges in Bulgaria: A Case Study, Structural Engineering International, 2020, 30(1), pp. 124–127, ISSN: 1016-8664 (Print) 1683-0350 (Online), DOI:10.1080/10168664.2019.1686337

[4] Georgiev L., Ivanov St., Longitudinal Stability Of Continuous Composite Steel-Concrete Superstructures For Railway Viaducts, X Jubilee International Scientific Conference „Civil Engineering Design and Construction" (Science and Practice), Sept. 20-22, 2018, Varna, Bulgaria, ISSN: 2603-4255, ISSN 2683-071X (online), pp.404-411 (in bulgarian)

[5] Nikolova, M., Georgiev, L., Problems and innovative solutions considering "open type" bridge deck structures in old steel riveted railway bridges, 2018 IOP Conference Series: Materials Science and Engineering; 9th INTERNATIONAL SYMPOSIUM ON STEEL BRIDGES, 10-11 September 2018, Prague; 2018 IOP Conf. Ser.: Mater. Sci. Eng. 419 012001, https://doi.org/10.1088/1757-899X/419/1/012001

[6] Alexander Jiponov, Lazar Georgiev, Condition assessment of several bridges in Sofia based on

visual inspection and as-built drawings, Structural Engineering for Future Societal Needs, IABSE CONGRESS GHENT 2021, pp.418-424, ISBN: 978-3-85748-176-5

[7] Lazar Georgiev, Vladimir Zhekov, Determination of a construction gauge allowing the movement of a railway vehicle for kinematic gauge GC, based on national construction gauge 1-CM2, AIP Conference Proceedings 2439, 020009 (2021); https://doi.org/10.1063/5.0071089, Proceedings of the technical and scientific conference ECO Varna, ISSN (Online) 2367- 6965, ISSN (Print) 2367- 6299

[8] Dulevski E. M., Georgiev L. D., Diagnostic And Strengthening Of An Old Riveted Bridge, 5th International Conference on Bridges Across the Danube, Volume II, Novi Sad, Serbia&Montenegro, 2004, ISBN 86-905637-2-5, pp.289-294

[9] Georgiev L., Ivanov St., BULGARIAN NATIONAL ANNEXES FOR BRIDGES – COMMENTS, ANNUAL of the UACEG, Sofia, Volume 51, Issue 4, pp.27-33, 2018, ISSN 1310-814X, ISSN 2534-9759 (in bulgarian)

[10] Jiponov A. (2020), A Technical solution for Frame Rail Viaduct that can guarantee Durability and Safety Exploitation, XI International Scientific Conference “Civil Engineering Design and Construction” (Science and Practice), September, 2020, Varna, Bulgaria, IOP Conference Series Materials Science and Engineering 951(1):012021, doi:10.1088/1757-899X/951/1/012021

[11] Jiponov A., Nikolov V. (2022), Extending the Lifecycle of Damaged Structure by Retrofitting New Bridge Design Concepts in Old Structures, Volume 200 LNCE, 1st Conference of the European Association on Quality Control of Bridges and Structures, EUROSTRUCT 2021, August 2021, pp.651–661

[12] https://bg.wikipedia.org/wiki/%D0%90%D0%B2%D1%80%D0%B0%D0%BC%D0%BE%D0%B2%D0%BE_(%D0%B6%D0%B5%D0%BB%D0%B5%D0%B7%D0%BE%D0%BF%D1%8A%D1%82%D0%BD%D0%B0_%D0%B3%D0%B0%D1%80%D0%B0)#/media/%D0%A4%D0%B0%D0%B9%D0%BB:Avramovo_railway_station_2021_01.jpg, 06.2022

AUTHOR INDEX